Vertebrate Zoology

Vertebrate Zoology

Toby Marquez

Callisto Reference

www.callistoreference.com

Callisto Reference,
118-35 Queens Blvd., Suite 400,
Forest Hills, NY 11375, USA

Visit us on the World Wide Web at:
www.callistoreference.com

ISBN: 978-1-64116-644-7 (Hardback)

Cataloging-in-Publication Data

Vertebrate zoology / Toby Marquez.
p. cm.
Includes bibliographical references and index.
ISBN 978-1-64116-644-7
1. Vertebrates. 2. Zoology. I. Marquez, Toby.
QL605 .V47 2022
596--dc23

TABLE OF CONTENTS

PREFACE

The purpose of this book is to help students understand the fundamental concepts of this discipline. It is designed to motivate students to learn and prosper. I am grateful for the support of my colleagues. I would also like to acknowledge the encouragement of my family.

The biological discipline that focuses on the study of vertebrate animals is known as vertebrate zoology. The animals with a backbone such as amphibians, reptiles, birds and mammals collectively called as tetrapods, are known as vertebrate animals. Vertebrates include various groups such as jawless fishes, jawed vertebrates that include the cartilaginous fishes, and bony fishes. Vertebrate zoology can be further divided into various sub-fields including ichthyology, mammalogy, ornithology, herpetology and batrachology. These sub-divisions respectively study the fishes, mammals, birds, reptiles and amphibians. Different approaches, evaluations and studies on vertebrate zoology have been included in this book. It attempts to understand the multiple branches that fall under the discipline of vertebrate zoology. Coherent flow of topics, student-friendly language and extensive use of examples make this book an invaluable source of knowledge.

A foreword for all the chapters is provided below:

Chapter - Introduction

The species of animals which fall under the subphylum vertebrata are called vertebrates. They are all chordates possessing a backbone. The study of vertebrate animals is known as vertebrate zoology. This chapter delves into the classes, anatomy, evolution and ecology of vertebrates to provide an easy understanding of the varied facets of vertebrate zoology.

Chapter - Phylum Protochordata

Organisms which are considered as lower chordates fall under the category of protochordata. They are also called Acraniata as they don't possess a true skull. They are classified as hemichordata, urochordata and cephalochordata. The topics elaborated in this chapter will help in gaining a better perspective about phylum protochordata.

Chapter - Anatomy of Vertebrates

The branch of biology that defines body structure and functions, such as respiration, circulation, digestions, etc. of all living organisms, is referred to as anatomy. Vertebrates are divided into pisces, reptiles, amphibians, aves and mammals. This chapter closely examines these anatomy of vertebrates to provide an extensive understanding of the subject.

Chapter - Reproductive System of Vertebrates

A system consisting of reproductive organs, necessary to carry-on the generation of any living species is referred to as reproductive system. There are two modes of reproduction, sexual and asexual. This chapter has been carefully written to provide an easy understanding of the varied facts of reproduction in pisces, reptiles, amphibians, aves and mammals.

Chapter - Adaptation in Vertebrates

The evolution of the living organism to better suit its environment and habitat is called adaptation. Adaptation in vertebrates includes structural and functional adaptations of pisces and mammals, behavior and morphological adaptations of reptiles, flight adaptation in aves, amphibians' adaptation to water balance and cold survival. This chapter discusses in detail these adaptation theories related to vertebrates.

Toby Marquez

Introduction

CHAPTER 1

The species of animals which fall under the subphylum vertebrata are called vertebrates. They are all chordates possessing a backbone. The study of vertebrate animals is known as vertebrate zoology. This chapter delves into the classes, anatomy, evolution and ecology of vertebrates to provide an easy understanding of the varied facets of vertebrate zoology.

Vertebrate Zoology is the study of animals with backbones. The Department is organized into four Divisions: Fishes, Amphibians and Reptiles, Birds, and Mammals.

The systematic and taxonomic research conducted in the department provides a solid foundation of understanding biodiversity that benefits our scientific colleagues, government agencies, conservation organizations, and individuals involved in fish and wildlife management.

The department holds the largest collection of vertebrate specimens in the world, including historically important collections from the nineteenth and early twentieth centuries.

VERTEBRATES

Vertebrate, also called Craniata are any animal of the subphylum Vertebrata, the predominant subphylum of the phylum Chordata. They have backbones, from which they derive their name. The vertebrates are also characterized by a muscular system consisting primarily of bilaterally paired masses and a central nervous system partly enclosed within the backbone.

The major groups of vertebrates include fishes, amphibians, reptiles, birds, and mammals.

The subphylum is one of the best known of all groups of animals. Its members include the classes Agnatha, Chondrichthyes, and Osteichthyes (all fishes); Amphibia (amphibians); Reptilia (reptiles); Aves (birds); and Mammalia (mammals).

General Features of Vertebrates

Although the vertebral column is perhaps the most obvious vertebrate feature, it was not present in the first vertebrates, which probably had only a notochord. The vertebrate has a distinct head, with a differentiated tubular brain and three pairs of sense organs (nasal, optic, and otic). The body is divided into trunk and tail regions. The presence of pharyngeal slits with gills indicates a relatively high metabolic rate. A well-developed notochord enclosed in perichordal connective tissue, with a tubular spinal cord in a connective tissue canal above it, is flanked by a number of segmented muscle masses. A sensory ganglion develops on the dorsal root of the spinal nerve, and segmental autonomic ganglia grow below the notochord. The trunk region is filled with a large, bilateral body cavity (coelom) with contained viscera, and this coelom extends anteriorly into the visceral arches. A digestive system consists of an esophagus extending from the pharynx to the stomach and a gut from the stomach to the anus. A distinct heart, anteroventral to the liver, is enclosed in a pericardial sac. A basic pattern of closed circulatory vessels is largely preserved in most living forms. Unique, bilateral kidneys lie retroperitoneally (dorsal to the main body cavity) and serve blood maintenance and excretory functions. Reproductive organs are formed from tissue adjacent to the kidneys; this original close association is attested by the tubular connections seen in males of living forms. The ducts of the excretory organs open through the body wall into a cloacal chamber, as does the anus of the digestive tract. Reproductive cells are shed through nearby abdominal pores or through special ducts. A muscular tail continues the axial musculature of the trunk.

Approximately 45,000 living species constitute the vertebrates. Species of several classes are found from the high Arctic or Antarctic to the tropics around the Earth; they are missing only from interior Antarctica and Greenland and from the North Polar ice pack. In size, vertebrates range from minute fishes to elephants and whales (of up to 100 tons), the largest animals ever to have existed. Vertebrates are adapted to life underground, on the surface, and in the air. They feed upon plants, invertebrate animals, and one another. Vertebrate faunas are important to humans for food and recreation.

AREAS OF STUDY

Anatomy of Vertebrates

Comparative vertebrate anatomy: The study of structure, of the function of structure, & of the range of variation in structure & function among vertebrates:

- Kingdom: Animal.
- Phylum: Chordata.
- Subphylum: Vertebrata.

Vertebrate Characteristics

- Notochord (at least in the embryo).

- Pharynx with pouches or slits in wall (at least in the embryo).
- Dorsal and hollow nervous system.
- Vertebral column.

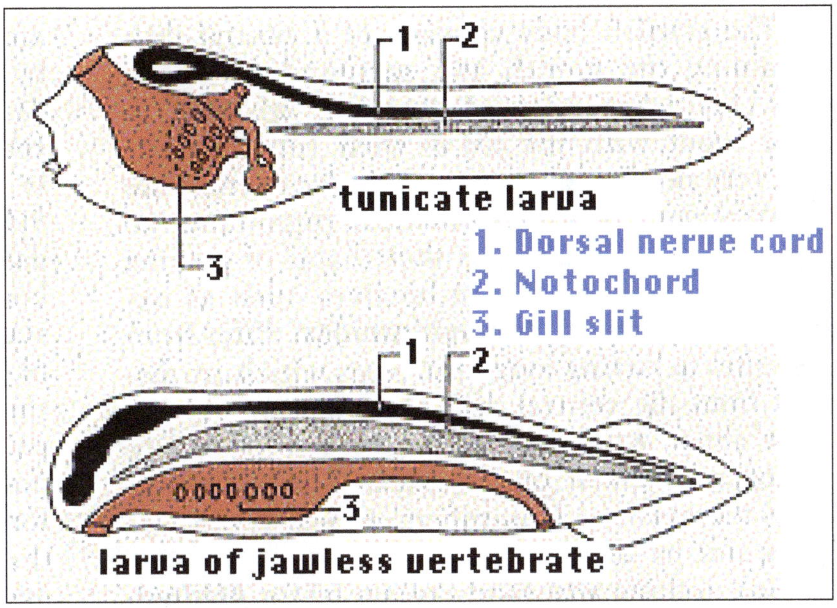

Notochord: Rod of living cells ventral to central nervous system & dorsal to alimentary canal.

Fate of notochord during development:

- Head region: Incorporated into floor of skull.
- Trunk & tail: Surrounded by cartilaginous or bony vertebrate (except in Agnathans).

Adults:

- Fishes & amphibians: Notochord persists the length of the trunk & tail but is constricted within the centrum of each vertebra.
- Reptiles, birds, & mammals: Notochord almost disappears during development (e.g., remains as a pulpy nucleus in the vertebrae of mammals).
- Protochordates: Notochord remains as the chief axial skeleton.
- Agnathans: Lateral neural cartilages are located on notochord lateral to the spinal cord.

Pharynx : Region of alimentary canal exhibiting pharyngeal pouches in embryo; pouches may open to the exterior as slits.

- Permanent slits: Adults that live in water & breathe via gills.
- Temporary slits: Adults live on land.

Dorsal, hollow central nervous system - consists of brain & spinal cord & contains a central cavity (called the neurocoel).

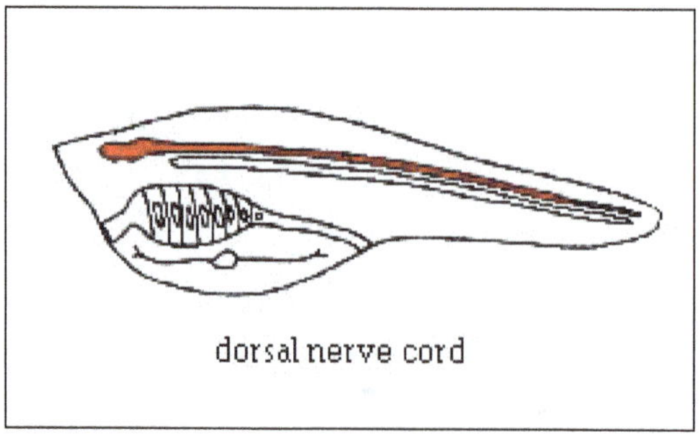

Vertebrate Beginnings

Among the oldest & best known = ostracoderms.

- Fishes that occurred in the late Cambrian period through the Devonian (about 400 - 525 million years before present).
- Had bony plates and scales (&, therefore, were easily fossilized).
- Jawless vertebrates called 'armored fishes'.

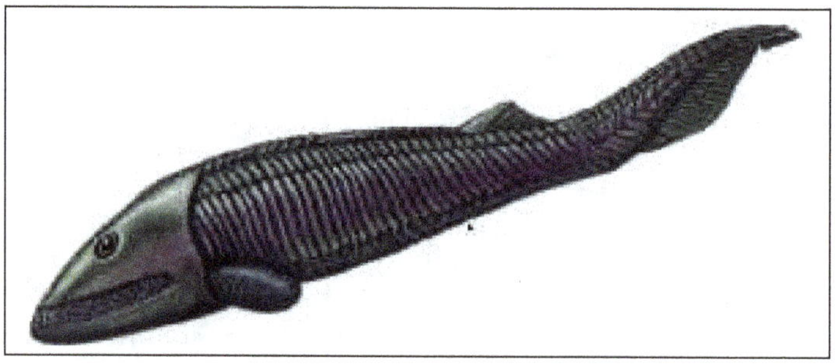

Before ostracoderms:

- Myllokunmingia fengjiaoa & Haikouichthys ercaicunensis - primitive fish that have many similarities to living hagfishes and are the oldest vertebrates (530 mybf) ever found.

Before Vertebrates:

- Cathaymyrus diadexus (literally the 'Chinese eel of good fortune') is not the fossil of an eel. At just 5 cm long, but 535 m.y. old, it is the earliest known chordate. Researchers think that Cathaymyrus is a fossil relative of modern lancelets (amphioxus).

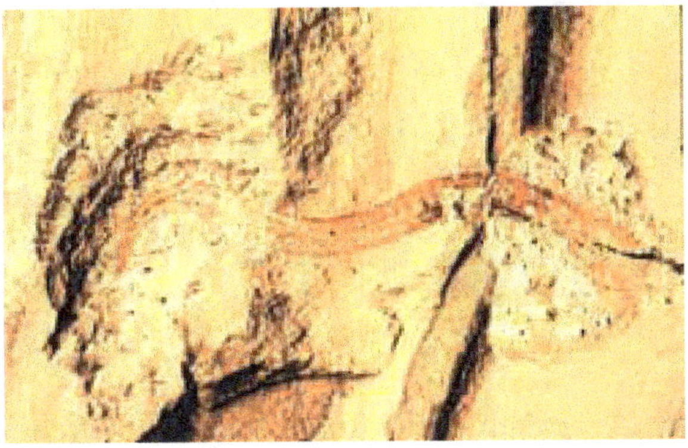

Non-vertebrate chordates still alive today include tunicates (or sea squirts; urochordates) & amphioxus (or branchiostoma). (cephalochordates).

Phylum Chordata - established in 1874 & included organisms with,

- Notochord,
- Pharyngeal pouches or slits,
- Dorsal, hollow nervous system,
- Cells that produce the hormone thyroxine.

Subphylum Urochordata = tunicates

Chordate 'ancestor' of vertebrates:

- Sessile (like adult tunicates).
- Tail Evolved as adaptation in larvae to increase mobility.
- 'Higher forms' - came about by retention of tail (neoteny).

Tunicate larva - also called 'sea squirt'.

- Notochord is confined to the tail.
- Notochord is lost during metamorphosis into sessile adult.
- Possess pharyngeal slits.

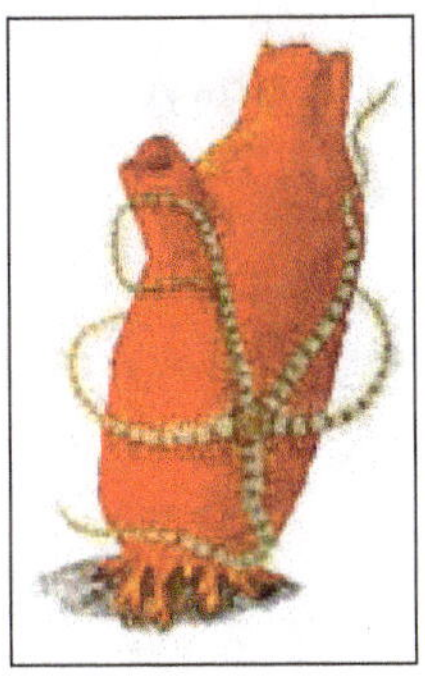

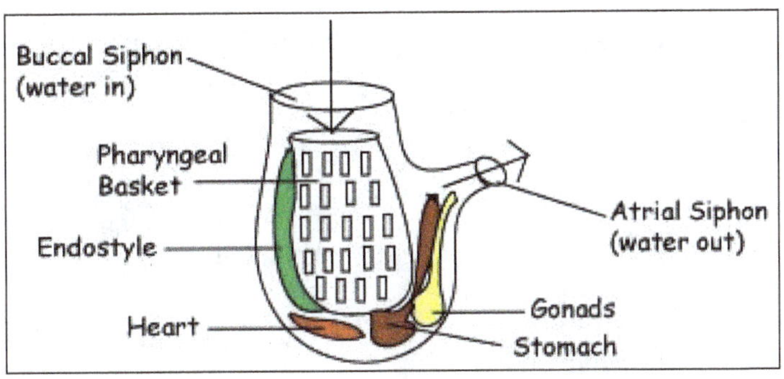

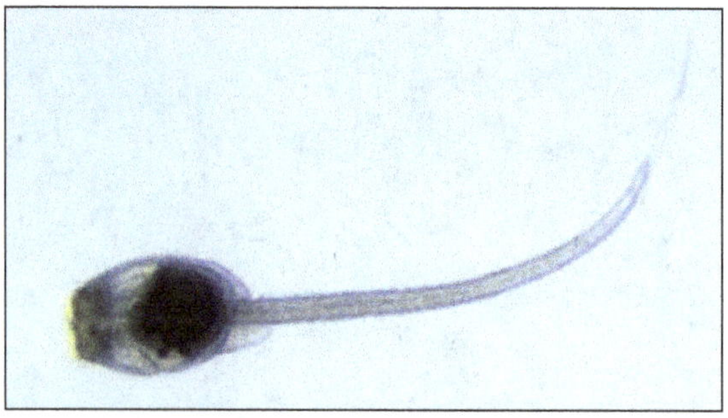

A 530 million-year-old (although perhaps as old as 560 million years) creature, *Cheungkongella ancestralis*, probably a tunicate, found in the Chengjiang fauna in China's southwest Yunnan Province, might be the earliest known fossil evidence of primitive chordates.

Subphylum Cephalochordata: Amphioxus (or Branchiostoma).

Vertebrate features:

- Notochord.
- Dorsal, hollow nervous system.
- Pharyngeal gill slits.
- 'Circulatory' system - vertebrate pattern with 'pumping vessels' (but no heart).

Hemichordates : Acorn worms.

Bateson added acorn worms to the phylum Chordata in 1884 because they have:

- A dorsal, hollow nervous system.
- Gill slits.
- A short diverticulum of the gut called the stomochord.

Present consensus: The stomochord is not homologous with the notochord and Hemichordates are placed in a separate phylum.

Possible invertebrate ancestors:

Annelid Worms

Evidence for:	Evidence against:
• Bilateral symmetry. • Segmented. • Central nervous system with brain and longitudinal nerve cord.	• Nerve cord is solid. • Nerve cord is ventral.

Echinodermata - Chordate characteristics include:

- Radial cleavage - blastomeres in adjacent tiers lie directly above one another (as opposed to spiral cleavage).
- Anus forms near or at blastopore (deuterostomous).
- Mesoderm arises as outpocketing of the gut wall.
- Indeterminate cleavage (i.e., fate of blastomeres isn't predetermined).

Phylum: Chordata.

Subphylum: Vertebrata.

Superclass - Pisces:

- Class Agnatha.

- Class Placodermii.
- Class Chondricthyes.
- Class Acanthodii.
- Class Osteichthyes.

Superclass - Tetrapoda:

- Class Amphibia.
- Class Reptilia.
- Class Aves.
- Class Mammalia.

Agnathans vs. Gnathostomes

- Semicircular canals:
    - Agnathans have 1 or 2.
    - Gnathostomes have 3.
- Jointed, paired lateral appendages:
    - Agnathans have none.
    - Gnathostomes do.
- Jaws:
    - Agnathans have none.
    - Gnathostomes do.

Class Agnatha

Orders:

- Osteostraci.
- Anaspida.
- Thelodonti.
- Galeaspida.
- Pituriaspida.
- Petromyzontia (lampreys).
- Myxinoidea (hagfishes).

Ostracoderms (Osteostraci, Anaspida, Heterostraci, & Coelolepid):

- Extinct Paleozoic (Cambrian to Devonian) jawless fish with an external skeleton of bone ('bony armor').
- Oldest known vertebrates.
- Many had flattened appearance (some may have been bottom-dwellers).

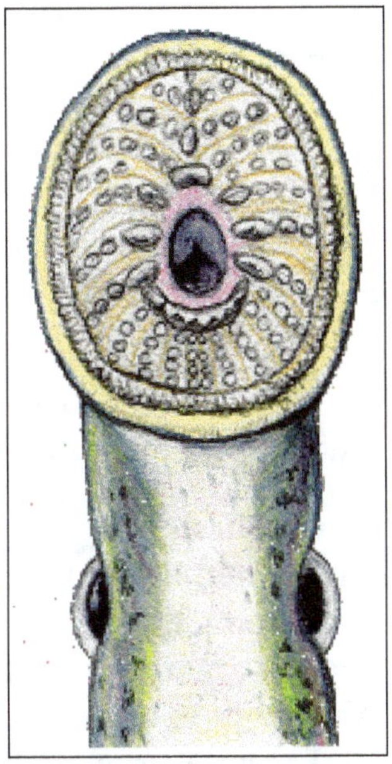

Cyclostomes (Petromyzontia and Myxinoidea)

- Lampreys: Parasitic with horny and rasping teeth.
- Hagfishes: Primarily scavengers.

Gnathostomes

- Acanthodians:
    - Earliest known gnathostomes (Silurian; about 440 mybp).
    - Probably related to modern bony fishes.
    - Small (less than 20 cm long) with large eyes.
    - Acanthodians most likely died out because of the rapidly increasing number of ray-finned fishes and sharks during the Permian.

The relationships of acanthodians to other vertebrates has been the subject of considerable debate. Early researchers considered them to be most closely related to the ray-finned fishes, but most scientists during the mid-20th Century considered acanthodians to have a closer affinity to the sharks. Opinion has now generally swung back in favor of a closer relationship with ray-fins, but this is far from universally accepted.

Class Placodermii

- Silurian (about 420 million years before present).
- Probably off the main line of vertebrate evolution.
- Many had bony dermal shields.
- Some were probably predators (with large, sharp 'tooth plates').

Placoderms were armored jawed fishes that first appeared about 420 million years ago (MYA) during the Silurian Period. They had diversified dramatically by the beginning of the Devonian and came to dominate most marine and freshwater ecosystems before becoming extinct at the end of that period (355 MYA). About 200 genera of placoderms have been discovered, with most of these occurring during the Devonian radiations. The rapid evolution and diversity of placoderms have made them useful in dating Devonian rocks. Placoderms (plated skin) were named for their heavy armor of dermal bone, which formed large shields on the head and thorax. The rest of their bodies was covered with small bony scales or was without dental armor. The head and trunk shields of most placoderms were articulated by bony joints. This joint apparently allowed the forward part of the skull to tilt up, increasing the gape. Placoderms lacked teeth, but biting or grinding structures are often be found in the dermal bones lining their mouths. Placoderms evolved into a variety of body forms in a relatively short time. Many were torpedo-shaped, but there were notable expections, including the flatten Phyllolepida and the bottom feeding Antiarchi. Most placoderms were less than 30 cm (2 feet) in length, but some members of the dinichthyids (= terrible fish) reached or exceeded 6 m (20 ft), making them the first giants of the vertebrate lineage.

Class Chondrichthyes - Cartilaginous Fishes

- Ancestors had bony skeletons so cartilaginous skeleton is specialized.
- Pelvic fins of males are modified as claspers.

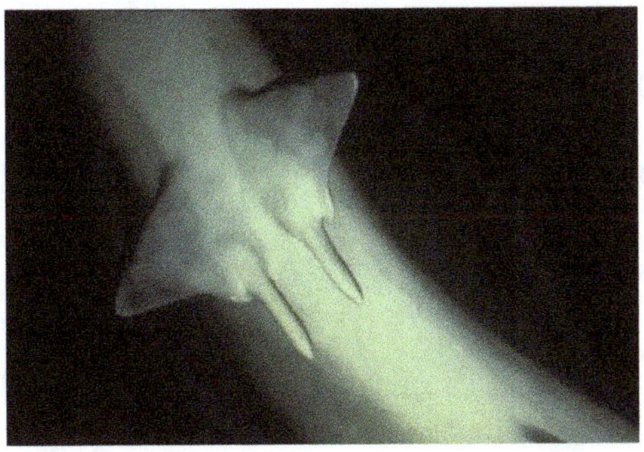

- Placoid scales.

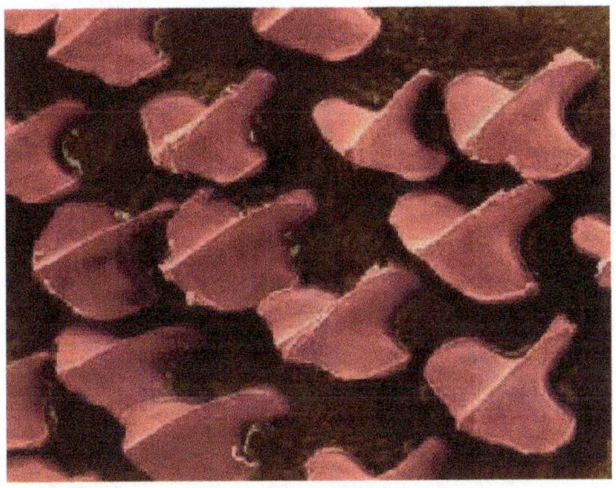

- Numerous today but more abundant in the past.

Subclass Elasmobranchii: most common cartilaginous fishes.

- Cladoselachii: Primitive sharks (300-400 mybp).
- Selachii: 'Modern' sharks.
- Batoidea: Rays and skates.

Elasmobranchs:

- 1st pharyngeal slit modified as a spiracle.
- Naked gill slits (no operculum).
- Mouth located ventrally.

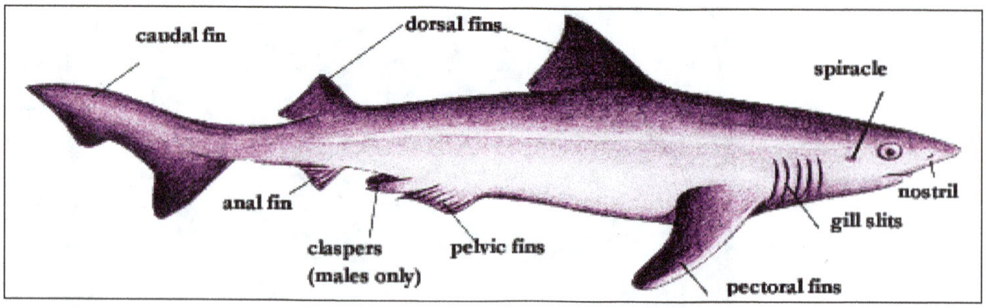

Subclass Holocephali:

- Chimaeriformes:
    - Marine.
    - Gill Slits Have A Fleshy Operculum & The Spiracle Is Closed.
    - Few Scales.
    - Common ancestor with sharks but an independent line.

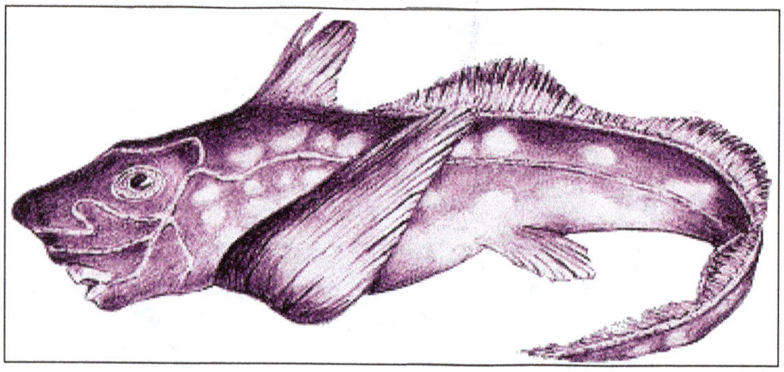

Class Osteichthyes - Bony Fishes

- Skeleton is partly or chiefly bone.
- Gill slits are covered by a bony operculum.
- Skin has scales with, typically, little bone.
- Most have a swim bladder.
- Ray-finned or lobe-finned.

Subclass Actinopterygii - ray-fins.

Superorder Chondrostei:

- Most primitive ray-fins.
- Chiefly paleozoic (300-400 mybp).
- Include present-day Sturgeons & Paddlefish.

Superorder Neopterygii.

Order Semionotiformes:

- Dominant Mesozoic fishes.
- Possess ganoid scales.
- Two extant genera:
    - Lepidosteus - Predatory; includes present-day gars.
    - Amia - Includes present-day bowfins (or dogfish).

Division Teleostei - modern ray-finned fishes

- Recent bony fishes.
- 95% of all living fish.
- About 40 living orders.
- Well-ossified skeleton.
- Cycloid & ctenoid scales (flexible & overlapping).
- Pelvic fins often located far forward.
- No spiracle.

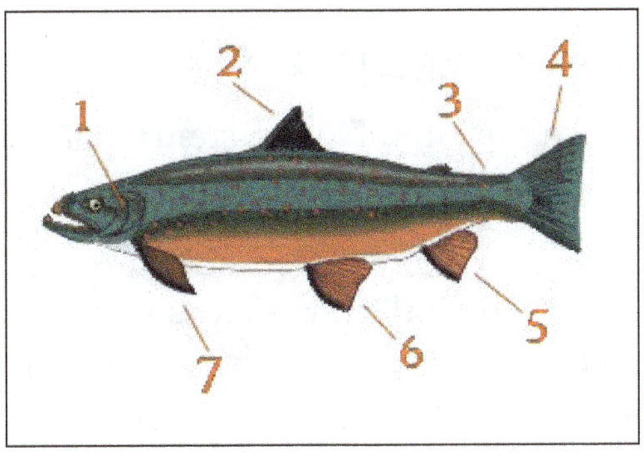

1 = operculum, 2 = dorsal fin, 3 = caudal peduncle (The narrow section of a fish's body directly anterior to the insertion of the tail but before the mid-body.), 4 = caudal fin or "tail", 5 = anal fins, 6 = pelvic fins, & 7 = pectoral fins.

Subclass Sarcopterygii - lobe-finned fishes:

- Crossopterygii: Chiefly Paleozoic except *Latimeria.*
    - Resemble early amphibians.
    - Skeleton of fin lobe corresponds closely to proximal skeletal elements of early tetrapod limbs.

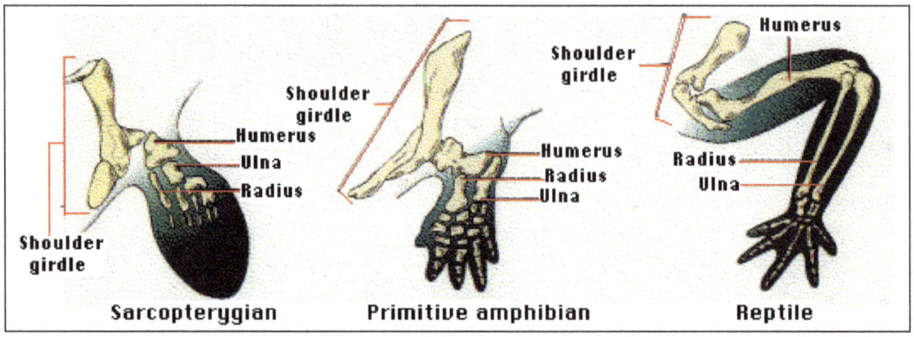

    - Skull similar to that of early amphibians.
    - Had swim bladders that may have been used as lungs.
- O. Dipnoi - lungfish (3 living genera; Africa, Australia, & South America):
    - African & South American species have inefficient gills & will drown if held under water.
    - Australian species (*Neoceratodus* spp.) relies on gills unless oxygen content of water is too low.

Class Amphibia

Oldest known = subclass Labyrinthodontia.

Fish-like features:

- Small bony scales in the skin.
- Fin-rays in the tail (for swimming).
- A skull similar to that of some Crossopterygians.
- A sensory canal system (like the lateral line system) that indicates a primarily aquatic existence.

Labyrinthodonts are distinguished by deeply folded structure of enamel and dentine layers in the teeth, that look like an intricate labyrinth in the cross section, hence the name of this group. Labyrinthodonts were probably similar to fishes in their mode of living. Labyrinthodonts, like fishes and most modern amphibians, laid eggs in the water, where their larvae developed into mature animals. All labyrinthodonts had special sense organs in the skin, that formed a system for perception of water fluctuations. Moreover, some of them possessed well developed gills. In contrast, many labyrinthodonts seemingly had primitive lungs. They could breath atmospheric air, that was a great advantage for residents of warm shoals with low oxygen levels in the water. The air was inflated into the lungs by contractions of a special throat sac. Primitive members of all labyrinthodont groups were probably true water predators, and only advanced forms that arose independently in different groups and times, gained an amphibious, semi-aquatic mode of living. Mature individuals of advanced labyrinthodonts could live on land, feeding mainly on insects and other small invertebrates. Well ossified robust skeletons in some Late Carboniferous and Early Permian labyrinthodonts prove their adaptation to the terrestrial mode of life. It suggests that amphibians had successfully 'organized' actual terrestrial assemblages prior to the wide expansion of reptiles.

The most diverse group of the labyrinthodonts was the batrachomorphs ('similar to a frog'). Though these animals looked more like crocodiles, they most probably gave rise to the order Anura, the amphibians without tails, which include, in particular, the modern frogs. Batrachomorphs appeared in the Late Devonian, but they had worldwide distribution in the continental shallow basins of the Permian (Platyoposaurus, Melosaurus) and Triassic Periods (Thoosuchus, Benthosuchus, Eryosuchus). Some batrachomorphs existed until the end of the Cretaceous.

Subclass Lepospondyli:

- Ancestry uncertain due to lack of fossil evidence.
- Probably on a 'side branch' of vertebrate evolution.

Subclass Lissamphibia - modern amphibians:

- Anura: Frogs & toads.
- Urodela: Tailed amphibians.
- Gymnophiona (apodans) - Wormlike, burrowing amphibians.

Modern amphibian characteristics:

- Aquatic larval stage with external gills.
- Middle ear cavity with ear ossicle (columella).
- No bony scales (except apodans).

Class Reptilia - The First Amniotes

- Scaly.
- Clawed.
- Large, yolk-laden, shell-covered eggs laid on land.

Stem reptiles = Cotylosaurs (about 300 mybp).

Reptile Subclasses:

- Anapsida.
    - Cotylosauria: Stem reptiles.
    - Chelonia: Turtles & tortoises.

      Unchanged for about 175 million years.

      Identified by bony dermal plates to which ribs & trunk vertebrae are fused.
- Lepidosauria:
    - Rhynchocephalia (Sphenodonta) - Only living representative is the Tuatara.
    - Squamata - Lizards, geckos, & snakes.
- Archosauria:
    - Thecodontia - stem archosaurs.
    - Pterosauria.
    - Saurischia - 2 major groups: sauropods & theropods.
    - Ornithischia (like Iguanodon).
    - Crocodilia.
- Euryapsida - Marine reptiles, includes the plesiosaurs & ichthyosaurs.

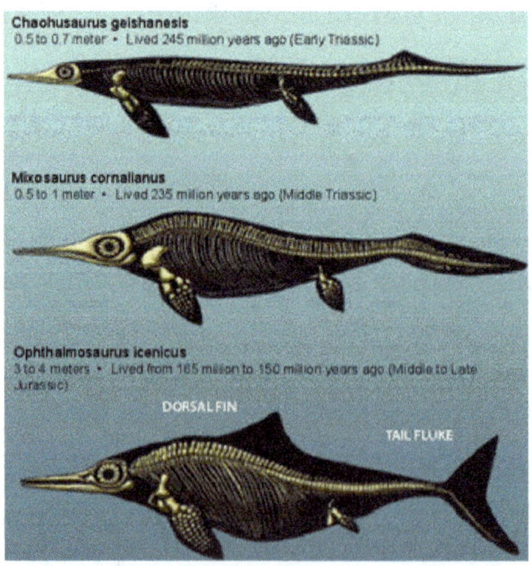

- Synapsida:
  - Pelycosauria - First stage in evolution to mammals.
  - Therapsida.

Saurischia (sawr-RIS-kee-ah) & Ornithischia are the two orders of dinosaurs, with the division based on the shape of the pelvic bone. The saurischian pubis (left) juts forward, and its ischium points backward. The ornithischian pubis and ischium (right) both point backward. The ornithischians were all herbivorous, and included some of the most interesting-looking dinosaurs. Ornithischian dinosaurs include three suborders: Ornithopoda, Marginocephalia and Thyreophora. The famous carnivorous dinosaurs were from the saurischian order, as were the largest herbivorous dinosaurs. The saurischian dinosaurs include two suborders: Theropoda and Sauropodomorpha.

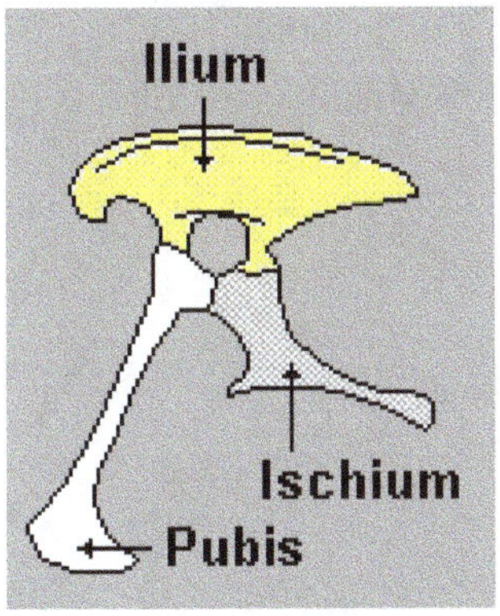

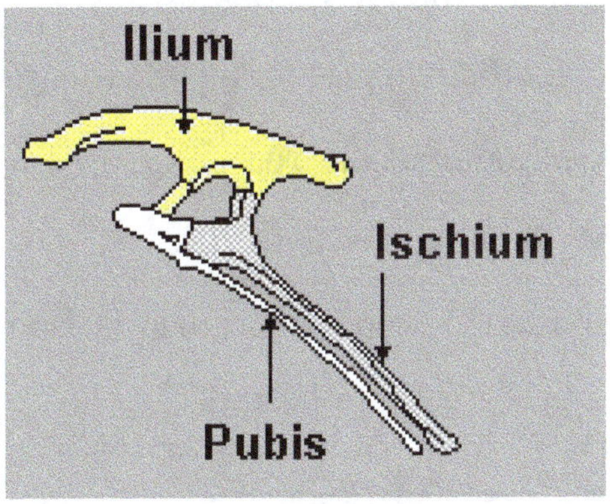

The first vertebrates to evolve true flight were the pterosaurs, flying archosaurian reptiles. After the discovery of pterosaur fossils in the 18th century, it was thought that pterosaurs were a failed experiment in flight; a humorous mishap; or that they were simply gliders, too weak to fly. More recent studies have revealed that pterosaurs were definitely proficient flyers, and were no evolutionary failure; as a group they lasted about 140 million years (about as long as birds have)! Pterosaurs are thought to be derived from a bipedal, cursorial (running) archosaur in the late Triassic period (about 225 million years ago). No other phylogenetic hypothesis has withstood examination; however, the early history of pterosaurs is not yet fully understood because of their poor fossil record in the Triassic period. We can infer that the origin of flight in pterosaurs fits the "ground up" evolutionary scenario, supported by the fact that pterosaurs had no evident arboreal adaptations.

The pterosaur wing was supported by an elongated fourth digit (imagine having a "pinky finger" several feet long, and using that to fly!). Pterosaurs had other morphological adaptations for flight as a keeled sternum for the attachment of flight muscles, a short and stout humerus (the first arm bone), and hollow but strong limb and skull bones. Pterosaurs also had modified scales that were wing-supporting fibers, and that possibly formed hairlike structures to provide insulation -- bird feathers are analogous to the wing fibers of pterosaurs, and both are thought to possibly have been evolved originally for the primary purpose of thermoregulation (which implies, but does not prove, that both pterosaurs and the earliest birds were endothermic).

Early pterosaurs (such as Dimorphodon) had long tails that assisted balance, but later pterosaurs had no tails, and may have been more adept flyers. The most derived pterosaurs, such as Pteranodon and Quetzalcoatlus, were so large that soaring was the only feasible option; these were the largest flyers ever to cast a shadow on the Earth's surface.

Reptile subclasses - classified in part according to presence or absence of temporal openings.

- Synapsid type: Mammal-like reptiles.
- Anapsid type: Stem reptiles & turtles.
- Diapsid type: Rhynchocephalians, lizards, & snakes.
- Euryapsid type: Extinct plesiosaurs.

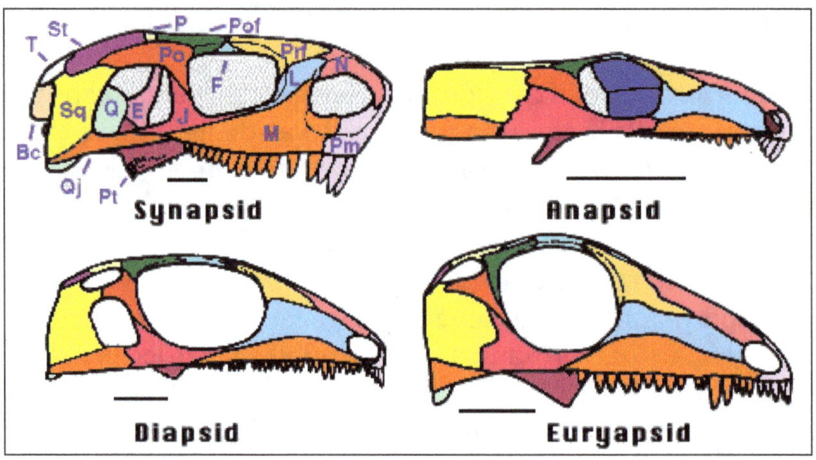

Temporal fenestration has long been used to classify amniotes. Taxa such as Anapsida, Diapsida, Euryapsida, and Synapsida were named after their type of temporal fenestration. Temporal fenestra are large holes in the side of the skull. The function of these holes has long been debated. Many believe that they allow muscles to expand and to lengthen. The resulting greater bulk of muscles results in a stronger jaw musculature, and the longer muscle fibers allow an increase in the gape.

Class Aves-birds

- May have arisen from an archosaurian reptile, perhaps a small bipedal dinosaur.
- Lost several dinosaur characteristics (e.g., long tail & teeth) but retained others (e.g., claws, scales, diapsid skull, single occipital condyle, perhaps and feathers).

Subclass Archaeornithes.

Genera: Archaeopteryx & Archaeornis.

Characteristics:

- Solid bones.
- Weakly developed keel &, probably, weakly developed flight muscles.

Subclass Neornithes.

Superorder Odontognathae:

- Extinct.
- Many features of modern birds (e.g., hollow bones & short tail).

Superorder Paleognathae:

- Ratites.
- Small wings but powerful leg muscles.

Superorder Neognathae: Birds adapted for sustained flight.

Modifications to reduce weight include:

- Loss of some bones.
- Pneumatic bones.
- Reduced tail.
- Loss of teeth.
- Loss of urinary bladder.

Class Mammalia

Characteristics:

- Hair.
- Mammary glands.
- 3 middle ear bones.
- 4 Muscular diaphragm.
- Sweat glands.
- Marrow within bones.
- 2 sets of teeth.
- Biconcave, enucleate red blood cells'.
- Well-developed cerebral cortex.

Subclass Prototheria: Egg-laying mammals.

- Monotremata-platypus + 2 spiny anteaters.
  - Lay eggs.
  - Testes within the abdominal cavity.
  - No pinna.
  - No corpus callosum.
  - Less stable body temperature.

Subclass Theria.

Infraclass Metatheria:

- Marsupialia: Pouched mammals; young born alive, but at a very immature stage
- Infraclass Eutheria: Placental mammals.

Evolution of Vertebrates

Origin of vertebrates is widely appreciated as one of the most formative events in evolutionary history. The gulf is so large between those blessed with a backbone and their spineless relatives which have not, that the invertebrate–vertebrate divide is one of the most fundamental in biology, discriminating teaching, research, journals and conferences. To some great extent, this may be perceived as entirely arbitrary and a consequence of undue anthropocentricism. And yet, the origin of vertebrates is associated with a fundamental embryological revolution and the doubling or quadrupling of the genome. Without doubt, these profound changes helped vertebrates get ahead and, what is more, to get a head.

A Perspective on Early Vertebrate Evolution based on the Living

To obtain an understanding of vertebrate origins, it is necessary to establish first their phylogenetic context standing of the anatomical characteristics of key fossil groups, especially the anaspids, thelodonts, galeaspids and pituriaspids; tests of the monophyly of traditional groups; and the application of divergence time methods that integrate not just molecular data from living species, but also morphological data and extinct species. The resulting framework will provide for rigorous tests of rates of character evolution and diversification, and of hypotheses of long-term trends in ecological evolution that themselves suffer for lack of quantitative functional tests. The fossil record has been silent on the nature of the transition from jawless vertebrates to the jawed vertebrates that have dominated communities since the middle Palaeozoic. Elucidation of this most formative of episodes likely rests with the overhaul of early vertebrate systematics that we propose, but perhaps more fundamentally with fossil grades that await discovery.

Their closest living relatives are the tunicates, together with which they comprise the clade Olfactores, to which the cephalochordates are a sister lineage. All three comprise the chordate phylum which, together with the echinoderms and hemichordates (Ambulacraria), and perhaps the acoelomorph flatworms and Xenoturbella, comprise the deuterostomes. Vertebrates inherit bilateral symmetry, a centralized nervous system and a perforated pharynx from the deuterostome crown ancestor, segmented musculature and a notochord from the chordate crown ancestor, and the rudiments of cranial sense organs and a fourth germ layer (neural crest) from the crown ancestor of Olfactores. These are major innovations, but they are merely the chassis onto which is bolted, in evolution and development, the fundamental novelties that distinguish vertebrates.

Vertebrate novelties can be discriminated as those features that vertebrates share, that are absent from their nearest living invertebrate relatives, the tunicates and the cephalochordates. The three principal lineages of vertebrates are the jawless hagfishes and lampreys, and the jawed gnathostomes. There has been a long-standing debate over the relationships of these three lineages, which impacts on this exercise, viz. whether the lampreys are more closely related to hagfishes (cyclostome monophyly) or gnathostomes (cyclostome paraphyly). This debate appears to have lurched in favour of cyclostome monophyly, revealing the ancestral crown vertebrate to have been more complex than either hagfishes or lampreys, both of which lineages have lost vertebrate characters.

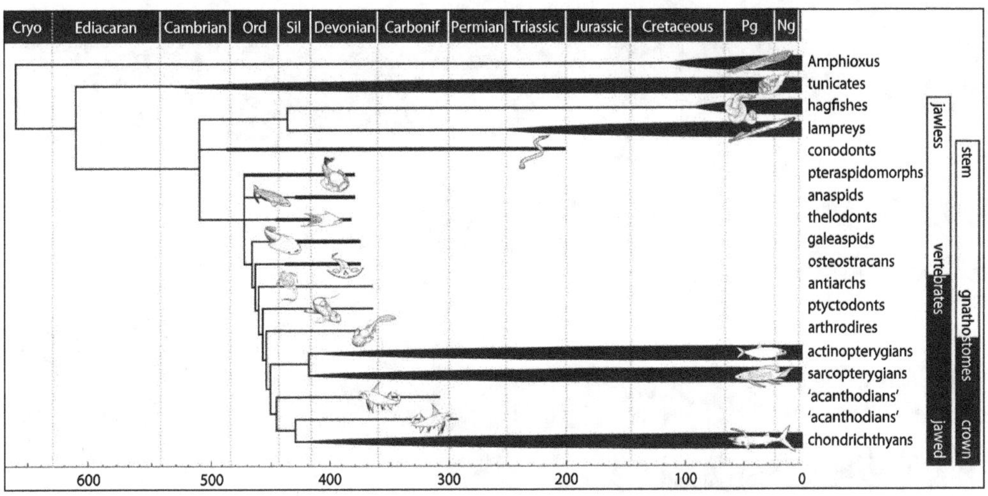

Time in millions of years before present (Ma).

Phylogeny of living chordates and a representative selection of extinct vertebrates plotted against geological time. Relationships are based upon Donoghue et al. The spindles (dark grey) reflect the crown group divergence of the respective lineages: amphioxus; tunicates; hagfishes and lampreys; chondrichthyans; and actinopterygians. Node dates, where applicable, correspond with divergence estimates from. Geological timescale based on Gradstein et al. Abbreviations: Cryo, Cryogenian; Ord, Ordovician; Sil, Silurian; Carbonif, Carboniferous; Pg, Palaeogene; Ng, Neogene. hagfishes or lampreys, both of which lineages have lost vertebrate characters.

Thus, vertebrate novelties include paired optic, otic and nasal organs, a sensory line system, multifarious neurons and glia, a differentiated brain, braincase, a sympathetic nervous system, diverse pigment cells, cartilage, connective tissue, an adaptive immune system, a post anal tail, fin rays, arcualia, and gills, among very many other characters. The majority of these features are not unrelated, and they can be accounted for by the elaboration of two embryological phenomena that evolved within stem Ol factores: neurogenic placodes, from which cranial sense organs develop, and neural crest, which is a population of migratory mulitpotent stem-like cells from which the novel neural, skeletal, pigment cells and their derivative tissues develop. The profound elaboration in the fate of neurogenic placodes and neural crest in vertebrates, in comparison with tunicates, has been explained as a consequence of the evolutionary assembly of regulatory interactions among the genes directing development. This will have responded to, and may perhaps have been facilitated by, the explosion in coding regulatory genes that occurred as a consequence of one or more whole-genome duplication events, as well as an independent but equally dramatic diversification in noncoding regulatory microRNA diversity, both of which occurred in the vertebrate stem lineage.

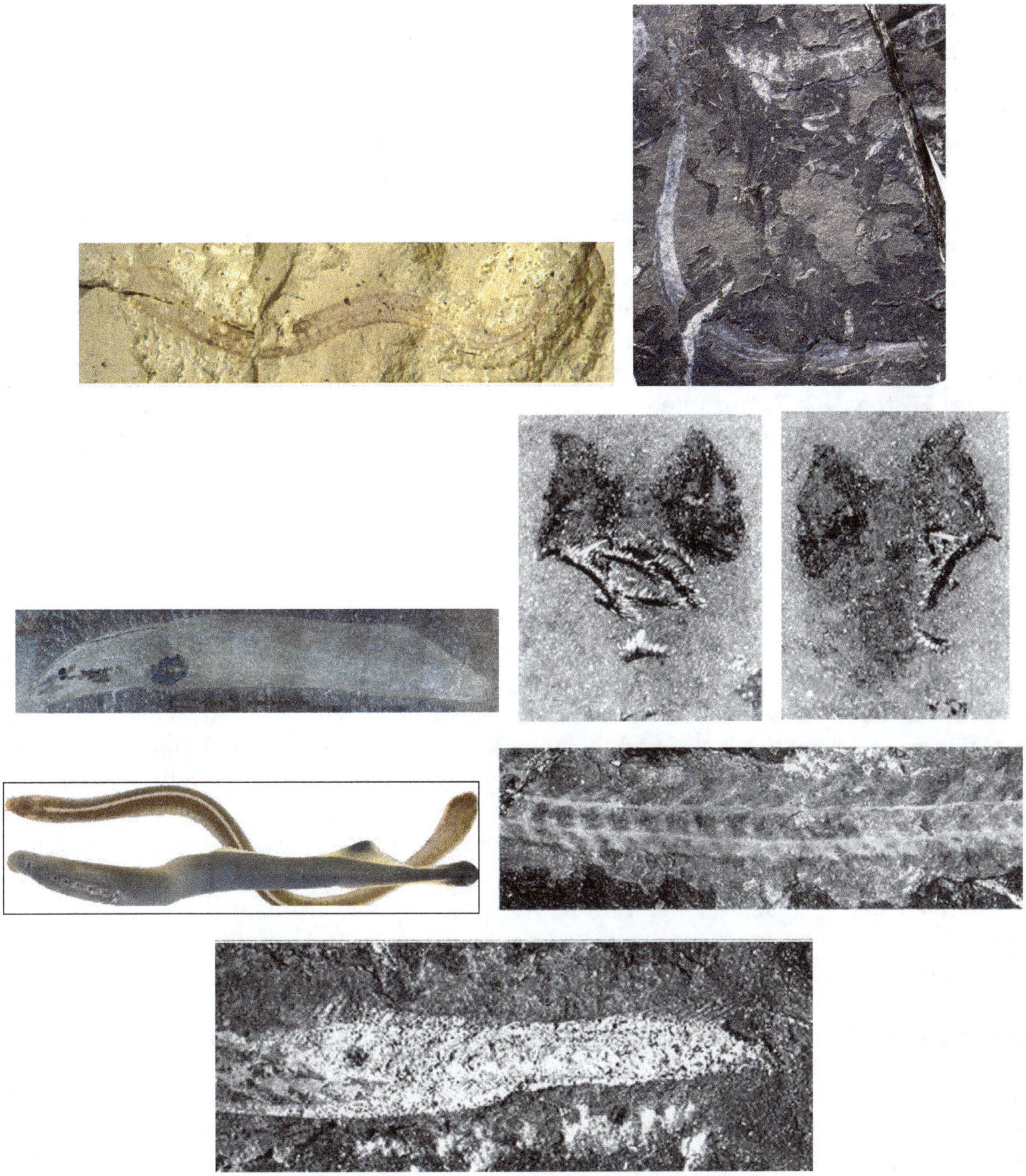

Soft-bodied invertebrate chordates and vertebrates.

A–B, the tunicate Ciona intestinalis adult (A) and larva (B). C, Shankouclava shankouense, a putative tunicate from the Cambrian Chengjiang Lagerstatten of Kunming, Yunnan Province, South China. D, Branchiostoma lanceolata adult male. E, Cathaymyrus diadexus, a putative cephalochordate from the Cambrian Chengjiang Lagerstatten of Kunming, Yunnan Province, South China. F, Myxinikela siroka, a putative hagfish from the Pennsylvanian (Carboniferous) Mazon Creek Lagerstatten of Illinois, USA. G, The inshore hagfish Eptatretus burgeri in dorsal aspect. H, the Arctic lamprey Lethenteron camtschaticum in lateral aspect. I, Mayomyzon pieckoensis, a lamprey from the Pennsylvanian (Carboniferous) Mazon Creek Lagerstatten of Illinois, USA. J, Haikouella lanceolata, a putative chordate that has been assigned to multifarious deuterostome (and nondeuterostome) clades, from the Cambrian Chengjiang Lagerstatten of Kunming, Yunnan Province, South China. K, Haikouichthys ercaicunensis, a total-group vertebrate from the Cambrian Chengjiang Lagerstatten of Kunming, Yunnan Province, South China. L–P, IGSE 13821 Clydagnathus winsorensis, a conodont (total-group vertebrate) from the Mississippian (early Carboniferous) of Granton, Scotland; M–N, part and counterpart of the head showing paired structures interpreted as eyes, as well as the characteristic phosphatic elements comprising the feeding apparatus; O, detail of trunk showing v-shaped structures interpreted as shrunken muscle blocks; P, detail of caudal fin with fin rays. Scale bar represents 25 mm (A); 150 lm (B); 3 mm (C); 6 mm (D–E); 11 mm (F); 24 mm (G); 45 mm (H); 1 mm (I, P); 3.5 mm (J); 4 mm (K); 2.5 mm (L); 0.5 mm (M–N); 1.5 mm (O). Abbreviations: EC, Early Life Research Center, Jinning, China; FMNH, Field Museum of Natural History; IGSE, Institute of Geological Science, Edinburgh, UK; NEIU, Northeastern Illinois University; NWU, Department of Geology, Northwest University, Xi'an, China; YKLP (formerly RCCBYU), Yunnan Key Laboratory for Palaeontology, Kunming, Yunna Province, China.

While vertebrate innovations are profound, most non- experts would be hard pressed to distinguish the living cyclostomes as vertebrates. For starters, they do not even have vertebrae (although they have rudiments in the form of arcualia. The majority of those features, like vertebrae, that are widely but incorrectly assumed to be general to all vertebrates, were acquired in the gnathostome lineage after it separated from cyclostomes. In some senses, this is an even more dramatic evolutionary episode than the origin of vertebrates more generally, as the inventory of features shared by all living jawed vertebrates, but absent from the ancestral crown vertebrate, is even more extensive. Gnathostome novelties include teeth comprised of dentine, enameloid and cellular bone, a mineralized dermal skeleton comprised of tooth-like scales, mineralized fin radials, paired pectoral and pelvic fins with both an endoskeleton and dermal fin rays, a mineralized braincase, a labyrinth associated with the oticorgan that has a horizontal semicircular canal, heterocercal tail, anal fin, differentiated gut, parasympathetic nervous system, paired external nostrils, externally open endolymphatic ducts, among very many other features. It is not surprising therefore that nonuniformitarian macroevolutionary mechanisms have been invoked to explain the transformation of the plesiomorphic body plan of vertebrates into that of gnathostomes.

Refining Vertebrate Evolutionary History with Insights from the Dead

Fossil evidence of the emergence of vertebrates is dominated by gnathostomes with mineralized skeletons, for obvious reasons. However, the discovery and investigation of Cambrian Burgess Shale type Lagerstatte has yielded the remains of entirely soft-bodied vertebrates, invertebrate chordates, and deuterostomes more generally. These have been recovered principally from the Chengjiang and Burgess Shale Lagerstatte. Cambrian fossils interpreted as members of known lineages include cephalochordates Cathaymyrus and Pikaia; tunicates Cheungkongella and Shankouclava, Zhongjianichthys and Zhongxiniscus. Fossils described from the Devonian and Carboniferous as hagfishes include Gilpichthys, and lampreys Hardistiella, Mayomyzon and Priscomyzon; Mesomyzon has been described as a lamprey from the Jurassic Jehol Biota of China. Interpreting the precise phylogenetic position and, therefore, the evolutionary significance of many of these fossils is problematic as the majority of chordate, Olfactores and vertebrate characters are developmental or cytological and therefore not usually fossilized even under conditions of exceptional preservation. Phylogenetic resolution is complicated further by the impact of decay on unmineralized vertebrates and chordates, which can affect the loss of anatomical characters in reverse phylogenetic order.

Yunnanozoon and Haikouella

Although the soft-bodied vertebrates described from fossil Lagerstatte are useful in that they provide minimum constraints on the age and palaeobiogeography of the lineages to which they have been attributed, they do not provide much additional insight into the evolutionary origin of vertebrates. This occurs principally because these fossils do not deviate materially from the anatomy of their living representatives and, hence, they have proven comparatively easy to classify. Perhaps more interesting, and concomitantly frustrating, are those soft-bodied fossils that have proven so much more difficult to classify, precisely because they deviate anatomically from living vertebrates and invertebrate chordates. Perhaps, the most interesting of these problematica are Yunnanozoon and Haikouella; which do not differ materially) that have been interpreted variably as craniates (vertebrates) with external gills, paired sense organs, a differentiated brain, a notochord against which segmental musculature appears to have been asymmetrically developed and a trunk enveloped in cuticle. This is not a combination of characters that anyone envisaged in a vertebrate, and therefore, Yunnanozoon constitutes a welcome test of hypotheses on the anatomical assembly of the vertebrate body plan. Or perhaps Yunnanozoon is such a radical challenge to expectation because interpretations of its anatomy and therefore its systematic classification are incorrect. And, indeed, the interpretations of almost every one of the yunnanozoans' preserved anatomical characters have been contested, and the organisms alternately classified as cephalochordates, stem chordates, crown hemichordates, stem ambulacrarians, neither chordates nor hemichordates, stem deuterostomes and stem bilaterians. Whichever, if any, of these systematic interpretations is correct, the yunnanozoans will impact upon our perspective of anatomical character evolution within the deuterostomes. For instance, only vertebrates among deuterostomes have true gills (as opposed to a perforated pharynx, yet trunk musculature is radially developed about the notochord in chordates and chordates do not have a cuticularized integument; indeed, cephalochordates lack an integument. However, until the phylogenetic position of the yunnanozoans is resolved, their evolutionary significance will remain moot.

Conodonts

Debate over the systematic position and therefore the evolutionary significance of conodonts is a much longerrunning parlour game. Known largely from their microscopic phosphatic tooth-like 'elements', conodonts have been interpreted as fossil plants, algae, all manner of invertebrates and, of course, vertebrates. The rules of the game were tightened upon discovery of their soft tissue anatomy, which revealed conodonts to have a notochord, segmental trunk musculature, a ray supported midline caudal fin, paired sensory organs and a short series of gill pouches. Given this suite of anatomical characters, conodonts were readily interpreted as chordates and even vertebrates, and within this phylogenetic context, the homologies of conodont and vertebrate skeletal tissues considered in more detail. Conodont elements are bicomponent, comprised of a hypermineralized coarsely crystalline crown tissue that often includes a homogenous vacuolated 'white matter', and a basal tissue that is laminated and or spheritic or includes ramifying tubules. The two tissues are arranged with the crown tissue capping the basal tissue, and incremental growth lines reflect appositional growth. Of all chordate tissues, conodont crown tissue compares best to enamel, and the basal tissue to dentine. These similarities appear even more significant given that in conodont elements, the crown and basal tissues are arranged topologically and grow in the same manner as enamel and dentine in vertebrate teeth and dermal denticles. Together, these hard and soft tissue homologies have supported a stem gnathostome affinity for conodonts.

However, these putative homologies have been contested, principally in terms of the homology of conodont and vertebrate skeletal tissues. Ultimately, conodont and vertebrate skeletal tissue homologies have been rejected because it has been demonstrated that the putative homologues were not present in ancestral conodonts, and therefore, they could not have been present in the last common ancestor of conodonts and other vertebrates. It does not follow, as Blieck, Turner and colleagues have argued that conodonts are anything other than vertebrates, an hypothesis that was substantiated explicitly on the nature of soft anatomical characters and not skeletal histological homologies. Indeed, conodonts were revealed to be stem vertebrates by Turner. Perhaps more importantly, with a mineralized skeleton that evolved in convergence rather than common descent with gnathostomes, conodonts do not contribute materially to our understanding of the evolutionary assembly of the vertebrate body plan.

Ostracoderms

Thus, for all of the excitement that fossils of soft-bodied vertebrates attract, the fossils that contribute the most to our understanding the evolutionary assembly of the vertebrate body plan are the boney 'ostracoderms'. These are an assemblage of apparently distinct clades that include both jawless and jawed vertebrates, all of which exhibit characters that are exclusive to gnathostomes among the living vertebrates. Chief among these characters is the presence of a mineralized dermal skeleton in which the canonical vertebrate skeletal tissues are first expressed in evolution, viz. enameloid, dentine and bone. It is this key apomorphy which transforms the fossil record of vertebrates from the exceptional to the routine, recording many subsequent innovations within early vertebrate evolution. Component ostracoderm lineages are related by degree to the living jawed vertebrates, the chondrichthyans and osteichthyans, betrayed by the hierarchically nested suites of characters that they possess.

Jawless and jawed ostracoderms and crown gnathostomes.

Figure indicates, A, Errivaspis waynensis, an pteraspidid heterostracan from the Lochkovian of Herefordshire, England. B, Birkenia elegans, an anaspid from the Llandovery of Scotland. C, Loganellia scotica, a thelodont from the early Silurian of Lesmahagow, Scotland. D–E, headshields of the galeaspids Nochelaspis maeandrine (D) and Duyunolepis paoyangensis (E) from the early Devonian of China, the latter preserving a natural endocast of an originally mineralized braincase. F, Zenaspis waynensis (NHMUK P24940), an osteostracan from the early Devonian of Herefordshire, England, preserved in dorsal aspect and preserving the dextral pectoral fin tucked between the cornual process and the trunk; G, Ctenurella gladbachensis, a pycnodont 'placoderm' from the late Devonian of Germany; H, Ptomacanthus anglicus, an early Devonian 'acanthodian' from Herefordshire, England, which has recently been assigned to both the gnathostome stem and crown. Scale bar represents 12 mm (A); 6 mm (B); 11 mm (C); 25.5 mm (D); 16 mm (E); 17.5 mm (F); 19.5 mm (G); 29 mm (H). Abbreviations: GLAHM, Hunterian Museum, Glasgow, UK; GMC, Geological Museum of China (Beijing); IVPP, Institute of Vertebrate Palaeontology and Palaeoanthropology, Beijing, China; MB, Museum fur Naturkunde in Berlin; NHMUK, Natural History Museum, London, UK.

Thus, for instance, the jawless pteraspidomorphs, including the heterostracans, are the first vertebrates with a mineralized dermal skeleton. The anaspids and thelodonts have such a skeleton, but also possess a differentiated gut. Galeaspids additionally possess a mineralized cartilaginous neurocranium that reveals details of the gross anatomy of the brain; in osteostracans, the neurocranium is comprised of perichondral bone, dermal bone is cellular, the tail fin is heterocecral, gill openings are slit shaped, and they possess paired pectoral fins, all features otherwise exclusive to jawed vertebrates. The apparently paraphyletic placoderms record the acquisition of pelvic appendages, mineralized vertebrae and fin radials, a jointed gill skeleton, and also jaws. Thus, the ostracoderms are stem gnathostomes, effectively recording the piecemeal assembly of the body plan shared by all crown gnathostomes.

The Timing and Tempo of Early Vertebrate Evolution

The timescale of early vertebrate evolution has been interpreted largely in the light of phylogenetically constrained stratigraphy, which reveals extensive ghost lineages in the known record, particularly for stem and, perhaps, crown gnathostomes in the Ordovician. Furthermore, simple statistical interpretation of stratigraphic occurrence data indicates that these inferences of the missing record are conservative Molecular clocks inevitably infer a Neoproterozoic origin of vertebrates; however, these divergence estimates vary by almost 180 million years. The apparent discordance with the vertebrate fossil record may seem unpalatable as, even when interpreted optimistically, the record of crown vertebrates cannot be extended beyond epoch 2 of the Cambrian and, more likely (because of equivocation over the stem versus crown classification of entirely soft-bodied vertebrates), may not be substantiable beyond the Middle Ordovician. Molecular estimates for more exclusive vertebrate clades are in much closer accord, both between divergence time studies, and with the fossil record, an inevitable consequence of the tighter temporal constraints imposed by a fossil record that increases in fidelity with the origin of boney skeletons. Indeed, it is very likely that the fossil record of early vertebrates provides a poor approximation of their evolutionary history. This inference is underlined by considering the fossil record of conodonts whose mineralized elements occur effectively continuously in all marine facies from Cambrian to Triassic; yet only a handful of fossils preserve their unmineralized anatomy, the oldest of which antedates the first appearance of their skeletal remains by some 60 million years. Fossil calibrations derived exclusively from taxa with poor preservation potential inevitably encompass greater uncertainty, typically described by broad prior probability densities, which provide less precise posterior age estimates. Intriguingly, Erwin et al. provided consistently younger divergence estimates for the major vertebrate nodes. This study encompasses a broad taxon sample of metazoan clades, some of which have a Precambrian fossil record, which effect to better constrain the age of deeper nodes within the tree and therefore better inform estimates of the age of vertebrate lineages, predicting a Late Ediacaran origin of crown Olfactores, a Middle Cambrian radiation of crown vertebrates and an earliest Devonian divergence of crown osteichthyan.

While there are some constraints on the timescale of early vertebrate evolution, our understanding of its tempo, in terms of the rate of phenotypic character evolution, remains crude. The genome of almost all living vertebrates preserves the phylogenetic footprint of at least two rounds of whole-genome duplication, once in the vertebrate stem lineage and again in the gnathostome stem lineage (or perhaps just twice in the vertebrate stem lineage); additionally, a third broadly inherited wholegenome duplication event occurred in the teleost stem lineage. The impact of these events upon phenotypic evolution is not clear; it has been argued that these events facilitated developmental and phenotypic evolution deterministically, or at least permissively; yet it remains possible that the phylogenetic footprint of genome duplication remains so clear because the gene regulatory controls on development were already sufficiently complex that their functions could be partitioned among duplicated genes, preventing their loss through drift (Donoghue and Purnell 2005). In this case, the fossil record indicates that in the assembly of the gnathostome bauplan, there was nothing unusual about the tempo and mode of phenotypic character evolution, as might be expected by the hypothesis of whole-genome duplication driving developmental evolution, although this may merely reflect the possibility that the first two whole-genome duplication events occurred within the vertebrate, and not the gnathostome, stem lineage. Thus, the evolutionary assembly of the gnathostome solution may be a permissive consequence of these events. Better constraints on genome evolution among chordates, provided by better assemblies of cyclostome genomes, are key to resolving this conundrum concerning the relationship between the evolution of phenotype and genotype.

It should go without saying that our knowledge of early vertebrate evolution relies not merely upon the anatomical interpretation of its fossil record but upon our understanding of the phylogenetic relationships among the fossil species, and among their living relatives. Early vertebrates were long an exemplar of phylogenetic rigour in animal systematics; indeed, they constituted the very battleground on which competing approaches to phylogenetic analysis were debated However, the scale of such phylogenetic studies has not extended much beyond these earliest incarnations, invariably considering either the internal relationships of component plesions or else the interrelationships among exemplars of plesions. Early vertebrates have yet to be subjected to rigorous tests of the monophyly of the plesions of stem gnathostomes recognized traditionally, which are constructs based on similarity. It is likely that these encompass both shared derived from shared primitive similarities.

Tests of the coherence of traditional plesions of jawed vertebrates have found placoderms and acanthodians wanting, now recognized as a series of successive sister lineages of stem gnathostomes and stem chondrichthyans, respectively. The putative clades of jawless stem gnathostomes have undergone no such revision and, indeed, where their monophyly has been tested, clades such as the thelodonts and anaspids have been found to be paraphyletic. The scale of the problem can, perhaps, be evaluated through a consensus network analysis of the trees derived from the study by Donoghue and Smith.

Consensus networks are a type of consensus tree that summarizes not only the common topologies present in the source trees (at a specified threshold level of occurrence), but also the conflicting topologies indicated by the data. Like regular cladograms, in consensus networks, the internal branches reflect relationships among the taxa and the nodes reflect possible common ancestors, and the branch weights reflect the number of input trees in which a given split is manifest. However, the consensus network also represents conflict in the data by also presenting conflicting relationships among the taxa. The more strictly bifurcating the cladogram, the less conflict present in the data. Donoghue and Smith recovered 37 most parsimonious trees through parsimony analysis of a data set that encompassed multiple representatives for each of the stem gnathostome plesions. A consensus network for splits with a weight threshold of 0.5 (i.e. including relationships supported by half or more of the source trees) is well resolved. However, when lower weight thresholds are considered, the strictly bifurcating nature of the consensus network breaks down, yielding increasingly net-like consensus trees at weight thresholds of 0.25 and 0.10. These indicate that, given the data, there is robust support for the relationships among living invertebrate chordates and vertebrates, for the monophyly and successive branching order of conodonts, anaspids, osteostracans and jawed vertebrates, but the relative relationships among heterostracans, thelodonts and galeaspids are less clear. Given the artificiality of the phena (higher taxa), the volume of missing data and the poor overlap in the classes of characters that can be coded for between living and fossil chordates, the degree of conflict in the phylogenetic relationships suggested by the data is concerning.

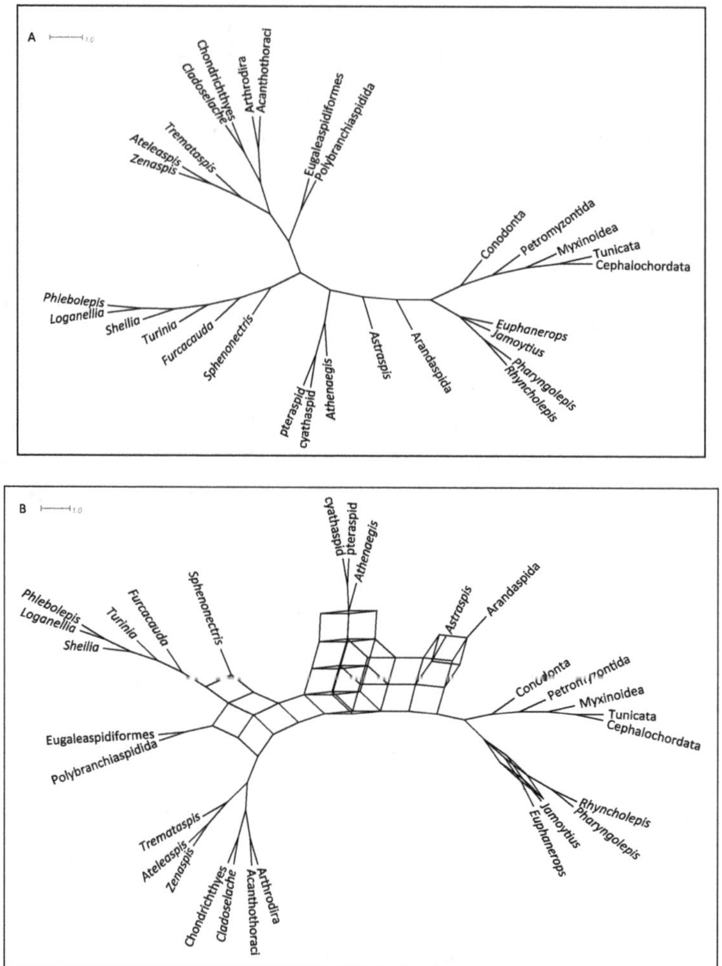

A serious attempt to test the monophyly of the stem gnathostome plesions could build on recent analyses to establish ingroup relationships within the Anaspida, Galeaspida, Heterostraci. However, there is a limit to the number of characters that can be extracted from the skeletons of these fossil fishes, particularly those that are currently resolved as branching early within the gnathostome stem, viz. anaspids, heterostracans and thelodonts, fossils of which comprise little more than scale-studded silhouettes. These data are not yet exhausted. There is little depth or breadth to our knowledge of dermoskeletal histology in any of the stem gnathostome plesions, and the existing data for groups such as the anaspids are very poorly understood. The external cranial anatomy of galeaspids has been well characterized, the detailed investigation of endocranial anatomy has but begun, and postcranial anatomy is virtually unknown. The pituriaspids, an apparently distinct plesion of stem gnathostomes, are known only from a small number of external and partial internal moulds. Nevertheless, it is likely that the core of the putative plesions of stem gnathostomes will be corroborated as clades, leaving a residue of lineages and taxa that remain phylogenetically unresolved for lack of data.

There is great room for improvement in our knowledge of the timing and tempo of early vertebrate evolution. The thorough phylogenetic review that we propose will serve as a foundation for a revised timescale that is based not merely on phylogenetically constrained stratigraphic occurrence but, rather, upon molecular divergence time estimates integrating both morphological and molecular data, and both fossil and living taxa. In this way, it is possible to date not merely the divergences among living and fossil species but, consequently, the rate of phenotypic character evolution and of diversification, facilitating more effective tests of the impact of whole-genome duplication in effecting the assembly of the vertebrate and gnathostome body plans and their ensuing taxic and ecological diversification.

The ecology of early vertebrates has been the topic of much speculation, based both on living and fossil species. For instance, the 'New Head Hypothesis' proposes that early vertebrate evolution is characterized by a trend towards increasingly active food acquisition. Yet the ecology of early vertebrates has been the subject of pitifully little rigorous attention. Conodonts are by far the best understood in terms of ecology, from functional analyses of their feeding elements using microwear, microstructural, occlusal kinematic and finite elements analyses, although they have been applied to few conodont species. Nevertheless, they establish methodological and experimental approaches that can be employed in studies of feeding ecology in other groups of early vertebrates, particularly the stem and crown gnathostomes. Indeed, microwear has been employed to test among hypotheses of micro- versus macrophagy in heterostracans; there is scope for further work on heterostracomorphs, and this approach could be extended readily to osteostracans, placoderms and acanthodians whose feeding plates are not uncommonly preserved. Similarly, disparity analyses of mandible characteristics have been employed to characterize the expansion in feeding ecologies during the initial radiation of jawed vertebrates. Speculative hypotheses of feeding function in these and other groups, including galeaspids, thelodonts and anaspids, can be tested using computed fluid dynamics. There is certainly no shortage of hypotheses in circulation, and the shift of focus from speculation based on vague analogy, to quantitative tests of these hypotheses, is long overdue.

Finally, although we raise serious concern about the validity of traditional groups of early vertebrates, stem gnathostomes most especially, the backbone of the relationships among heterostracans, galeaspids, osteostracans and jawed vertebrates will surely withstand all scrutiny because it is founded on the nested hierarchy of characters that evidence the evolutionary assembly of the gnathostome body plan. It is concerning, however, that although the fossil record evidences the acquisition of the fundamental characters that distinguish living jawed from jawless vertebrates, it, or our understanding of it, is so far silent on the evolutionary origin of jaws themselves. Illumination may yet be found through further anatomical and phylogenetic scrutiny of the fossils that occupy museum drawers, but the known facies-driven gaps in the fossil record of early vertebrate evolution along with phylogenetic–biogeographic inconsistencies, and the discovery of cryptic taxa like the pituriaspids, all combine to suggest that fundamental insights into this evolutionary episode reside in rock, awaiting discovery.

Ecology of Vertebrates Nutrition

Vertebrate Nutritional Needs

All vertebrates require certain amounts of energy, protein and 'essential' nutrients. Dietary sources of energy include carbohydrates, protein and fat. Dietary protein is necessary for satisfying an animal's protein requirements. Certain dietary nutrients are classified as 'essential' because they are required for synthesis or functioning of crucial chemical compounds, yet they cannot be synthesised by the organism in sufficient amounts to satisfy requirements. Thus, essential nutrients must come from diet. For most birds and mammals, about 38 nutrients are required in the diet including 10 essential amino acids, 1–3 essential fatty acids, 13 vitamins and at least 14 essential minerals.

Energy Requirements

The energy requirements of the animal and the energy density of the food primarily determine the amount of foodeaten by an animal. Energy requirements of an animal change to some extent with environmental temperature depending on the thermoregulatory capacity of the animal. Energy requirements increase with activity and during periods of production (e.g. growth, storage and reproduction). However, metabolism and energy use continues even when an animal is at rest. Resting metabolic rate of a bird or mammal can be predicted from its body mass using the following equations.

$$\text{Passerine birds}: Y = 114.8(X)^{0.726}$$

$$\text{NonPasserine birds}: Y = 73.5(X)^{0.734}$$

$$\text{Eutherian (or placeental) mammal}: Y = 57.2(X)^{0.716}$$

$$\text{Marsupial mammals}: Y = 46.6(X)^{0.737}$$

In the above equations, *Y* is resting metabolic rate in kilocalories per day, and *X* is body weight in kilograms. The similar exponents suggest a common allometric relationship between metabolic rate and body size in these vertebrates. The nutritional implication of this allometric relationship (with an exponent of about 0.72) is that larger animals have lower energy requirements per unit body weight than smaller animals. In addition, the higher value of the multiplication constant in birds compared to mammals demonstrates that birds have higher energetic costs for a given body size than mammals. For example, using the equations above, a 2kg nonpasserine bird such as a goose would require 122 kcal day^{-1} to satisfy the energy costs of resting metabolic rate compared to only 94 kcal day^{-1} (or about 25% less) for a similar-sized placental mammal such as a snowshoe hare.

The additional energy required to fuel the costs of thermoregulation, activity and production are also important in determining the total energy requirements of an animal. Each of these costs can be estimated separately and then summed to estimate total energy requirements for the animal, although most such estimates use captive animals in various degrees of restraint, which poses many problems. Recently, nutritional ecologists have used isotopes of hydrogen and oxygen to label the metabolic water of free living animals. Using the relative change over time in concentration of hydrogen and oxygen isotopes in the metabolic water of the animal, an estimate of the total energy requirements of a free-living animal is obtained (called 'field metabolic rate'). Studies using this so-called 'doubly labelled water method' have revolutionised our understanding of energy requirements of wild and captive animals.

Protein and Amino Acid Requirements

Proteins are amino acids bound together by peptide bonds. Dietary protein can be used as a source of body protein, fat or carbohydrate whereas dietary fat or carbohydrate alone cannot be used to increase the net amount of body protein. Hence, insufficient dietary protein has serious consequences because of inadequate synthesis of body protein. Of the 20 or so common amino acids, animals without significant fermentative capacity usually require 10 essential amino acids including arginine, histidine, isoleucine, leucine, threonine, lysine, methionine, phenylalanine, tryptophan and valine. Cats have lost the ability to synthesise taurine and so they also require this amino acid in their diet. Animals with significant fermentative capacity probably have similar amino acid requirements as non - fermenting animals. However, the protein produced by gut microbes during fermentation and subsequently digested and absorbed by the herbivore effectively eliminates the need for dietary sources of essential amino acids.

Protein requirements depend on the physiological state of the animal. For example, protein requirements are highest in growing animals and generally decline with age. Protein requirements of adults increase during productive periods such as egg-laying in birds or during lactation in mammals. The quantity of protein required by an animal also depends on the energy density of the diet and the amino acid composition of dietary protein. If there is a shortage of dietary fat or carbohydrate, then some protein must be used to satisfy energy requirements so that all dietary protein is not available to satisfy protein requirements of the animal. For animals without significant fermentative capacity, amino acid composition of dietary protein must provide a balanced profile of essential amino acids relative to need so that normal rates of protein synthesis are maintained. If dietary availability of an essential amino acid is deficient, then rate of protein synthesis will decline and other amino acids not used in protein synthesis will quickly be catabolised.

Carbohydrate and Fat Requirements

Carbohydrates and fats are the primary energy sources in most vertebrate diets. Dietary requirements for specific carbohydrates and fats are less exacting than for proteins. Carbohydrates are a biochemically diverse group of compounds that vary widely in how completely they can be digested by vertebrates. For example, the most abundant forms of carbohydrate in plants are cellulose and starch. Vertebrates lack cellulase, the digestive enzyme that digests cellulose, and so cellulose is either not utilised at all or must be digested by gut microbes during fermentation.

Vertebrate animals can grow normally on a diet without carbohydrates provided sufficient triglycerides are available for satisfying energy requirements. In contrast, a few specific fats are required in the diet of vertebrates. Three fatty acids (linoleic, linolenic and arachidonic acid) are usually categorised as essential for vertebrates, although current evidence suggests that linoleic acid is the only essential fatty acid in birds. Essential fatty acids are important as precursors for prostaglandin synthesis and as structural components in cell membranes.

Vitamins and Mineral Requirements

Vitamins and minerals are often referred to as micronutrients because they are required in small quantities compared to macronutrients such as protein, carbohydrates and fats. Animals require at least 14 minerals, including macroelements (e.g. calcium, phosphorus, sodium, potassium, magnesium, chlorine and sulfur), which are needed in relatively large amounts (milligrams per gram of food intake), and trace elements, which are needed in relatively small amounts (e.g. iron, zinc, manganese, copper, molybdenum, iodine, selenium, cobalt, fluoride and chromium).

Most vertebrates without significant fermentative capacity require 13 vitamins: fat-soluble vitamins include vitamins A, D, E and K; water-soluble vitamins include thiamin (B_1), riboflavin (B_2), pyridoxal (B_6), nicotinamide (B_7), pantothenic acid, biotin, folic acid, cyanocobalamin (B_{12}) and choline. Fat soluble vitamins can usually be stored in the animal so that deficiencies are extremely rare in free living individuals. Excesses of fat-soluble vitamins, especially vitamins A and D, can be extremely toxic. In contrast, most water-soluble vitamins cannot be stored and excesses are excreted, although some vitamin B_{12} and riboflavin can be stored in the liver. Thus, a regular dietary supply of water soluble vitamins is needed to satisfy vitamin requirements of vertebrates. Water-soluble vitamins primarily function as precursors for coenzymes. Deficiencies include growth retardation, dermatitis, poor feathering or fur and reduced reproductive output.

In addition to these 13 vitamins required of most vertebrates, vitamin C is required in the diet of many vertebrates including all fish, and some mammals (e.g. bats, guinea-pigs and some primates and cetaceans). Other vertebrates synthesise vitamin C in the kidney (e.g. amphibians, reptiles, monotremes and most nonpasserine birds) or in the liver (e.g. most mammals). About one-half of passerine birds can synthesise vitamin C in the liver whereas other passerine birds are unable to synthesise this vitamin and thus must satisfy requirements entirely from dietary sources.

Gut microbes synthesise some vitamins. For example, bacteria in the digestive tract of mammals synthesise vitamins K and B_{12}. Such symbiotic synthesis of vitamins is particularly important for foregut fermenting herbivores such as ruminants because it makes these animals independent of a dietary source of these vitamins. Although gut microbes in hindgut fermenting herbivores may also synthesise vitamins, coprophagy may be necessary for the herbivore to utilise these vitamins produced by microbes in the hindgut.

Procuring Food

Much of the daily activity of vertebrate animals is associated with obtaining an adequate quantity and quality of food. Vertebrates use a diversity of methods for obtaining food. The largest living vertebrate, the blue whale, and other plankton-eating whales use filtering devices called baleens to filter large volumes of water and capture the small plankton in hair-like filaments on horny plates. Other filter-feeding vertebrates include many pelagic fish (e.g. herring and mackerel), large sharks (e.g. basking and whale sharks) and even a few birds (e.g. petrels, flamingos and many ducks).

The majority of vertebrates procure food that is much larger than plankton and these larger masses of food must be captured and manipulated differently from small particles such as plankton. Most mammals mechanically break down food using teeth, whereas birds have no teeth and use a specialised portion of their stomach, the gizzard, for this purpose. Bill morphology in birds, and tooth and jaw structure in all other vertebrate classes are impressively diverse and provide excellent examples of evolutionary design in relation to procuring food. For example, dagger like bills of herons capture fast-moving aquatic prey, serrated bills of mergansers seize and hold fish, spatula like bills of spoonbills and shovellers filter small aquatic invertebrates, raptorial bills of falcons, hawks and eagles shred prey, chisel-shaped bills of woodpeckers uncover wood boring insects and conical bills of many songbirds crack open seeds. Adaptations of the teeth, jaws and jaw musculature are closely related to the specific styles of feeding in nonavian vertebrates and especially in mammals. For example, the structure of molar teeth includes blade-like molars in carnivores that shear, molars with rounded cusps in omnivores that crush and grind and molars with ridged cusps in herbivores that finely section and grind plants. Similarly, chisel-like incisors in rodents and rabbits are used for gnawing, dagger-like canines of most carnivorous mammals are used for piercing and tearing food. The arrangement and types of teeth in mammals allow them to masticate or chew their food, an ability relatively unique among vertebrates.

Muscular, Acidic Stomach in Macrophagous Food Processing

Most vertebrate food consists of proteins, fats and carbohydrates, which are large molecules that must be broken down into simpler units so they can be absorbed and utilised by the vertebrate (Karasov and Martinez del Rio, In vertebrates, this breakdown takes place primarily in the stomach and midgut using both mechanical and chemical processes.

Mechanical processes in the stomach involve contraction of circular and longitudinal muscle layers in the stomach wall that mixes the food, saliva and stomach secretions. The stomach initiates chemical digestion of protein by secreting enzymes such as pepsinogen and hydrochloric acid. The stomach of most birds is divided into a proventriculus, where chemical digestion is initiated, and a gizzard, where ingested food is mechanically mixed and ground (replacing the function of teeth). Often small, hard objects such as pebbles are swallowed by herbivorous birds to aid in the grinding of food in the gizzard.

Specificity of Vertebrate Digestive Enzymes

Before protein, fats and carbohydrates can be absorbed and utilised, they must be broken down into simpler building blocks of amino acids, esters of fatty acids and glycerol, and simple sugars (e.g. glucose, sucrose), respectively. In vertebrates, the chemical breakdown of ingested food involves digestive enzymes that increase the rate of hydrolysis, adding H^+ to one part of the large food molecule and OH^- to the other part. Hydrolysis of chemical bonds divides the larger food molecules into their simpler building blocks that can then be absorbed.

Digestive enzymes are proteins produced by cells of the stomach, intestine and pancreas that regulate the rate of chemical breakdown of food compounds. Like all enzymes, digestive enzymes are to some degree specific for a certain substrate or reactant molecule, and the activity of digestive enzymes is sensitive especially to temperature and pH. Some digestive enzymes hydrolyse certain chemical bonds regardless of their location in a molecule and so may act on many different substrates. For example, trypsin and chymotrypsin catalyse the hydrolysis of peptide bonds in any protein in which the carboxyl group is part of certain amino acids. Other digestive enzymes are much more specific and hydrolyse certain chemical bonds in one location in certain molecules. For example, sucrase (sucrose a-glucosidase) catalyses hydrolysis of the disaccharide sucrose into glucose and fructose, and cannot hydrolyse the same chemical bond in other disaccharides such as lactose and maltose (these disaccharides are digested by their own specific enzymes, lactase (ß – D –galactosidase) and maltase (a-glucosidase).

Proteases

Each of the three major types of foodstuffs has a corresponding group of enzymes associated with its digestion: proteases for digestion of protein, carbohydrases for digestion of carbohydrates and lipases for digestion of fats. Proteases include endopeptidases that catalyse hydrolysis of peptide bonds located well within the protein molecule, and exopeptidases that catalyse hydrolysis of peptide bonds near the end of a peptide chain. Thus, digestion of food protein involves endopeptidases such as pepsin, trypsin and chymotrypsin, which break large peptide chains into shorter polypeptide segments, and then exopeptidases such as carboxypeptidase and aminopeptidase, which break down the shorter polypeptide segments into free amino acids and di- or tripeptides.

To avoid digestion of their own body protein, vertebrates produce many proteolytic enzymes as zymogens, inactive forms of the digestive enzyme that become active at specific (usually low) pH or by interaction with another specific enzyme. For example, protein digestion usually begins in the stomach with secretion of pepsinogen, which is later converted to pepsin in the highly acidic environment of the stomach.

Carbohydrases

Carbohydrases include polysaccharidases (e.g. amylases), which hydrolyse the glycosidic bonds of long-chain carbohydrates such as starch, and glycosidases (e.g. sucrase, lactase and maltase) which hydrolyse a specific glycosidic bond in certain disaccharidases. Amylase is secreted by the vertebrate pancreas and in the saliva and is located primarily in the stomach. In contrast, most glycosidases are located in the intestinal mucosa.

Lipases

Lipases, which are responsible for most of the chemical digestion of fats, act only at a lipid–water interface. Fats are water-insoluble and are rendered water-soluble by emulsification, a process involving the mixing of fat with detergents such as bile salts, secreted by the liver, and lecithin to form small droplets called micelles. Formation of micelles is necessary for fat digestion because it increases the surface area available for pancreatic lipase digestion. Lipases secreted by the pancreas break down triglycerides into fatty acids, monoglycerides and diglycerides that can be absorbed by the intestine.

Modulation of Digestive Enzyme Activity in Relation to Diet

Activity of digestive enzymes in vertebrates, including pancreatic proteases, amylase and lipase as well as intestinal aminopeptidase, sucrase and maltase, generally changes in proportion to the amount of dietary substrate (i.e. protein-digesting enzymes increase with their respective dietary protein substrate, carbohydrate-digesting enzymes increase with their respective dietary carbohydrate substrate). Although this generalisation holds for laboratory rats, a fish (Tilapia mossambica), and poultry, the few wild birds studied to date exhibit somewhat different patterns of modulation in digestive enzymes. Specifically, wild birds fed diets with higher carbohydrate concentrations did not increase their digestive disaccharidases, whereas birds fed diets with higher protein concentrations increased their aminopeptidase-N activity. This pattern of modulation of digestive peptidases but not disaccharidases holds across bird species that are dietary generalists and specialists, although more comparative studies are needed to determine if dietary flexibility is generally unrelated to this type of digestive plasticity.

Nutrient Absorption

Chemical digestion of proteins, fats and carbohydrates along with absorption of the end products of digestion primarily occurs in the small intestine. In the anterior small intestine (the duodenum) secretions from the liver and pancreas are delivered. In the posterior small intestine (the jejunum and ileum) digestion is completed and absorption takes precedence. Products of digestion are absorbed across the intestinal microvilli of the apical membrane of the absorptive cells and then move via blood circulation into the animal's tissues and cells.

Mechanisms of Absorption

Absorption involves several processes including passive diffusion, facilitated diffusion and active transport. Fat-soluble nutrients such as fatty acids and monoglycerides passively diffuse across the lipid bilayer intestinal membrane. Once fatty acids and glycerol are inside the absorptive cell they are reconstructed into triglycerides, collected together with other cellular lipids to form tiny droplets called chylomicrons and then exported to the blood by exocytosis. Water and small, water-soluble nutrients such as certain sugars and alcohols passively diffuse through water-filled pores in the intestinal membrane. Such passive diffusion of nutrients uses the energy of the concentration gradient and requires no cellular metabolic energy. Although all nutrients are probably absorbed passively to some extent, many nutrients are absorbed against a concentration gradient and this process, called active transport, requires cellular metabolic energy. Certain sugars and amino acids are actively transported across the intestinal membrane.

Modulation of Nutrient Absorption Rate in Relation to Diet

Absorption rate of dietary nutrients in the intestine of vertebrates provides strong evidence for adaptive modulation of digestive features in response to diet change. Nutrient transport rates for glucose, vitamins and amino acids increases with dietary substrate in mammals. However, no modulation of nutrient transport rates for glucose has been observed in the four species of bird studied to date. The absence of modulation of mediated glucose transport in birds may occur because birds rely less on active transport for absorption of glucose and more on passive absorption of glucose.

Microbial Symbionts in Digestion

Cellulose is the most abundant compound in the biosphere and is the primary structural material in plants. However, no vertebrate produces cellulose-digesting enzymes (called cellulases). Digestion of cellulose and other complex carbohydrates in plants by vertebrates is accomplished by microbes (primarily bacteria and some protozoa). The evolution of enlarged portions of the gut to house symbiotic microbes was crucial in the radiation of herbivorous mammals because it provided the appropriate conditions for microbial fermentation to have significant nutritional benefits. Digestion of cellulose is only one of many nutritionally important functions of gut microbes. Microbes ferment dietary carbohydrate and hydrolyse dietary lipids into absorbable nutrients, use nitrogenous compounds including dietary protein to synthesise microbial protein and synthesise vitamins.

Fermentation Sites in the Gut

Fermentation of plant material occurs in the hindgut and/or foregut of vertebrate herbivores. Large (>50 kg) hindgut fermenters including horses, rhinos and elephants typically have enlarged colons, whereas small (>10 kg) hindgut fermenters including rabbits, most rodents, some monkeys and most herbivorous birds and reptiles typically have enlarged caeca where plant material is microbially digested. Marsupial herbivores include groups such as the kangaroos and wallabies (Macropodidae), which have their primary site for microbial fermentation in complex, large stomachs, but also include groups such as the wombats (Vombatidae), brushtail possums (Phalangeridae), koala (Phascolarctidae) and gliders (Pseudocheiridae), which have their primary site for microbial fermentation in the hindgut (i.e. colon, caeca or both).

Fermentation of plant material in the foregut occurs in enlarged compartment(s) anterior to the true digestive stomach, although these structures are often referred to collectively as a multichambered stomach. Mammals in the suborder Ruminantia (e.g. deer, elk, giraffe, bison, sheep and cattle) and in the order Tylopoda (camels, llama, alpaca and vicuana) have multichambered stomachs. The hoatzin (Opisthocomus hoatzin) is the only bird in which portions of the foregut are enlarged and primarily used for fermentation of plant material.

Nutritional Importance of Fermentation and Rumination

In vertebrates with multichambered stomachs, fermentation occurs in the first division of the stomach (consisting of the rumen and reticulum in ruminants) and involves the anaerobic conversion of carbohydrates to short-chain fatty acids (SCFAs). SCFAs and amino acids produced by fermentation are absorbed into the bloodstream from the rumen fluid. Partially digested food from the first division of the stomach is regurgitated, chewed and then swallowed and passed into the second division of the stomach (consisting of the omasum and abomasum in ruminants). Hydrolysis and enzymatic digestion of the ruminated food (especially protein in the microbes) occurs in this second division of the stomach. Foregut fermenters that ruminate maximise the breakdown and absorption of nutrients from food because the undigested particles can be regurgitated and remasticated until sufficiently broken down.

Foregut fermentation allows the products of fermentation to be efficiently digested and absorbed because the fermentation site is anterior to the intestine where digestion and absorption occurs. SCFAs produced by microbes are the primary source of energy for ruminants (contributing as much as 80% of maintenance energy requirements (Stevens and Hume, 1995)). Digestion of microbial protein provides an important source of amino acids for ruminants. Microbes can synthesise protein from inorganic nitrogen compounds and from urea, a normal waste product of protein metabolism in the liver of herbivores. In many foregut-fermenting animals and especially in ruminants, urea reenters the foregut directly and in part via saliva and is resynthesised into microbial protein. Recycling of urea is especially important for herbivorous animals eating food with low protein.

Foregut fermenters, in general, and ruminants, in particular, are more efficient at utilising dietary nutrients than hindgut fermenters when feeding on moderately fibrous diets because of their ability to maximise digestion of plant material. Foregut fermenters also have an advantage over hindgut fermenters when eating low-protein or low-quality protein foods because they are better able to recycle urea nitrogen and digest microbial protein that usually has all the essential amino acids.

Since fermentation in the hindgut occurs posterior to the primary site of absorption (i.e. the intestine), hydrolysis of microbial protein and absorption of the amino acids is relatively less efficient in hindgut compared to foregut fermenters. Many small hindgut fermenting vertebrates (e.g. rodents, rabbits and hares), however, produce a special kind of faeces from the contents of their caecum and then ingest this faeces (called coprophagy). Coprophagy is nutritionally important because it allows recovery of otherwise lost protein, vitamins and other fermentation products. The nutritional significance of coprophagy is evident when animals are prevented from ingesting caecal faeces. Coprophagic animals that are prevented from ingesting caecal faeces develop vitamin deficiencies (especially vitamin K) and utilise dietary protein and fibre less efficiently.

If primarily high fibre forage is available, then hindgut fermenting herbivores are favoured over foregut-fermenting herbivores because they can increase their intake and passage rate and so obtain adequate nutrients despite the fast passage rates and large amounts of undigested residue. Foregut-fermenting herbivores have difficulty eating high fibre forage because they selectively retain the large fibre particles in the foregut for so long that the rate of delivery of fermentation products is too slow. If primarily low fibre forage is available, then direct digestion of food is usually more efficient than fermentative digestion.

Diet Diversity and the Evolution of Phenotypic Plasticity

Physiological adaptation can be deduced from comparative studies of physiological traits (e.g. organ size and function, nutrient transport rates and metabolic rate) in vertebrate taxa with different life styles. However, many such physiological traits exhibit considerable variability in ecological time both within and among individuals, which may make it difficult to detect important adapted patterns. The form of phenotypic variation that involves a single genotype producing different phenotypes in response to variation in some environmental variable is called 'phenotypic plasticity'. Rapid reversible changes in organ size and digestive processes provide examples of phenotypic flexibility in that they may flexibly respond to changes in diet. Such phenotypic flexibility in physiological traits may itself be a critical component of the adaptive repertoire of animals that may influence diet diversity, niche width, feeding rate, and thus the acquisition of energy and essential nutrients.

CLASSES IN VERTEBRATES

Fishes

Fishes constitute the non-tetrapod vertebrates, meaning that they do not have four legs or limbs, although they generally have two pairs of fins and several unpaired fins that allow them to move efficiently through water. They are almost all entirely aquatic, although there are a few species that can survive out of the water for short periods of time. There are approximately 28,000 fish species subdivided into five different classes. Examples of different fish species are shown in figure.

The fishes include species that are common human food sources such as salmon, trout, bass, and even sharks and rays. Their habitats cover all aquatic regions on Earth including oceans, rivers, lakes, and other freshwater sources. Most fishes use gills to obtain oxygen from their aquatic environments and have scales covering their bodies. Although the majority of fish species are jawed fish, hagfish of the class Myxini and lampreys of the class Hyperoartia are jawless. The classes of Myxini, Hyperoartia, and Chondrichthyes all possess endoskeletons that consist of cartilage. Fishes in the classes Actinopterygii, and Sarcopterygii are considered the "bony fishes" because they have endoskeletons consisting primarily of bone.

Amphibians

Amphibians are cold-blooded (ectothermic) vertebrate tetrapods that generally respire air as adults but usually require an aquatic (freshwater) environment for reproduction. The term tetrapod, refers to four-legged/limbed animals that generally have two pairs of limbs: one pair on the anterior, or front, end of the animal and one pair on the posterior, or tail, end of the animal. Other vertebrate tetrapods include reptiles, birds, and mammals. There are roughly 6,000 existing amphibian species. They include frogs, toads, and salamanders, and several examples are shown in figure.

Almost all amphibian species produce eggs that cannot survive outside of water, and this is the reason for their reproductive dependence on an aquatic environment. Amphibians usually undergo indirect development via aquatic larval forms called tadpoles or polliwogs that respire using gills. During metamorphosis, the larvae develop limbs and replace their gills with lungs to become semi-terrestrial adult animals.

Reptiles

Reptiles are also cold-blooded (ectothermic) vertebrate tetrapods, but they differ from amphibians in a number of important ways. Unlike amphibians, reptiles are amniotes that are not dependent on water in order to reproduce. Amniotes lay eggs that are surrounded by an extensive amniotic membrane to facilitate gas exchange and waste processing in a terrestrial environment. A solid eggshell surrounds the membrane-enclosed embryos and provides protection to the developing organism. Another difference between reptiles and amphibians is that reptiles generally undergo direct development that does not include an aquatic larval stage. All reptiles use lungs to breath, although there are some species that are also able to absorb oxygen through their skin.

All of these evolutionary developments have allowed reptiles to live an exclusively terrestrial existence. Despite these adaptations, there are some species that spend most of their lives in water such as aquatic turtles. In these species, the limbs have been modified to have a more fin-like structure for swimming.

There are approximately 8,000 species of reptiles living today that occupy all continents on Earth except Antarctica. These species include lizards, snakes, and turtles, and examples are shown in figure below. There are some species of reptiles, such as snakes, that have lost their limbs during the course of evolution. However, they are still considered tetrapods because they evolved from a tetrapod ancestor.

The diversity of reptile species. (a) Caimen lizard. (b) Florida redbelly turtle. (c) Texas coral snake.

Birds

Birds are warm-blooded (endothermic), egg-laying vertebrate tetrapods. They are characterized by the presence of feathers and wings. Almost all birds are capable of flight. The few species that are not, such as ostriches and penguins, evolved from flying ancestor species but lost the capability over the course of their evolution. There are approximately 10,000 existing species of birds that live in nearly all habitats on Earth. This is the largest number of species of any of the tetrapod classes. Because birds are endothermic, they are not dependent on environmental temperatures to maintain both their internal body temperature and activity levels; their ectothermal tetrapod counterparts, the reptiles and amphibians, are dependent on the environmental temperatures. This probably explains birds' vast range over the planet, including both extremely cold and extremely warm regions such as the North and South poles as well as desert climates. Examples of several different species of birds are shown in figure below.

Mammals

Mammals are warm-blooded (endothermic) vertebrates that differ in many ways from members of other vertebrate classes. Mammals have hair on their skin, three bones in the middle ear, a four-chambered heart, and a region of the brain called a neocortex that is involved in higher neurological functions such as spatial reasoning, conscious thought, and, in some species (ours), language. There are roughly 5,400 species of mammals including our species, *Homo sapiens*. Figure below shows a number of examples of mammalian species.

Mammals derive their name from the fact that they produce milk to feed their young using mammary glands. Nearly all mammals are viviparous and give birth to live young, although there is one group, the monotremes, that lays eggs, which undergo external development.

- Vertebrates can be subdivided into five major groups: fishes, amphibians, reptiles, birds, and mammals.
- Fishes are almost all entirely aquatic, although there are a few species that can survive out of water for short periods of time.
- Amphibians are ectothermic tetrapod vertebrate that generally respire air as adults but usually require an aquatic (freshwater) environment for reproduction.
- Reptiles are amniotes that are not dependent on water in order to reproduce.
- Birds are endothermic, egg-laying tetrapod vertebrates. They are characterized by the presence of feathers and wings.
- Mammals have hair on their skin, three bones in the middle ear, a four-chambered heart, and a region of the brain called a neocortex.

Various species of mammals. (a) Bottlenose dolphin. (b) Brown rat. (c) Baboon. Notice the extreme diversity in the different body forms of mammals. Mammals are almost all terrestrial species but also include a few aquatic species such as dophins.

References

- Vvertebrate-zoology, research: naturalhistory.si.edu, Retrieved 31 July, 2019
- Vertebrate, animal: britannica.com, Retrieved 10 May, 2019
- Ecology-of-Vertebrate-Nutrition-23208498: academia.edu, Retrieved 02 January, 2019
- Molecular phylogeny and divergence times of deuterostome animals. Molecular Biology and Evolution, 22, 2275–2284
- Two rounds of whole genome duplication in the ancestral vertebrate. PLoS Biology, 3, e314

Phylum Protochordata

CHAPTER 2

Organisms which are considered as lower chordates fall under the category of protochordata. They are also called Acraniata as they don't possess a true skull. They are classified as hemichordata, urochordata and cephalochordata. The topics elaborated in this chapter will help in gaining a better perspective about phylum protochordata.

The organisms belonging to the Protochordata are generally known as the lower chordates. They don't form a "proper" taxonomic group and are only classified as such for convenience purposes. However, they do form a major division of Chordata. They are also known as Acraniata because they lack a true skull. They are divided into three sub-phyla- Hemichordata, Urochordata, Cephalochordata.

Characteristics of Protochordata

- They are generally found in marine water.
- Their body is bilaterally symmetrical, triploblastic, and coelomated.
- At a certain stage of their lives, their body develops a long, rod-like structure for support called notochord.
- They exhibit organ system level of organization.
- Eg., Herdmania, Amphioxus.

ANATOMY AND CLASSIFICATION

Anatomy

The glass catfish (*Kryptopterus vitreolus*) is one of the few chordates with a visible backbone. The spinal cord is housed within its backbone.

Chordates form a phylum of animals that are defined by having at some stage in their lives all of the following anatomical features:

- A notochord, a fairly stiff rod of cartilage that extends along the inside of the body. Among the vertebrate sub-group of chordates the notochord develops into the spine, and in wholly aquatic species this helps the animal to swim by flexing its tail.
- A dorsal neural tube. In fish and other vertebrates, this develops into the spinal cord, the main communications trunk of the nervous system.
- Pharyngeal slits. The pharynx is the part of the throat immediately behind the mouth. In fish, the slits are modified to form gills, but in some other chordates they are part of a filter-feeding system that extracts particles of food from the water in which the animals live.
- Post-anal tail. A muscular tail that extends backwards behind the anus.
- An endostyle. This is a groove in the ventral wall of the pharynx. In filter-feeding species it produces mucus to gather food particles, which helps in transporting food to the esophagus. It also stores iodine, and may be a precursor of the vertebrate thyroid gland.

There are soft constraints that separate chordates from certain other biological lineages, but are not part of the formal definition:

- All chordates are deuterostomes. This means that, during the embryo development stage, the anus forms before the mouth.
- All chordates are based on a bilateral body plan.
- All chordates are coelomates, and have a fluid filled body cavity called a coelom with a complete lining called peritoneum derived from mesoderm.

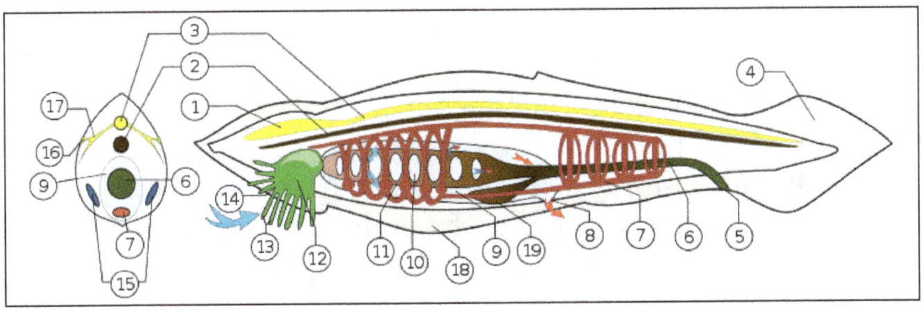

Anatomy of the cephalochordate *Amphioxus*.

In the figure above:

- Bulge in spinal cord ("brain").
- Notochord.
- Dorsal nerve cord.
- Post-anal tail.
- Anus.

- Digestive canal.
- Circulatory system.
- Atriopore.
- Space above pharynx.
- Pharyngeal slit (gill).
- Pharynx.
- Vestibule.
- Oral cirri.
- Mouth opening.
- Gonads (ovary / testicle).
- Light sensor.
- Nerves.
- Metapleural fold.
- Hepatic caecum (liver-like sack).

SUBPHYLA OF CHORDATES

Craniate

A craniate is a member of the Craniata (sometimes called the Craniota), a proposed clade of chordate animals with a skull of hard bone or cartilage. Living representatives are the Myxini (hagfishes), Hyperoartia (including lampreys), and the much more numerous Gnathostomata (jawed vertebrates). Formerly distinct from vertebrates by excluding hagfish, molecular and anatomical research in the 21st century has led to the reinclusion of hagfish, making living craniates synonymous with living vertebrates.

The clade was conceived largely on the basis of the Hyperoartia (lampreys and kin) being more closely related to the Gnathostomata (jawed vertebrates) than the Myxini (hagfishes). This, combined with an apparent lack of vertebral elements within the Myxini, suggested that the Myxini were descended from a more ancient lineage than the vertebrates, and that the skull developed before the vertebral column. The clade was thus composed of the Myxini and the vertebrates, and any extinct chordates with skulls.

However recent studies using molecular phylogenetics has contradicted this view, with evidence that the Cyclostomata (Hyperoartia and Myxini) is monophyletic; this suggests that the Myxini are degenerate vertebrates, and therefore the vertebrates and craniates are cladistically equivalent, at least for the living representatives. The placement of the Myxini within the vertebrates has been further strengthened by recent anatomical analysis, with vestiges of a vertebral column being discovered in the Myxini.

Characteristics

In the simplest sense, craniates are chordates with well-defined heads, thus excluding members of the chordate subphyla Tunicata (tunicates) and Cephalochordata (lancelets), but including Myxini, which have cartilaginous skulls and tooth-like structures composed of keratin. Craniata also includes all lampreys and armoured jawless fishes, armoured fish, sharks, skates, and rays, and teleostomians: spiny sharks, bony fish, lissamphibians, temnospondyls and protoreptiles, sauropsids and mammals. The craniate head consists of a brain, sense organs, including eyes, and a skull.

In addition to distinct crania (sing. *cranium*), craniates possess many derived characteristics, which have allowed for more complexity to follow. Molecular-genetic analysis of craniates reveals that, compared to less complex animals, they developed duplicate sets of many gene families that are involved in cell signaling, transcription, and morphogenesis.

In general, craniates are much more active than tunicates and lancelets and, as a result, have greater metabolic demands, as well as several anatomical adaptations. Aquatic craniates have gill slits, which are connected to muscles and nerves that pump water through the slits, engaging in both feeding and gas exchange (as opposed to lancelets, whose pharyngeal slits are used only for suspension feeding). Muscles line the alimentary canal, moving food through the canal, allowing higher craniates such as mammals to develop more complex digestive systems for optimal food processing. Craniates have cardiovascular systems that include a heart with two or more chambers, red blood cells, and oxygen transporting hemoglobin, as well as kidneys.

Systematics and Taxonomy

Linnaeus used the terms *Craniata* and *Vertebrata* interchangeably to include lampreys, jawed fishes, and terrestrial vertebrates (or tetrapods). Hagfishes were classified as Vermes, possibly representing a transitional form between 'worms' and fishes.

Dumeril grouped hagfishes and lampreys in the taxon *Cyclostomi*, characterized by horny teeth borne on a tongue-like apparatus, a large notochord as adults, and pouch-shaped gills (Marspibranchii). Cyclostomes were regarded as either degenerate cartilaginous fishes or primitive vertebrates. Cope coined the name *Agnatha* ("jawless") for a group that included the cyclostomes and a number of fossil groups in which jaws could not be observed. Vertebrates were subsequently divided into two major sister-groups: the Agnatha and the Gnathostomata (jawed vertebrates). Stensiö suggested that the two groups of living agnathans (i.e. the cyclostomes) arose independently from different groups of fossil agnathans.

Løvtrup argued that lampreys are more closely related to gnathostomes based on a number of uniquely derived characters, including:

- Arcualia (serially arranged paired cartilages above the notochord).
- Extrinsic eyeball muscles.

- Radial muscles in the fins.
- A closely set atrium and ventricle of the heart.
- Nervous regulation of the heart by the vagus nerve.
- A typhlosole (a spirally coiled valve of the intestinal wall).
- True lymphocytes.
- A differentiated anterior lobe of the pituitary gland (adenohypophysis).
- Three inner ear maculae (patches of acceleration sensitive 'hair cells' used in balance) organized into two or three vertical semicircular canals.
- Neuromast organs (composed of vibration sensitive hair cells) in the laterosensory canals.
- An electroreceptive lateral line (with voltage sensitive hair cells).
- Electrosensory lateral line nerves.
- A cerebellum, i.e. the multi-layered roof of the hindbrain with unique structure (characteristic neural architecture including direct inputs from the lateral line and large output Purkinje cells) and function (integrating sensory perception and coordinating motor control).

In other words, the cyclostome characteristics (e.g. horny teeth on a "tongue", gill pouches) are either instances of convergent evolution for feeding and gill ventilation in animals with an eel-like body shape, or represent primitive craniate characteristics subsequently lost or modified in gnathostomes. On this basis Janvier proposed to use the names *Vertebrata* and *Craniata* as two distinct and nested taxa.

Validity

The validity of the taxon "Craniata" was recently examined by using mtDNA sequence data, concluding that Myxini is more closely related to Hyperoartia than to Gnathostomata - i.e., that modern jawless fishes form a clade called Cyclostomata. The argument is that, if Cyclostomata is indeed monophyletic, Vertebrata would return to its old content (Gnathostomata + Cyclostomata) and the name Craniata, being superfluous, would become a junior synonym.

The new evidence removes support for the hypothesis for the evolutionary sequence by which (from among tunicate-like chordates) first the hard cranium arose as it is exhibited by the hagfishes, then the backbone as exhibited by the lampreys, and then finally the hinged jaw that is now ubiquitous. In 2010, Philippe Janvier stated:

> "Although I was among the early supporters of vertebrate paraphyly, I am impressed by the evidence provided by Heimberg et al. and prepared to admit that cyclostomes are, in fact, monophyletic. The consequence is that they may tell us little, if anything, about the dawn of vertebrate evolution, except that the intuitions of 19th century zoologists were correct in assuming that these odd vertebrates (notably, hagfishes) are strongly degenerate and have lost many characters over time".

Tunicate

A tunicate is a marine invertebrate animal, a member of the subphylum Tunicata. It is part of the Chordata, a phylum which includes all animals with dorsal nerve cords and notochords. The subphylum was at one time called Urochordata, and the term urochordates is still sometimes used for these animals. They are the only chordates that have lost their myomeric segmentation, with the possible exception of the seriation of the gill slits.

Some tunicates live as solitary individuals, but others replicate by budding and become colonies, each unit being known as a zooid. They are marine filter feeders with a water-filled, sac-like body structure and two tubular openings, known as siphons, through which they draw in and expel water. During their respiration and feeding, they take in water through the incurrent (or inhalant) siphon and expel the filtered water through the excurrent (or exhalant) siphon. Most adult tunicates are sessile, immobile and permanently attached to rocks or other hard surfaces on the ocean floor; others, such as salps, doliolids and pyrosomes, swim in the pelagic zone of the sea as adults.

Various species of the subphylum tunicata are commonly known as ascidians, tunicates, sea pork, sea livers, or sea tulips.

The earliest probable species of tunicate appears in the fossil record in the early Cambrian period. Despite their simple appearance and very different adult form, their close relationship to the vertebrates is evidenced by the fact that during their mobile larval stage, they possess a notochord or stiffening rod and resemble a tadpole. Their name derives from their unique outer covering or "tunic", which is formed from proteins and carbohydrates, and acts as an exoskeleton. In some species, it is thin, translucent, and gelatinous, while in others it is thick, tough, and stiff.

Taxonomy

Clavelina moluccensis, the bluebell tunicate.

Botrylloides violaceus showing oral tentacles at openings of buccal siphons.

About 2,150 species of tunicate exist in the world's oceans, living mostly in shallow water. The most numerous group is the ascidians; fewer than 100 species of these are found at depths greater than 200 m (660 ft). Some are solitary animals leading a sessile existence attached to the seabed, but others are colonial and a few are pelagic. Some are supported by a stalk, but most are attached directly to a substrate, which may be a rock, shell, coral, seaweed, mangrove root, dock, piling, or ship's hull. They are found in a range of solid or translucent colours and may resemble seeds, grapes, peaches, barrels, or bottles. One of the largest is a stalked sea tulip, *Pyura pachydermatina*, which can grow to be over 1 metre (3.3 ft) tall.

The Tunicata were established by Jean-Baptiste Lamarck in 1816. In 1881, Francis Maitland Balfour introduced a second name for the same group, "Urochorda", to emphasize the affinity of the group to other chordates. No doubt largely because of his influence, various authors supported the term, either as such, or as "Urochordata", but this usage is invalid because "Tunicata" has precedence, and grounds for superseding the name never existed. Accordingly, the current (formally correct) trend is to abandon the name Urochorda or Urochordata in favour of the original Tunicata, and the name Tunicata is almost invariably used in modern scientific works. It is accepted as valid by the World Register of Marine Species but not by the Integrated Taxonomic Information System.

Various common names are used for different species. Sea tulips are tunicates with colourful bodies supported on slender stalks. Sea squirts are so named because of their habit of contracting their bodies sharply and squirting out water when disturbed. Sea liver and sea pork get their names from the resemblance of their dead colonies to pieces of meat.

Classification

Tunicates are more closely related to craniates, (including hagfish, lampreys, and jawed vertebrates) than to lancelets, echinoderms, hemichordates, *Xenoturbella* or other invertebrates.

The clade consisting of tunicates and vertebrates is called Olfactores.

The Tunicata contain roughly 3,051 described species, traditionally divided into these classes:

- Ascidiacea (Aplousobranchia, Phlebobranchia, and Stolidobranchia).

- Thaliacea (Pyrosomida, Doliolida, and Salpida).
- Appendicularia (Larvacea).

Members of the Sorberacea were included in Ascidiacea in 2011 as a result of rDNA sequencing studies. Although the traditional classification is provisionally accepted, newer evidence suggests the Ascidiacea are an artificial group of paraphyletic status.

The following cladogram is based on the phylogenomic study of Delsuc and colleagues.

Tuncicata

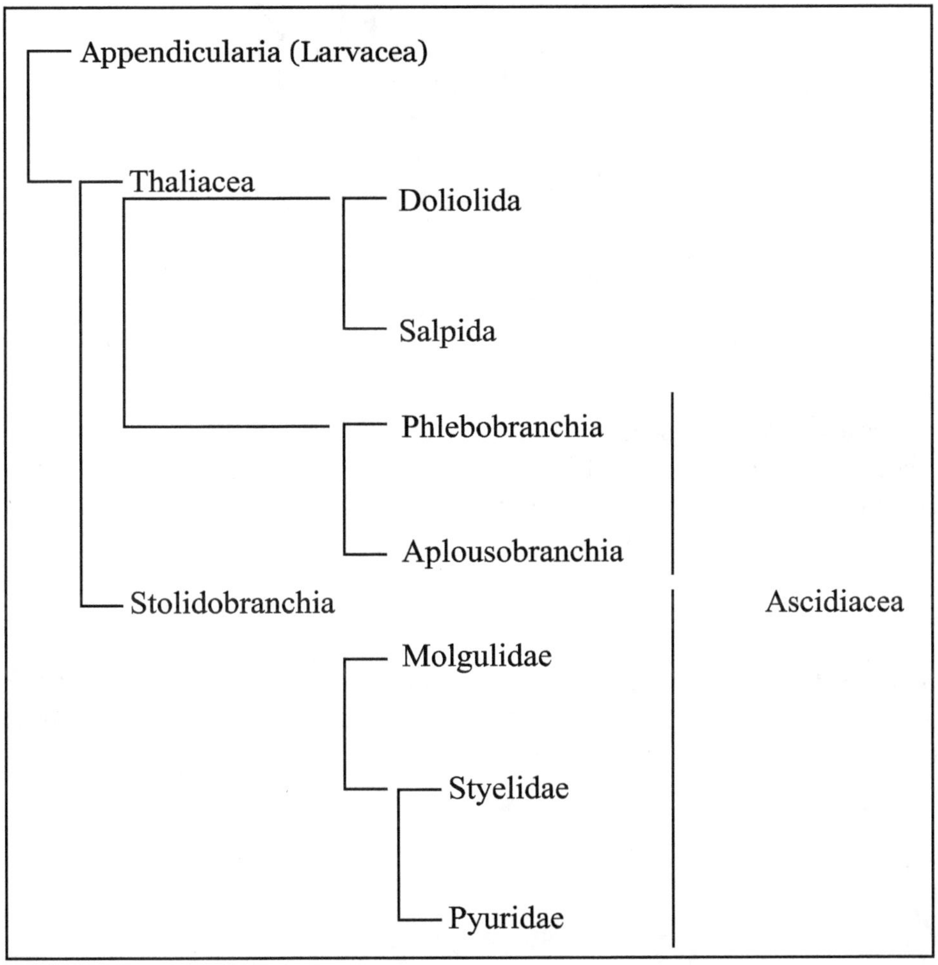

Fossil Record

Undisputed fossils of tunicates are rare. The best known and earliest unequivocally identified species is *Shankouclava shankouense* from the Lower Cambrian Maotianshan Shale at Shankou village, Anning, near Kunming (South China). There is also a common bioimmuration, (*Catellocaula vallata*), of a possible tunicate found in Upper Ordovician bryozoan skeletons of the upper midwestern United States.

Three enigmatic species were also found from the Ediacaran period – *Ausia fenestrata* from the

Nama Group of Namibia, the sac-like *Yarnemia acidiformis*, and one from a second new *Ausia*-like genus from the Onega Peninsula of northern Russia, *Burykhia hunti*. Results of a new study have shown possible affinity of these Ediacaran organisms to the ascidians. *Ausia* and *Burykhia* lived in shallow coastal waters slightly more than 555 to 548 million years ago, and are believed to be the oldest evidence of the chordate lineage of metazoans. The Russian Precambrian fossil *Yarnemia* is identified as a tunicate only tentatively, because its fossils are nowhere near as well-preserved as those of *Ausia* and *Burykhia*, so this identification has been questioned.

The star-shaped holes (*Catellocaula vallata*) in this Upper Ordovician bryozoan may represent a tunicate preserved by bioimmuration in the bryozoan skeleton.

Fossils of tunicates are rare because their bodies decay soon after death, but in some tunicate families, microscopic spicules are present, which may be preserved as microfossils. These spicules have occasionally been found in Jurassic and later rocks, but, as few palaeontologists are familiar with them, they may have been mistaken for sponge spicules.

Hybridization Studies

A multi-taxon molecular study in 2010 proposed that sea squirts are descended from a hybrid between a chordate and a protostome ancestor. This study was based on a quartet partitioning approach designed to reveal horizontal gene transfer events among metazoan phyla.

Anatomy

Body Form

Colonial tunicate with multiple openings in a single tunic.

Colonies of tunicates occur in a range of forms, and vary in the degree to which individual organisms, known as zooids, integrate with one another. In the simplest systems, the individual animals are widely separated, but linked together by horizontal connections called stolons, which grow along the seabed. Other species have the zooids growing closer together in a tuft or clustered together and sharing a common base. The most advanced colonies involve the integration of the zooids into a common structure surrounded by the tunic. These may have separate buccal siphons and a single central atrial siphon and may be organized into larger systems, with hundreds of star-shaped units. Often, the zooids in a colony are tiny but very numerous, and the colonies can form large encrusting or mat-like patches.

Body Structure

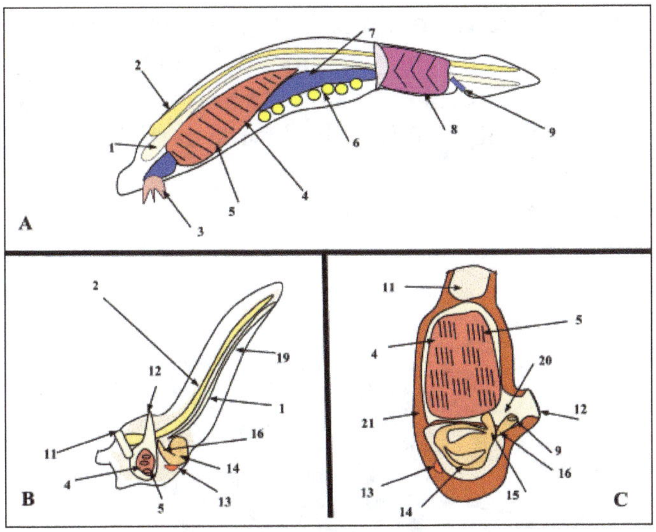

A. Lancelet, B. Larval tunicate, C. Adult tunicate. 1. Notochord, 2. Nerve chord, 3. Buccal cirri, 4. Pharynx, 5. Gill slit, 6. Gonad, 7. Gut, 8. V-shaped muscles, 9. Anus, 11. Inhalant syphon, 12. Exhalant syphon, 13. Heart, 14. Stomach, 15. Esophagus, 16. Intestines, 17. Tail, 18. Atrium, 19. Tunic.

By far the largest class of tunicates is the Ascidiacea. The body of an ascidiacean is surrounded by a test or tunic, from which the subphylum derives its name. This varies in thickness between species but may be tough, resembling cartilage, thin and delicate, or transparent and gelatinous. The tunic is composed of proteins and complex carbohydrates, and includes tunicin, a variety of cellulose. The tunic is unique among invertebrate exoskeletons in that it can grow as the animal enlarges and does not need to be periodically shed. Inside the tunic is the body wall or mantle composed of connective tissue, muscle fibres, blood vessels, and nerves. Two openings are found in the body wall: the buccal siphon at the top through which water flows into the interior, and the atrial siphon on the ventral side through which it is expelled. A large pharynx occupies most of the interior of the body. It is a muscular tube linking the buccal opening with the rest of the gut. It has a ciliated groove known as an endostyle on its ventral surface, and this secretes a mucous net which collects food particles and is wound up on the dorsal side of the pharynx. The gullet, at the lower end of the pharynx, links it to a loop of gut which terminates near the atrial siphon. The walls of the pharynx are perforated by several bands of slits, known as stigmata, through which water escapes into the surrounding water-filled cavity, the atrium. This is criss-crossed by various rope-like mesenteries which extend from the mantle and provide support for the pharynx, preventing it from collapsing, and also hold up the other organs.

The Thaliacea, the other main class of tunicates, is characterised by free-swimming, pelagic individuals. They are all filter feeders using a pharyngeal mucous net to catch their prey. The pyrosomes are bioluminous colonial tunicates with a hollow cylindrical structure. The buccal siphons are on the outside and the atrial siphons inside. About 10 species are known, and all are found in the tropics. The 23 species of doliolids are small, mostly under 2 cm (0.79 in) long. They are solitary, have the two siphons at opposite ends of their barrel-shaped bodies, and swim by jet propulsion. The 40 species of salps are also small, under 4 cm (1.6 in) long, and found in the surface waters of both warm and cold seas. They also move by jet propulsion, and often form long chains by budding off new individuals.

A third class, the Larvacea (or Appendicularia), is the only group of tunicates to retain their chordate characteristics in the adult state, a product of extensive neoteny. The 70 species of larvaceans superficially resemble the tadpole larvae of amphibians, although the tail is at right angles to the body. The notochord is retained, and the animals, mostly under 1 cm long, are propelled by undulations of the tail. They secrete an external mucous net known as a house, which may completely surround them and is very efficient at trapping planktonic particles.

Physiology and Internal Anatomy

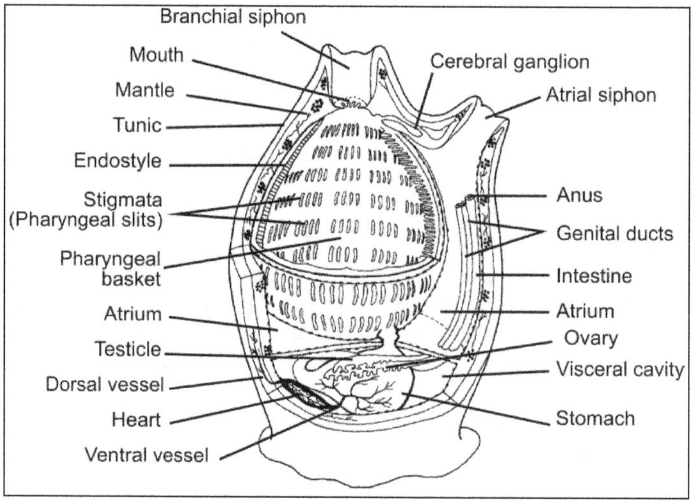

Internal anatomy of a generalised tunicate.

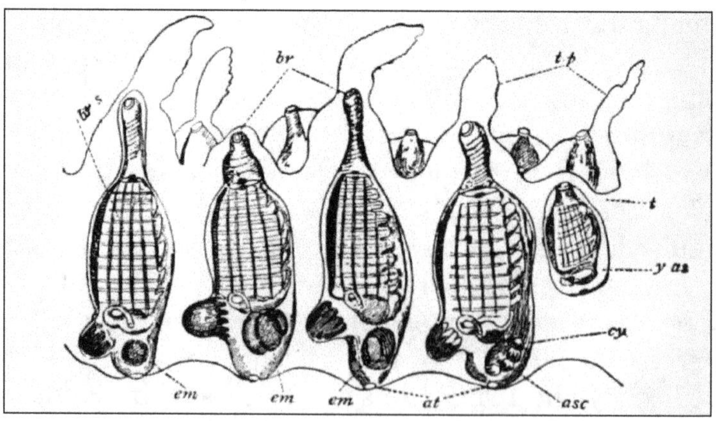

Section through the wall of a pyrosoma showing ascidiozooids; br) buccal siphon; at) atrial siphon; tp) process of the test; br s) pharynx.

Like other chordates, tunicates have a notochord during their early development, but which is lost by the time they have completed their metamorphosis. As members of the Chordata, they are true Coelomata with endoderm, ectoderm, and mesoderm, but they do not develop very clear coelomic body cavities, if any at all. Whether they do or not, by the end of their larval development, all that remain are the pericardial, renal, and gonadal cavities of the adults. Except for the heart, gonads, and pharynx (or branchial sac), the organs are enclosed in a membrane called an epicardium, which is surrounded by the jelly-like mesenchyme. Tunicates begin life in a mobile larval stage that resembles a tadpole. A minority of species, those in the Larvacea, retain the general larval form throughout life, but most Tunicata very rapidly settle down and attach themselves to a suitable surface, later developing into a barrel-like and usually sedentary adult form. The Thaliacea, however, are pelagic throughout their lives and may have complex lifecycles.

Tunicates have a well-developed heart and circulatory system. The heart is a double U-shaped tube situated just below the gut. The blood vessels are simple connective tissue tubes, and their blood has several types of corpuscle. The blood may appear pale green, but this is not due to any respiratory pigments, and oxygen is transported dissolved in the plasma. Exact details of the circulatory system are unclear, but the gut, pharynx, gills, gonads, and nervous system seem to be arranged in series rather than in parallel, as happens in most other animals. Every few minutes, the heart stops beating and then restarts, pumping fluid in the reverse direction.

Tunicate blood has some unusual features. In some species of Ascidiidae and Perophoridae, it contains high concentrations of the transitional metal vanadium and vanadium-associated proteins in vacuoles in blood cells known as vanadocytes. Some tunicates can concentrate vanadium up to a level ten million times that of the surrounding seawater. It is stored in a +3 oxidation form that requires a pH of less than 2 for stability, and this is achieved by the vacuoles also containing sulfuric acid. The vanadocytes are later deposited just below the outer surface of the tunic, where their presence is thought to deter predation, although it is unclear whether this is due to the presence of the metal or low pH. Other species of tunicates concentrate lithium, iron, niobium, and tantalum, which may serve a similar function. Other tunicate species produce distasteful organic compounds as chemical defenses against predators.

Oikopleura cophocerca in its “house”. Arrows indicate water movement and (x) the lateral reticulated parts of the house.

Tunicates lack the kidney-like metanephridial organs typical of deuterostomes. Most have no excretory structures, but rely on the diffusion of ammonia across their tissues to rid themselves of nitrogenous waste, though some have a simple excretory system. The typical renal organ is a mass of large clear-walled vesicles that occupy the rectal loop, and the structure has no duct. Each vesicle is a remnant of a part of the primitive coelom, and its cells extract nitrogenous waste matter from circulating blood. They accumulate the wastes inside the vesicles as urate crystals, and do not have any obvious means of disposing of the material during their lifetimes.

Adult tunicates have a hollow cerebral ganglion, equivalent to a brain, and a hollow structure known as a neural gland. Both originate from the embryonic neural tube and are located between the two siphons. Nerves arise from the two ends of the ganglion; those from the anterior end innervate the buccal siphon and those from the posterior end supply the rest of the body, the atrial siphon, organs, gut and the musculature of the body wall. There are no sense organs but there are sensory cells on the siphons, the buccal tentacles and in the atrium.

Blue sea squirts from the genus *Rhopalaea*.

Tunicates are unusual among animals in that they produce a large fraction of their tunic and some other structures in the form of cellulose. The production in animals of cellulose is so unusual that at first some researchers denied its presence outside of plants, but the tunicates were later found to possess a functional cellulose synthesizing enzyme, encoded by a gene horizontally transferred from a bacterium. When, in 1845, Carl Schmidt first announced the presence in the test of some ascidians of a substance very similar to cellulose, he called it "tunicine", but it is now recognized as cellulose rather than any alternative substance.

Fluorescent-colored sea squirts, *Rhopalaea crassa*.

Didemnum molle.

Feeding

Nearly all tunicates are suspension feeders, capturing planktonic particles by filtering sea water through their bodies. Ascidians are typical in their digestive processes, but other tunicates have similar systems. Water is drawn into the body through the buccal siphon by the action of cilia lining the gill slits. To obtain enough food, an average ascidian needs to process one body-volume of water per second. This is drawn through a net lining the pharynx which is being continuously secreted by the endostyle. The net is made of sticky mucus threads with holes about 0.5 μm in diameter which can trap planktonic particles including bacteria. The net is rolled up on the dorsal side of the pharynx, and it and the trapped particles are drawn into the oesophagus. The gut is U-shaped and also ciliated to move the contents along. The stomach is an enlarged region at the lowest part of the U-bend. Here, digestive enzymes are secreted and a pyloric gland adds further secretions. After digestion, the food is moved on through the intestine, where absorption takes place, and the rectum, where undigested remains are formed into faecal pellets or strings. The anus opens into the dorsal or cloacal part of the peribranchial cavity near the atrial siphon. Here, the faeces are caught up by the constant stream of water which carries the waste to the exterior. The animal orientates itself to the current in such a way that the buccal siphon is always upstream and does not draw in contaminated water.

Clavelina robusta (black and white) and *Pycnoclavella flava* (orange) showing siphons.

Some ascidians that live on soft sediments are detritivores. A few deepwater species, such as *Megalodicopia hians*, are sit-and-wait predators, trapping tiny crustacea, nematodes, and other small invertebrates with the muscular lobes which surround their buccal siphons. Certain tropical species in the family Didemnidae have symbiotic green algae or cyanobacteria in their tunics, and one of these symbionts, *Prochloron*, is unique to tunicates. Excess photosynthetic products are assumed to be available to the host.

Life Cycle

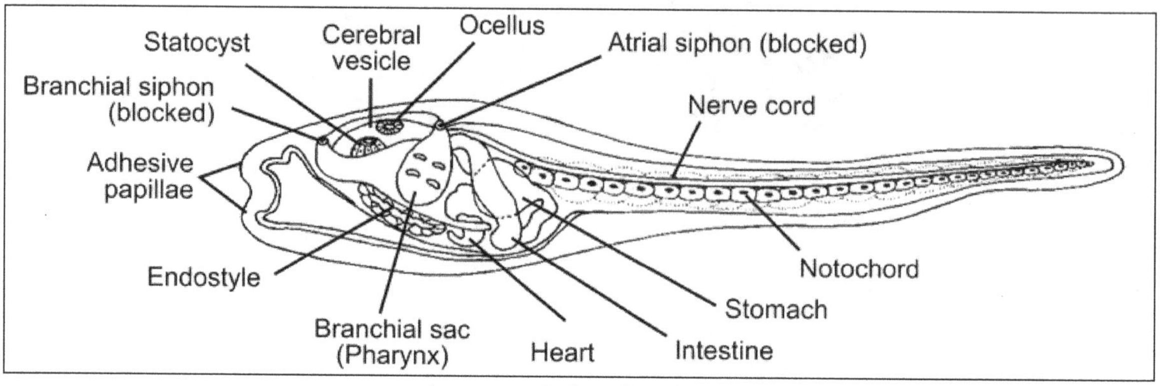

Anatomy of a larval tunicate.

Ascidians are almost all hermaphrodites and each has a single ovary and testis, either near the gut or on the body wall. In some solitary species, sperm and eggs are shed into the sea and the larvae are planktonic. In others, especially colonial species, sperm is released into the water and drawn into the atria of other individuals with the incoming water current. Fertilization takes place here and the eggs are brooded through their early developmental stages. Some larval forms appear very much like primitive chordates with a notochord (stiffening rod) and superficially resemble small tadpoles. These swim by undulations of the tail and may have a simple eye, an ocellus, and a balancing organ, a statocyst.

When sufficiently developed, the larva of the sessile species finds a suitable rock and cements itself in place. The larval form is not capable of feeding, though it may have a rudimentary digestive system, and is only a dispersal mechanism. Many physical changes occur to the tunicate's body during metamorphosis, one of the most significant being the reduction of the cerebral ganglion, which controls movement and is the equivalent of the vertebrate brain. From this comes the common saying that the sea squirt "eats its own brain".However, the adult does possess a cerebral ganglion which may even be larger than in the embryonic stage, so the scientific validity of this joke is questionable. In some classes, the adults remain pelagic (swimming or drifting in the open sea), although their larvae undergo similar metamorphoses to a higher or lower degree. Colonial forms also increase the size of the colony by budding off new individuals to share the same tunic.

Pyrosome colonies grow by budding off new zooids near the posterior end of the colony. Sexual reproduction starts within a zooid with an internally fertilized egg. This develops directly into an oozooid without any intervening larval form. This buds precociously to form four blastozooids which become detached in a single unit when the oozoid disintegrates. The atrial siphon of the oozoid becomes the exhalent siphon for the new, four-zooid colony.

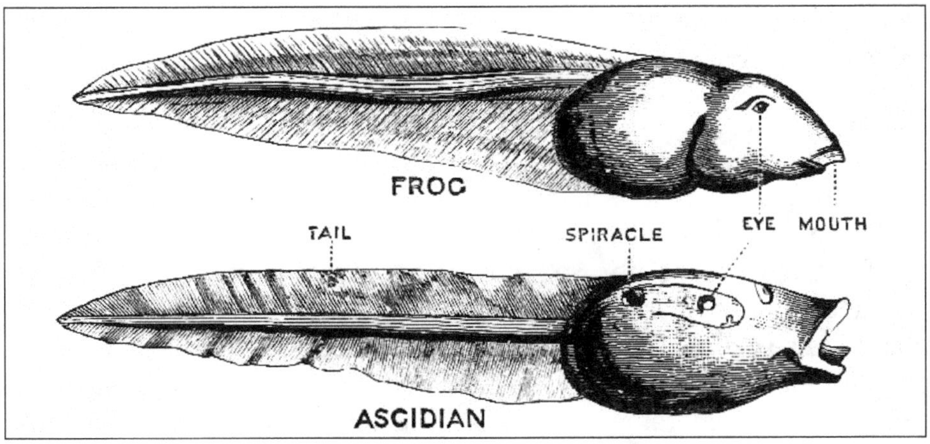

A 1901 comparison of frog tadpole and a tunicate larva.

Doliolids have a very complex life cycle that includes various zooids with different functions. The sexually reproducing members of the colony are known as gonozooids. Each one is a hermaphrodite with the eggs being fertilised by sperm from another individual. The gonozooid is viviparous, and at first, the developing embryo feeds on its yolk sac before being released into the sea as a free-swimming, tadpole-like larva. This undergoes metamorphosis in the water column into an oozooid. This is known as a "nurse" as it develops a tail of zooids produced by budding asexually. Some of these are known as trophozooids, have a nutritional function, and are arranged in lateral rows. Others are phorozooids, have a transport function, and are arranged in a single central row. Other zooids link to the phorozooids, which then detach themselves from the nurse. These zooids develop into gonozooids, and when these are mature, they separate from the phorozooids to live independently and start the cycle over again. Meanwhile, the phorozooids have served their purpose and disintegrate. The asexual phase in the lifecycle allows the doliolid to multiply very rapidly when conditions are favourable.

Salps also have a complex lifecycle with an alternation of generations. In the solitary life history phase, an oozoid reproduces asexually, producing a chain of tens or hundreds of individual zooids by budding along the length of a stolon. The chain of salps is the 'aggregate' portion of the lifecycle. The aggregate individuals, known as blastozooids, remain attached together while swimming and feeding and growing larger. The blastozooids are sequential hermaphrodites. An egg in each is fertilized internally by a sperm from another colony. The egg develops in a brood sac inside the blastozooid and has a placental connection to the circulating blood of its "nurse". When it fills the blastozooid's body, it is released to start the independent life of an oozooid.

Larvaceans only reproduce sexually. They are protandrous hermaphrodites, except for *Oikopleura dioica* which is gonochoric, and a larva resembles the tadpole larva of ascidians. Once the trunk is fully developed, the larva undergoes "tail shift", in which the tail moves from a rearward position to a ventral orientation and twists through 90° relative to the trunk. The larva consists of a small, fixed number of cells, and grows by enlargement of these rather than cell division. Development is very rapid and only takes seven hours for a zygote to develop into a house-building juvenile starting to feed.

During embryonic development, tunicates exhibit determinate cleavage, where the fate of the cells is set early on with reduced cell numbers and genomes that are rapidly evolving. In contrast, the amphioxus and vertebrates show cell determination relatively late in development and cell cleavage is indeterminate. The genome evolution of amphioxus and vertebrates is also relatively slow.

Promotion of Out-crossing

Ciona intestinalis (class Ascidiacea) is a hermaphrodite that releases sperm and eggs into the surrounding seawater almost simultaneously. It is self-sterile, and thus has been used for studies on the mechanism of self-incompatibility. Self/non-self-recognition molecules play a key role in the process of interaction between sperm and the vitelline coat of the egg. It appears that self/non-self recognition in ascidians such as *C. intestinalis* is mechanistically similar to self-incompatibility systems in flowering plants. Self-incompatibility promotes out-crossing, and thus provides the adaptive advantage at each generation of the masking of deleterious recessive mutations (i.e. genetic complementation) and the avoidance of inbreeding depression.

Botryllus schlosseri (class Ascidiacea) is a colonial tunicate, a member of the only group of chordates that are able to reproduce both sexually and asexually. *B. schlosseri* is a sequential (protogynous) hermaphrodite, and in a colony, eggs are ovulated about two days before the peak of sperm emission. Thus self-fertilization is avoided, and cross-fertilization is favored. Although avoided, self-fertilization is still possible in *B. schlosseri*. Self-fertilized eggs develop with a substantially higher frequency of anomalies during cleavage than cross-fertilized eggs (23% vs. 1.6%).Also a significantly lower percentage of larvae derived from self-fertilized eggs metamorphose, and the growth of the colonies derived from their metamorphosis is significantly lower. These findings suggest that self-fertilization gives rise to inbreeding depression associated with developmental deficits that are likely caused by expression of deleterious recessive mutations.

A Model Tunicate

Oikopleura dioica (class Appendicularia) is a semelparous organism, reproducing only once in its lifetime. It employs an original reproductive strategy in which the entire female germ-line is contained within an ovary that is a single giant multinucleate cell termed the "coenocyst". *O. dioica* can be maintained in laboratory culture, and is of growing interest as a model organism because of its phylogenetic position within the closest sister group to vertebrates.

Lancelet

The lancelets, also known as amphioxi (singular, amphioxus), consist of about 30-35 species of "fish-like" benthic filter feeding chordates in the order Amphioxiformes. They are the modern representatives of the subphylum Cephalochordata. Lancelets closely resemble, and are believed to be related to, 530-million-year-old *Pikaia*, fossils of which are known from the Burgess Shale. Zoologists are interested in them because they provide evolutionary insight into the origins of vertebrates. Lancelets contain many organs and organ systems that are closely related to those of modern fish, but in more primitive form. Therefore, they provide a number of examples of possible evolution exaptation. For example, the gill-slits of lancelets are used for feeding only, and not for respiration. The circulatory system carries food throughout their body, but does not have red blood cells or hemoglobin for transporting oxygen. Lancelet genomes hold clues about the early evolution of vertebrates, by comparing genes from lancelets with the same genes in vetebrates, the change in gene function as vetebrates evolved can be discovered. The genome of a few species in the genus *Branchiostoma* have been sequenced: *B. floridae, B. belcheri, B. lanceolatum.*

In Asia, lancelets are harvested commercially as food for humans and as domesticated animals. In Japan, amphioxus (*B. belcheri*) has been listed in the registry of "Endangered Animals of Japanese Marine and Fresh Water Organisms.

Ecology

Habitat

Lancelets are distributed in shallow subtidal sand flats in temperate (as far north as Norway), subtropical and tropical seas around the world. The only exception is *Asymmetron inferum*, a species known from the vicinity of whale falls at a depth of about 225 m (738 ft).Although they are able to swim, adult amphioxi are mostly benthic. They live in sandy bottoms whose granulometry depends on the species and the site, and they are usually found half-buried in sand. When disturbed, they quickly leave their burrow, swim a short distance, and then rapidly burrow again, posterior end first, into the sand. Adults (*B. floridae*) can tolerate salinities as low as 6% and temperatures from 3 to 37°C.

Feeding

Their habitat preference reflects their feeding method: they only expose the front end to the water and filter-feed on plankton by means of a branchial ciliary current that passes water through a mucous sheet. *Branchiostoma floridae* is capable of trapping particles from microbial to small phytoplankton size, while *B. lanceolatum* preferentially traps bigger particles (>4µm).

Reproduction and Spawning

Lancelets are gonochoric animals, i.e. having two sexes, and reproduce via external fertilization. They only reproduce during their spawning season, which varies slightly between species - usually corresponding to spring and summer months. All lancelets species spawn shortly after sunset, either synchronously (e.g. *Branchiostoma floridae*, about once every 2 weeks during spawning season) or asynchronously (*Branchiostoma lanceolatum*, gradual spawning through the season.

Nicholas and Linda Holland were the first researchers to describe a method of obtaining amphioxus embryos by induction of spawning in captivity and in vitro fertilization. Spawning can be artificially induced in the lab by electric or thermal shock.

Anatomy

Depending on the exact species involved, the maximum length of lancelets is typically 2.5 to 8 cm (1.0–3.1 in). *Branchiostoma belcheri* and *B. lanceolatum* are among the largest. Except for the size, the species are very similar in general appearance, differing mainly in the number of myotomes and the pigmentation of their larvae. They have a translucent, somewhat fish-like body, but without any paired fins or other limbs. A relatively poorly developed tail fin is present, so they are not especially good swimmers. While they do possess some cartilage-like material stiffening the gill slits, mouth, and tail, they have no true skeleton.

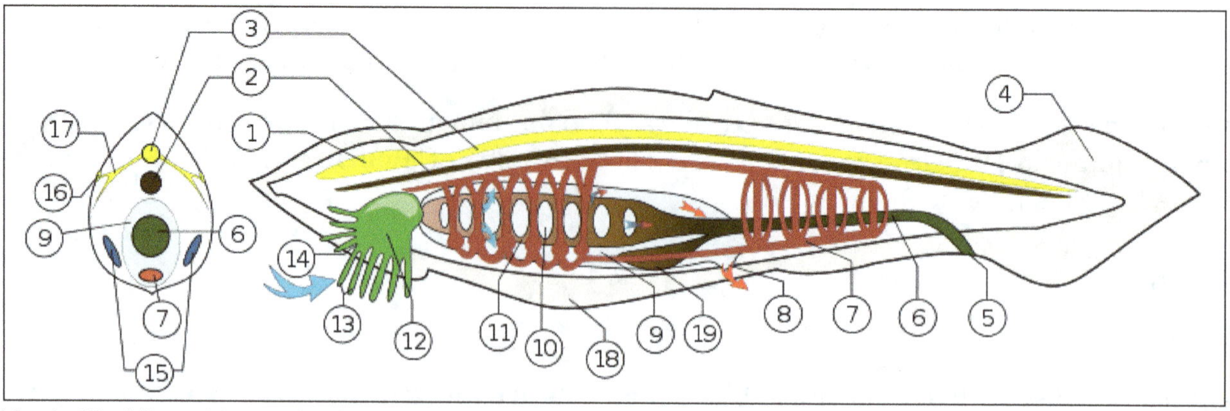

(1) brain-like blister (2) notochord, (3) dorsal nerve cell, (4) tail, (5) anus, (6) food canal, (7) blood system, (8) abdominal porus, (9) overpharynx lacuna, (10) gill-slit, (11) pharynx, (12) mouth lacuna, (13) cirri, (14) mouth gap, (15) gonads (ovary or testicle), (16) light sensor, (17) nerves, (18) abdominal ply, (19) hepatic caecum.

Nervous System

In common with vertebrates, lancelets have a hollow nerve cord running along the back, pharyngeal slits and a tail that runs past the anus. Also like vertebrates, the muscles are arranged in blocks called myomeres.

Unlike vertebrates, the dorsal nerve cord is not protected by bone but by a simpler notochord made up of a cylinder of cells that are closely packed to form a toughened rod. The lancelet notochord, unlike the vertebrate spine, extends into the head. This gives the subphylum its name (κεφαλή, *kephalē* means 'head'). The nerve cord is only slightly larger in the head region than in the rest of the body, so that lancelets do not appear to possess a true brain. However, developmental gene expression and transmission electron microscopy indicate the presence of a diencephalic forebrain, a possible midbrain, and a hindbrain but recent studies involving a comparison with vertebrates indicates that the vertebrate thalamus, pretectum, and midbrain domains jointly correspond to a single amphioxus region, which has been termed Di-Mesencephalic primordium (DiMes).

Visual System

Lancelets have four known kinds of light-sensing structures: Joseph cells, Hesse organs, an unpaired anterior eye and lamellar body, all of which utilize opsins as light receptors. All of these organs and structures are located in the neural tube, with the frontal eye at the front, followed by the lamellar body, the Joseph cells, and the Hesse organs.

Joseph Cells and Hesse Organs

Joseph cells are bare photoreceptors surrounded by a band of microvilli. These cells bear the opsin melanopsin. The Hesse organs (also known as dorsal ocelli) consist of a photoreceptor cell surrounded by a band of microvilli and bearing melanopsin, but half enveloped by a cup-shaped pigment cell. The peak sensitivity of both cells is ~470 nm (blue).

Both the Joseph cells and Hesse organs are in the neural tube, the Joseph cells forming a dorsal column, the Hesse organs in the ventral part along the length of the tube. The Joseph cells extend from the caudal end of the anterior vesicle (or cerebral vesicle) to the boundary between myomeres 3 and 4, where the Hesse organs begin and continue nearly to the tail.

Frontal Eye

The frontal eye consists of a pigment cup, a group of putative photoreceptor cells (termed *Row 1*), three rows of neurons (*Rows 2–4*), and glial cells. The frontal eye, which expresses the PAX6 gene, has been proposed as the homolog of vertebrate paired eyes, the pigment cup as the homolog of the RPE (retinal pigment epithelium), the putative photoreceptors as homologs of vertebrate rods and cones, and Row 2 neurons as homologs of the retinal ganglion cells.

The pigment cup is oriented concave dorsally. Its cells contain the pigment melanin.

The putative photoreceptor cells, Row 1, are arranged in two diagonal rows, one on either side of the pigment cup, symmetrically positioned with respect to the ventral midline. The cells are flask-shaped, with long, slender ciliary processes (one cilium per cell). The main bodies of the cells lie outside of the pigment cup, while the cilia extend into the pigment cup before turning and exiting. The cells bear the opsin *c-opsin 1*, except for a few which carry *c-opsin 3*.

The Row 2 cells are serotonergic neurons in direct contact with Row 1 cells. Row 3 and 4 cells are also neurons. Cells of all four rows have axons that project into the left and right ventrolateral nerves. For Row 2 neurons, axon projections have been traced to the tegmental neuropil. The tegmental neuropil has been compared with locomotor control regions of the vertebrate hypothalamus, where paracrine release modulates locomotor patterns such as feeding and swimming.

Feeding and Digestive System

Lancelets are inactive filter feeders, spending most of the time half-buried in sand with only their frontal part protruding. They eat a wide variety of small planktonic organisms, such as bacteria, fungi, diatoms, dinoflagellates and zooplankton, and they will also take detritus. Little is known about the diet of the lancelet larvae in the wild, but captive larvae of several species can be maintained on a diet of phytoplankton, although this apparently is not optimal for *Asymmetron lucayanum*.

Lancelets have oral cirri, thin tentacle-like strands that hang in front of the mouth and act as sensory devices and as a filter for the water passing into the body. Water passes from the mouth into the large pharynx, which is lined by numerous gill-slits. The ventral surface of the pharynx contains a groove, called the endostyle, which, connected to a structure known as Hatschek's pit, produces a film of mucus. Ciliary action pushes the mucus in a film over the surface of the gill slits, trapping suspended food particles as it does so. The mucus is collected in a second, dorsal groove and passed back to the rest of the digestive tract. Having passed through the gill slits, the water enters an atrium surrounding the pharynx, then exits the body via the atriopore.

Both adults and larvae exhibit a "cough" reflex to clear the mouth or throat of debris or items too large to swallow. In larvae the action is mediated by the pharyngeal muscles while in the adult animal it is accomplished by atrial contraction.

The remainder of the digestive system consists of a simple tube running from the pharynx to the anus. The hepatic caecum, a single blind-ending caecum, branches off from the underside of the gut, with a lining able to phagocytize the food particles, a feature not found in vertebrates. Although it performs many functions of a liver, it is not considered a true liver but a homolog of the vertebrate liver.

Other Systems

Lancelets have no respiratory system, breathing solely through their skin, which consists of a simple epithelium. Despite the name, little if any respiration occurs in the "gill" slits, which are solely devoted to feeding. The circulatory system does resemble that of primitive fish in its general layout, but is much simpler, and does not include a heart. There are no blood cells, and no hemoglobin.

The excretory system consists of segmented "kidneys" containing protonephridia instead of nephrons, and quite unlike those of vertebrates. Also unlike vertebrates, there are numerous, segmented gonads.

Model Organism

Lancelets became famous in the 1860s when Ernst Haeckel ("the German Darwin") began promoting them as a model for the ancestor of all vertebrates. By 1900 lancelets had become a model organism. By the mid-20th century they had fallen out of favor for a variety of reasons, including a decline of comparative anatomy and embryology, and due to the belief that lancelets were more derived than they appeared, e.g., the profound asymmetry in the larval stage.

With the advent of molecular genetics lancelets are once again regarded as a model of vertebrate ancestors, and are used again as a model organism.

As a result of their use in science, methods of keeping and breeding lancelets in captivity have been developed for several of the species, initially the European *Branchiostoma lanceolatum*, but later also the West Pacific *Branchiostoma belcheri* and *Branchiostoma japonicum*, the Gulf of Mexico and West Atlantic *Branchiostoma floridae* and the circumtropical (however, genetic evidence suggest the Atlantic and Indo-Pacific populations should be recognized as separate) *Asymmetron lucayanum*. They can reach an age of up to 7–8 years.

Taxonomy

The lancelet is a small, translucent, fish-like animal that is one of the closest living invertebrate relative of the vertebrates.

The Cephalochordata is traditionally seen as a sister subphylum to the vertebrates, with which it is grouped together into a clade (sometimes called Notochordata) which in turn is the sister group to the simpler still Urochordata. Newer research suggests this is not the case. The Cephalochordata is the most basal subphylum of the chordates, while the sister group of the vertebrates are the tunicates, previously known as the urochordates. However, older molecular studies placed cephalochordates nearer to vertebrates, and "[m]ost authors regard amphioxus as the closest relative of the Vertebrata on the basis of 10–15 [morphological] features that are not seen in tunicates".

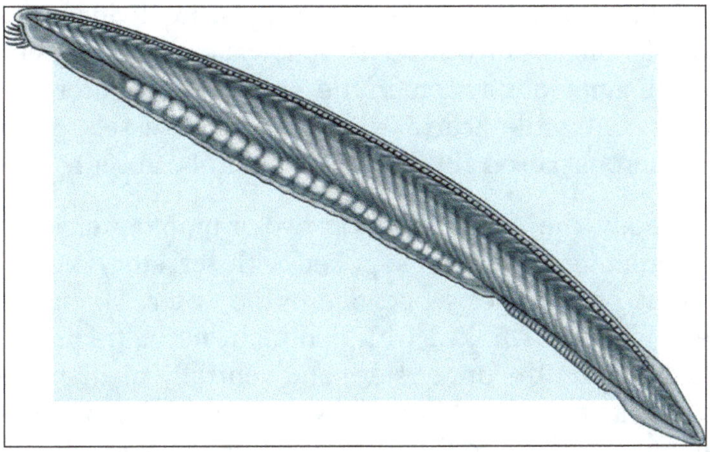

Among the three extant (living) genera, *Asymmetron* is basal and it is placed in its own family. Genetic studies have come to separate conclusions on their divergence, with some suggesting that *Asymmetron* diverged from other lancelets more than 100 million years ago and others (with higher statistic support) that it occurred about 42 million years ago. The two remaining genera, *Branchiostoma* and *Epigonichthys* diverged from each other about 36 million years ago. Despite this deep separation, hybrids between *Asymmetron lucayanum* and *Branchiostoma floridae* are viable (among the deepest split species known to be able to produce such hybrids).

CLOSEST NONCHORDATE RELATIVES

Hemichordate

Hemichordate, any of a group of wormlike marine invertebrates closely related to both chordates and echinoderms and usually considered to constitute a phylum, the Hemichordata. The term Hemichordata—from the word hemi, meaning "half," and chorde, meaning "string," thus, "half-chordate"—was first proposed because the buccal diverticulum, a tubular outgrowth from the mouth cavity forward into the proboscis, or "snout," resembled a rudimentary notochord—the dorsal, or back-side, supporting axis of the more primitive vertebrates. This theory has since been rejected, however, because it has been determined that the diverticulum bears little resemblance in origin and function to the vertebrate notochord. Molecular studies indicate that hemichordates are more closely related to echinoderms than to chordates. About 90 species have been described.

The Hemichordata consist of three classes: Enteropneusta, Pterobranchia, and Planctosphaeroidea. Enteropneusts, or acorn worms (about 70 species), are solitary, wormlike, bilaterally symmetrical animals, often brilliantly coloured. They are known as acorn worms because of the appearance of the proboscis and collar. Pterobranchs (about 20 species) are minute, colonial, tube-building forms. Planctosphaeroidea are known only from a few floating larvae. Enteropneusts are common in the intertidal zones from the White Sea and Greenland south to New Zealand and the Cape of Good Hope; they are found offshore to depths of 400 metres (about 1,300 feet) or more. They vary in size from a few centimetres long (Saccoglossus pygmaeus of the North Sea) to two metres (about seven feet) or more (Balanoglossus gigas of Brazilian coastal waters).

The adult hemichordate body and body cavities, or coeloms, are divided into three basic parts: the proboscis, collar, and trunk. There is no central nervous system: nerve tissue is concentrated in the collar, which is linked with a nervous system in the epidermis, or outer covering. The circulatory system usually includes a contractile heartlike vesicle, blood vessels, and sinuses. The pharynx may be perforated by numerous paired gill slits, or they may be absent.

The second region of the body, the collar, may bear two or more tentacle-like plumes, which may have a double row of ciliated tentacles well supplied with secretory cells. The tentacles are special adaptations for feeding on particles suspended in the water. The network of nerve cells and fibres lying within the epidermis is linked with two main nerve tracts that lie dorsally median (i.e., toward the body midline on the upper side) and ventrally median (on the lower side). The dorsal side of the collar has a neurochord formed by an inpocketing of the epidermis; it may have a central lumen, or cavity, that opens to the exterior anteriorly and posteriorly, or it may have a series of lacunae, or spaces. The neurochord contains large nerve cells, extensions of which reach almost to the tip of the proboscis and into the ventral nerve cord. These cells probably facilitate rapid responses such as abrupt contractions of the anterior trunk when the proboscis is touched. The general body surface is innervated by a primitive receptor system, which consists of scattered sensory cells. There is no well-defined centre of stimuli and responses.

All hemichordates are suspension feeders, capturing minute animals and algae as they drift by in the water. Acorn worms trap such organisms on their proboscis, whereas pterobranchs use their tentacles. Many hemichordates have a larval stage in their life cycle; the larvae, called tornariae, swim using microscopic hairs called cilia.

References

- Protochordata, biology: byjus.com, Retrieved 12 March, 2019
- Valentine, J.W. (2004). On the Origin of Phyla. Chicago: University Of Chicago Press. p. 7 The phylum chordata includes three subphylums: Cephalochoradta, Urochordata, Vertebras. It is determined that urochordates are closer to vertebrates that cephalochordates are due to their chordate–like features as a larval form. ISBN 978-0-226-84548-7
- Singh, T. R.; Tsagkogeorga, G.; Delsuc, F.; Blanquart, S.; Shenkar, N.; Loya, Y.; Douzery, E. J.; Huchon, D. (2009). "Tunicate mitogenomics and phylogenetics: peculiarities of the Herdmania momus mitochondrial genome and support for the new chordate phylogeny". BMC Genomics. 10: 534. doi:10.1186/1471-2164-10-534. PMC 2785839. PMID 19922605
- Barnes, M.K.S (7 June 2015). Tyler-Walters, H.; K. Hiscock (eds.). "Lancelet (Branchiostoma lanceolatum)". Marine Life Information Network: Biology and Sensitivity Key Information Reviews. Retrieved 7 January 2018
- Hemichordate, animal: britannica.com, Retrieved 17 June, 2019

Anatomy of Vertebrates

CHAPTER 3

The branch of biology that defines body structure and functions, such as respiration, circulation, digestions, etc. of all living organisms, is referred to as anatomy. Vertebrates are divided into pisces, reptiles, amphibians, aves and mammals. This chapter closely examines these anatomy of vertebrates to provide an extensive understanding of the subject.

Comparative anatomy, the comparative study of the body structures of different species of animals in order to understand the adaptive changes they have undergone in the course of evolution from common ancestors. The field is largely confined to the study of the vertebrate animals.

Modern comparative anatomy dates from the work of Pierre Belon, who in 1555 showed that the skeletons of humans and birds are constructed of similar elements arranged in the same way. From this humble beginning, knowledge of comparative anatomy advanced rapidly in the 18th century with the work of the Count de Buffon and Louis-Jean-Marie Daubenton, who compared the anatomies of a wide range of animals.

In the early 19th century, Baron Georges Cuvier placed the field on a more scientific basis by asserting that animals' structural and functional characteristics result from their interaction with their environment. Cuvier also rejected the 18th-century notion that the members of the animal kingdom are arranged in a single linear series from the simplest up to humans. Instead Cuvier arranged all animals into four large groups (vertebrates, mollusks, articulates, and radiates) according to body plan. Another great figure in the field was the mid-19th-century British anatomist Sir Richard Owen, whose vast knowledge of vertebrate structure did not prevent him from opposing Charles Darwin's theory of evolution. Darwin made extensive use of comparative anatomy in advancing his theory, and it in turn revolutionized the field by explaining the structural differences between species as arising out of their evolutionary descent by natural selection from a common ancestor.

Since Darwin's time, the study of comparative anatomy has centred largely on body structures that are homologous—i.e., ones in different species that have the same evolutionary origin regardless of their present-day function. Such structures may look quite different and perform different tasks, but they can still be traced back to a common structure in an animal that was ancestral to both.

The forelimbs of humans, birds, crocodiles, bats, dolphins, and rodents have been modified by evolution to perform different functions, but they are all evolutionarily traceable to the fins of crossopterygian fishes, in which that basic arrangement of bones was first established. Analogous structures, by contrast, may resemble each other because they perform the same function, but they have different evolutionary origins and often a different structure, the wings of insects and of birds being a prime example of this.

ANATOMY OF PISCES

Respiration in Fishes

Most fish exchange gases using gills on either side of the pharynx (throat). Gills are tissues which consist of threadlike structures called filaments. These filaments have many functions and "are involved in ion and water transfer as well as oxygen, carbon dioxide, acid and ammonia exchange. Each filament contains a capillary network that provides a large surface area for exchanging oxygen and carbon dioxide. Fish exchange gases by pulling oxygen-rich water through their mouths and pumping it over their gills. In some fish, capillary blood flows in the opposite direction to the water, causing countercurrent exchange. The gills push the oxygen-poor water out through openings in the sides of the pharynx.

Gills inside the head of a tuna. The head is snout-down, with the view looking towards the mouth. On the right are the detached gills.

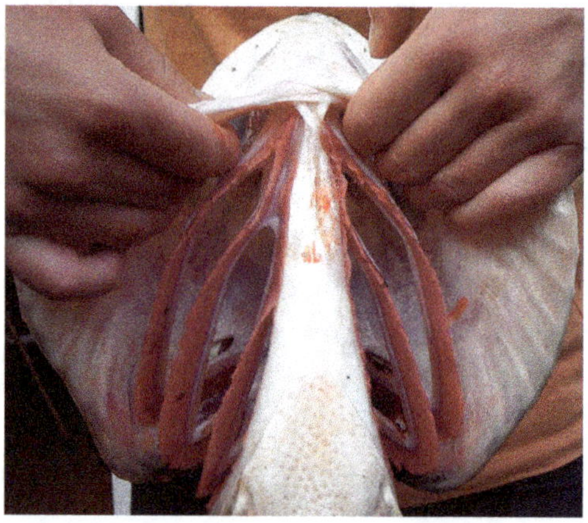

Gill arches bearing gills in a pike.

Fish from multiple groups can live out of the water for extended time periods. Amphibious fish such as the mudskipper can live and move about on land for up to several days, or live in stagnant or otherwise oxygen depleted water. Many such fish can breathe air via a variety of mechanisms. The skin of anguillid eels may absorb oxygen directly. The buccal cavity of the electric eel may breathe air. Catfish of the families Loricariidae, Callichthyidae, and Scoloplacidae absorb air through their digestive tracts. Lungfish, with the exception of the Australian lungfish, and bichirs have paired lungs similar to those of tetrapods and must surface to gulp fresh air through the mouth and pass spent air out through the gills. Gar and bowfin have a vascularized swim bladder that functions in the same way. Loaches, trahiras, and many catfish breathe by passing air through the gut. Mudskippers breathe by absorbing oxygen across the skin (similar to frogs). A number of fish have evolved so-called accessory breathing organs that extract oxygen from the air. Labyrinth fish (such as gouramis and bettas) have a labyrinth organ above the gills that performs this function. A few other fish have structures resembling labyrinth organs in form and function, most notably snakeheads, pikeheads, and the Clariidae catfish family.

Breathing air is primarily of use to fish that inhabit shallow, seasonally variable waters where the water's oxygen concentration may seasonally decline. Fish dependent solely on dissolved oxygen, such as perch and cichlids, quickly suffocate, while air-breathers survive for much longer, in some cases in water that is little more than wet mud. At the most extreme, some air-breathing fish are able to survive in damp burrows for weeks without water, entering a state of aestivation (summertime hibernation) until water returns.

Air breathing fish can be divided into *obligate* air breathers and *facultative* air breathers. Obligate air breathers, such as the African lungfish, are obligated to breathe air periodically or they suffocate. Facultative air breathers, such as the catfish *Hypostomus plecostomus*, only breathe air if they need to and can otherwise rely on their gills for oxygen. Most air breathing fish are facultative air breathers that avoid the energetic cost of rising to the surface and the fitness cost of exposure to surface predators.

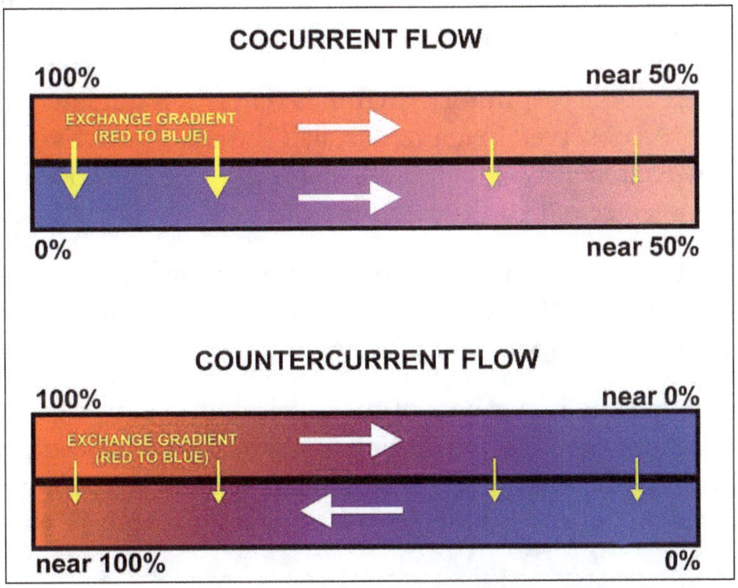

Cocurrent and countercurrent flow exchange systems- Red represents a higher value (e.g. of temperature or the partial pressure of a gas) than blue so the property being transported in the channels flows from red to blue. In fish a countercurrent flow (lower diagram) of blood and water in the gills is used to extract oxygen from the environment.

All basal vertebrates breathe with gills. The gills are carried right behind the head, bordering the posterior margins of a series of openings from the esophagus to the exterior. Each gill is supported by a cartilaginous or bony gill arch. The gills of vertebrates typically develop in the walls of the pharynx, along a series of gill slits opening to the exterior. Most species employ a countercurrent exchange system to enhance the diffusion of substances in and out of the gill, with blood and water flowing in opposite directions to each other, which increases the efficiency of oxygen-uptake from the water. Fresh oxygenated water taken in through the mouth is uninterruptedly "pumped" through the gills in one direction, while the blood in the lamellae flows in the opposite direction, creating the countercurrent blood and water flow, on which the fish's survival depends.

The gills are composed of comb-like filaments, the gill lamellae, which help increase their surface area for oxygen exchange. When a fish breathes, it draws in a mouthful of water at regular intervals. Then it draws the sides of its throat together, forcing the water through the gill openings, so that it passes over the gills to the outside. The bony fish have three pairs of arches, cartilaginous fish have five to seven pairs, while the primitive jawless fish have seven. The vertebrate ancestor no doubt had more arches, as some of their chordate relatives have more than 50 pairs of gills.

Higher vertebrates do not develop gills, the gill arches form during fetal development, and lay the basis of essential structures such as jaws, the thyroid gland, the larynx, the *columella* (corresponding to the stapes in mammals) and in mammals the malleus and incus. Fish gill slits may be the evolutionary ancestors of the tonsils, thymus gland, and Eustachian tubes, as well as many other structures derived from the embryonic branchial pouches.

Scientists have investigated what part of the body is responsible for maintaining the respiratory rhythm. They found that neurons located in the brainstem of fish are responsible for the genesis of the respiratory rhythm. The position of these neurons is slightly different from the centers of respiratory genesis in mammals but they are located in the same brain compartment, which has caused debates about the homology of respiratory centers between aquatic and terrestrial species. In both aquatic and terrestrial respiration, the exact mechanisms by which neurons can generate this involuntary rhythm are still not completely understood.

Another important feature of the respiratory rhythm is that it is modulated to adapt to the oxygen consumption of the body. As observed in mammals, fish "breathe" faster and heavier when they do physical exercise. The mechanisms by which these changes occur have been strongly debated over more than 100 years between scientists. The authors can be classified in 2 schools:

- Those who think that the major part of the respiratory changes are pre-programmed in the brain, which would imply that neurons from locomotion centers of the brain connect to respiratory centers in anticipation of movements.
- Those who think that the major part of the respiratory changes result from the detection of muscle contraction, and that respiration is adapted as a consequence of muscular contraction and oxygen consumption. This would imply that the brain possesses some kind of detection mechanisms that would trigger a respiratory response when muscular contraction occurs.

Many now agree that both mechanisms are probably present and complementary, or working alongside a mechanism that can detect changes in oxygen and/or carbon dioxide blood saturation.

Bony Fish

In bony fish, the gills lie in a branchial chamber covered by a bony operculum. The great majority of bony fish species have five pairs of gills, although a few have lost some over the course of evolution. The operculum can be important in adjusting the pressure of water inside of the pharynx to allow proper ventilation of the gills, so that bony fish do not have to rely on ram ventilation (and hence near constant motion) to breathe. Valves inside the mouth keep the water from escaping.

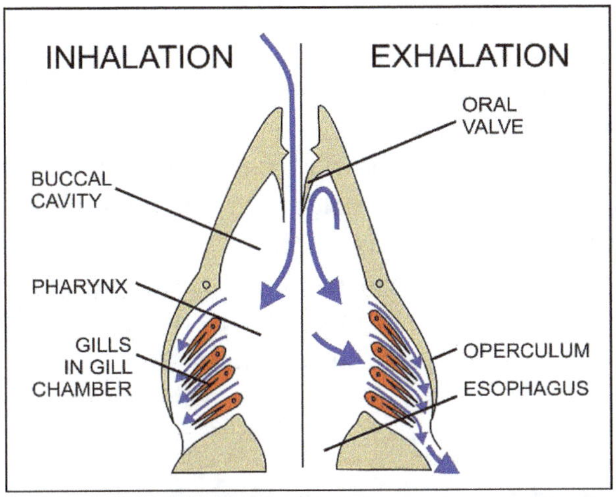

Respiratory mechanism in bony fishThe fish draws oxygen-rich water in through the mouth (left). It then pumps it over gills so oxygen enters the bloodstream, and allows oxygen-depleted water to exit through the gill slits (right).

The gill arches of bony fish typically have no septum, so that the gills alone project from the arch, supported by individual gill rays. Some species retain gill rakers. Though all but the most primitive bony fish lack a spiracle, the pseudobranch associated with it often remains, being located at the base of the operculum. This is, however, often greatly reduced, consisting of a small mass of cells without any remaining gill-like structure.

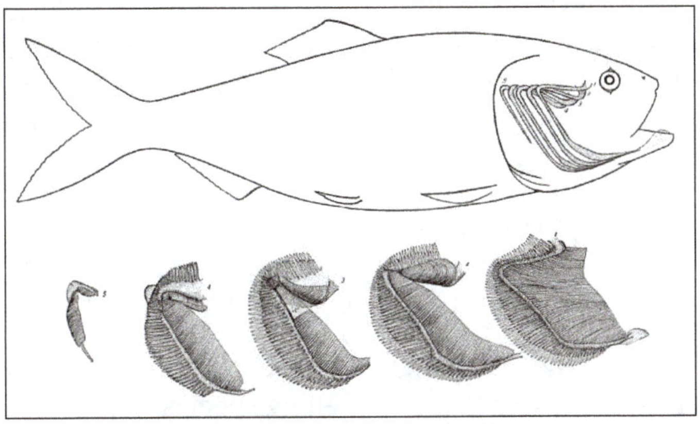

Most bony fish have five gills.

Marine teleosts also use gills to excrete electrolytes. The gills' large surface area tends to create a problem for fish that seek to regulate the osmolarity of their internal fluids. Saltwater is less dilute than these internal fluids, so saltwater fish lose large quantities of water osmotically through their gills. To regain the water, they drink large amounts of seawater and excrete the salt. Freshwater is more dilute than the internal fluids of fish, however, so freshwater fish gain water osmotically through their gills.

In some primitive bony fishes and amphibians, the larvae bear external gills, branching off from the gill arches. These are reduced in adulthood, their function taken over by the gills proper in fishes and by lungs in most amphibians. Some amphibians retain the external larval gills in adulthood, the complex internal gill system as seen in fish apparently being irrevocably lost very early in the evolution of tetrapods.

Cartilaginous Fish

Like other fish, sharks extract oxygen from seawater as it passes over their gills. Unlike other fish, shark gill slits are not covered, but lie in a row behind the head. A modified slit called a spiracle lies just behind the eye, which assists the shark with taking in water during respiration and plays a major role in bottom–dwelling sharks. Spiracles are reduced or missing in active pelagic sharks. While the shark is moving, water passes through the mouth and over the gills in a process known as "ram ventilation". While at rest, most sharks pump water over their gills to ensure a constant supply of oxygenated water. A small number of species have lost the ability to pump water through their gills and must swim without rest. These species are *obligate ram ventilators* and would presumably asphyxiate if unable to move. Obligate ram ventilation is also true of some pelagic bony fish species.

The respiration and circulation process begins when deoxygenated blood travels to the shark's two-chambered heart. Here the shark pumps blood to its gills via the ventral aorta artery where it branches into afferent brachial arteries. Reoxygenation takes place in the gills and the reoxygenated blood flows into the efferent brachial arteries, which come together to form the dorsal aorta. The blood flows from the dorsal aorta throughout the body. The deoxygenated blood from the body then flows through the posterior cardinal veins and enters the posterior cardinal sinuses. From there blood enters the heart ventricle and the cycle repeats.

Sharks and rays typically have five pairs of gill slits that open directly to the outside of the body, though some more primitive sharks have six or seven pairs. Adjacent slits are separated by a cartilaginous gill arch from which projects a long sheet-like septum, partly supported by a further piece of cartilage called the *gill ray*. The individual lamellae of the gills lie on either side of the septum. The base of the arch may also support gill rakers, small projecting elements that help to filter food from the water.

A smaller opening, the spiracle, lies in the back of the first gill slit. This bears a small *pseudobranch* that resembles a gill in structure, but only receives blood already oxygenated by the true gills. The spiracle is thought to be homologous to the ear opening in higher vertebrates.

Most sharks rely on ram ventilation, forcing water into the mouth and over the gills by rapidly swimming forward. In slow-moving or bottom dwelling species, especially among skates and rays, the spiracle may be enlarged, and the fish breathes by sucking water through this opening, instead of through the mouth.

Chimaeras differ from other cartilagenous fish, having lost both the spiracle and the fifth gill slit. The remaining slits are covered by an operculum, developed from the septum of the gill arch in front of the first gill.

Lampreys and Hagfish

Lampreys and hagfish do not have gill slits as such. Instead, the gills are contained in spherical pouches, with a circular opening to the outside. Like the gill slits of higher fish, each pouch contains two gills. In some cases, the openings may be fused together, effectively forming an operculum. Lampreys have seven pairs of pouches, while hagfishes may have six to fourteen, depending on the species. In the hagfish, the pouches connect with the pharynx internally. In adult lampreys, a separate respiratory tube develops beneath the pharynx proper, separating food and water from respiration by closing a valve at its anterior end.

Blood Circulation in Fishes

Transition from aquatic to terrestrial life during the Paleozoic era was one of the monumental steps in the evolution of vertebrates. Recent findings of new fossil records have substantially expanded our knowledge of osteological transformation during the fish-tetrapod transition that occurred in the Devonian period and inspired some new ideas about environmental settings in which the transition occurred. Nevertheless, those findings can give little insight into the question of how the cardiorespiratory system was transformed during the invasion of land by vertebrates. It is widely assumed that Devonian sarcopterygians that gave rise to tetrapods possessed the lungs and used them for aerial gas exchange while they were still inhabitants of the aquatic realm. Subsequently, the lungs have become the major aerial gas exchange organ in tetrapods and birds, with varying contributions of skin as a supplementary (in many amphibians) or in some exceptional cases, exclusive gas exchange surface. In addition to the sarcopterygian-amphibian lineage, air breathing has independently arisen numerous times in the evolutionary history of bony fishes. Many of the so-called 'primitive' bony fishes have a lung or lung-like gas bladder, such as polypterids, gars, bowfin and lungfishes. More advanced teleostean forms have developed a vast variety of air-breathing organs in different parts of the body, such as different segments of the alimentary canal (pharynx, esophagus, stomach or intestine), various types of outgrowth from the canal (e.g., respiratory swimbladders or suprabranchial chambers), specialized structures developing from the gills (e.g., labyrinthine organs) and the skin.

The circulatory system of the Devonian vertebrates perhaps underwent considerable modifications in con junction with the transition from aquatic to aerial mode of gas exchange, although this can- not be traced back from the fossil records but can only be inferred from comparison among extant animals. In all extant fishes, gills are placed upstream of and connected in series to the systemic bed in the vascular circuit. Fish hearts are undivided, consisting of a single atrium and a single ventricle that contain only systemic venous blood and impart kinetic energy only once during sequential perfusion of the gills and the systemic beds. This type of circulation is called a single circulatory system. On the other hand, in adult mammals and birds the pulmonary circuit is completely separated from the systemic circuit. The hearts of adult mammals and birds are completely divided into the right and left sides, in which the right atrium and ventricle deliver O2-poor systemic venous blood to the pulmonary arteries with a relatively low pressure, whilst the left chambers eject O2- rich pulmonary venous blood into a single systemic aorta with a much higher pressure to perfuse the systemic vascular bed. Thus, in adult mammals and birds the heart pressurizes the blood twice, once for pulmonary circulation and once again for systemic ejection, to complete a cycle through the body. This type of circulation is called a double or dual circulatory system. As the type and location of aerial respiratory organs varies widely among airbreathing fishes, so do their circulatory systems. It is usually considered that bony fishes except dipnoans have not evolved any vascular specialization to allow the separation of two streams of blood through the central vascular pathways and therefore, the mixing of deoxygenated systemic venous blood and oxygenated effluent blood from the air-breathing organ is inevitable, although such a statement is in fact not always based on physiological determinations of blood gases and flow patterns on fully recovered, normally behaving animals.

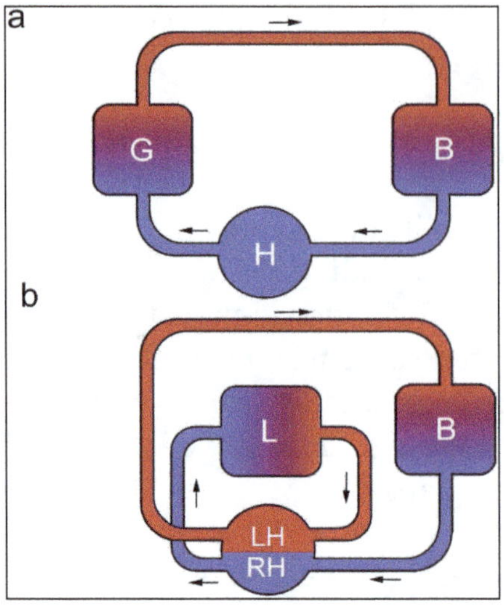

Cardiovascular systems in fish (a) and adult birds and mammals (b). Red and blue vessels convey oxygen-rich arterialized blood and oxygen-poor venous blood, respectively. B, body; G, gills; H, heart; L, lungs; LH, left side of the heart; RH, right side of the heart. Arrows indicate directions of blood flow.

Phylogenetic evolution from single circulation of fishes towards double circulation of tetrapods remains one of the central issues in zoology; however, it has not drawn as much interest in recent years as in a few decades ago.

Taxonomy and Natural History of Selected Airbreathing Fishes

Eel Gobies

Eel gobies are cryptic, inconspicuous fish inhabiting the coastal and estuarine waters of the Indo-West Pacific region. Eel gobies belong to the subfamily Amblyopinae (Gobiidae), which includes 23 species in 12 genera. Because of their cryptic nature, not much information is available on the natural history of eel gobies. Hora described that Taenioides rubicundus floated its head with gill chambers inflated at the surface of an aquarium containing foul water. Some 70 years later, the air-breathing capacity of another eel goby, Odontamblyopus lacepedii, was experimentally verified by Gonzales et al. Unlike syntopic mudskippers (Beleophthalmus pectinirostris and Periophthalmus modestus), O. lacepedii will not emerge from water during low tide, but remain inside burrows they excavate in mudflats. The burrows have two to seven openings on the mudflat surface and show a complex branching pattern, which probably varies with season. Water filling the burrows is severely hypoxic (mean oxygen pressure (Po_2) 18 torr: 1 torr = 1 mmHg = 0.1333 kPa) as in mudskipper burrows and O. lacepedii was found to switch from water breathing to air breathing by gulping air shortly after the burrow openings emerged in low tide.

Mudskippers

Mudskippers typically refers to four genera of oxudercine gobies (Gobiidae: Oxudercinae) including Boleophthalmus, Periophthalmodon, Periophthalmus, and Scartelaos, but may also include other oxudercine gobies exhibiting an amphibious lifestyle (e.g., Pseudapocryptes). Species of Periophthalmus and Periophthalmodon are usually more terrestrial, whilst Boleophthalmus and Scartelaos are more aquatic; neither of the latter two ventures far from water pools on an exposed mudflat during low tide and both appear to reside in their burrows during high tide, when the burrow openings are covered by the tide. Of the four mudskipper genera, species of Periophthalmodon and Periophthalmus are purely carnivorous, whereas Boleophthalmus species are exclusively herbivorous. Less is known about the feeding habits of Scartelaos, but Milward reported them to be omnivorous. The distribution of mudskippers centers in the tropical Indo-Pacific. The known localities of Boleophthalmus and Scartelaos species are similar, ranging from the Persian Gulf throughout the East Asian countries and south to the tropical regions of Australia. The distribution of Periophthalmodon species is more restricted to Southeast Asia and mid-northern Australia. Periophthalmus species show the widest distribution among mudskippers, from the west coast of Africa to the Polynesian islands. Mudskippers have a high commercial value in some Asian countries (China, Korea and Vietnam) and are produced by aquaculture.

Swamp Eels

Similar to oxudercine gobies, swamp eels show a wide spectrum of transition from aquatic to amphibious lifestyle within one taxonomic group. Swamp eels belong to the family Synbranchidae, which consists of four genera (Macrotrema, one species; Ophisternon, six; Synbranchus, two; Monopterus, eight: note that Amphipnous is now regarded as the subgenus of Monopterus, Nelson). Liem reported that Macrotrema caligans is purely aquatic and exclusively water-breathing, Ophisternon and Synbranchus are facultative air-breathers and Monopterus is an obligate air-breathing fish, even though we are unaware of any publication on the respiratory physiology of Macrotrema caligans or Ophisternon species. Swamp eels are widely distributed in tropical and subtropical regions from West Africa, Asia, Australia, Mexico and Central and Southern America. These fishes often occur in swamps or marshy areas, where burrowing and amphibious habits are commonly displayed. Swamp eels lack paired fins and the dorsal and anal fins are reduced to skin folds, which are generally confluent with the much reduced caudal fin. Another peculiar feature of swamp eels is that their gill opening is confined to a single slit to the ventral body wall, except Macrotrema, in which the gill opening is more extensive and reaches to the lateral sides of the body. Monopterus and Ophisternon are consumed in Asian countries such as China, Indonesia and Vietnam and are produced by aquaculture.

Snakeheads

Snakeheads are freshwater air-breathing fishes of the family Channidae, to which two genera (Channa, 26 species and Parachanna, three species) belong . Snakeheads inhabit stagnant freshwater ponds and creeks and their distribution is mainly limited to Asia (Channa) and Africa (Parachanna). Snakeheads are purely carnivorous, feeding on small fish, frogs, crayfish and other aquatic animals. Channa argus is the most northern species among the member of Channidae, inhabiting Northeast China, Siberia, and Korea, but has been recently found to populate the United States and reproduce there. Some snakeheads attain a body length of 1.2 m. Snakeheads have a high commercial value in Southeast Asian countries where the fish are produced by aquaculture.

Lungfishes

Lungfishes assume a key position in the research into the evolution from aquatic to terrestrial life in vertebrates. Together with the coelacanth, lungfishes are the only living sarcopterygian (lobe-finned) fishes and are probably the closest living relatives of the tetrapods. The extant lungfishes include three genera, Neoceratodus (one species), Protopterus (four species) and Lepidosiren (one species) inhabiting freshwaters of Australia, Africa and South America, respectively. All lungfishes have a lung and use it for breathing air but the relative importance of aerial gas exchange in respiration varies between the genera. Thus, Australian lungfish is a facultative air-breather practicing air breathing only during swimming activity or when the water is hypoxic. In contrast, African and South American lungfishes are obligate air-breathers and must breathe air even in normoxic water, although this has been questioned recently for Protopterus aethiopicus on the basis of behavioral monitoring by radio telemetry. It remains to be seen whether the study site of Mlewa et al, where the water seems to be rarely hypoxic, is typical for P. aethiopicus. There is no evidence for active emersion from the water for any lungfish species. On the other hand, it is well known that African and South American lungfishes both aestivate in mud during droughts. Protopterus spp. secrete mucus to form an underground cocoon, whereas Lepidosiren does not. Neoceratodus does not aestivate. African lungfishes are valued food fish and some major populations are under a threat of over-exploitation.

Anatomy of the Cardiorespiratory System in Air Breathing Fishes

Eel Gobies

The gills of Odontamblyopus lacepedii appear to be well-suited for aquatic respiration and lack any morphological specializations for aerial gas exchange; each gill arch contains two rows of gill filaments, which in turn bear welldeveloped secondary lamellae on either side, as in ordinary water-breathing teleosts. The eel goby lacks a specialized accessory respiratory organ for air breathing but has developed a dense capillary network in the inner opercular epithelia. The cardiovascular anatomy of O. lacepedii basically retains the typical circulatory pattern of teleosts. A single ventral aorta emanates from the bulbus arteriosus, giving rise to four pairs of afferent branchial arteries, each of which subdivides into afferent filamental arteries. The afferent filamental arteries supply blood to secondary lamellae. After traversing the gills, oxygenated blood is collected into four pairs of the efferent branchial arteries to perfuse the systemic circuit. There is no vascular shunt connecting the afferent and efferent sides of the gills. Respiratory capillaries on the inner epithelia of the opercular cavity are supplied from the effluent blood from the 1st and 2nd gill arches and drained by the anterior cardinal vein, which also drains venous blood from the head. The anterior cardinal vein unites with the posterior cardinal vein (returning venous blood from the trunk) to form the ductus Cuvieri, which opens into the sinus venosus together with the hepatic vein. Thus, O_2-rich blood from the air-breathing organ will probably mix with systemic venous blood before and during its passage through the heart.

Mudskippers

Similar to eel gobies, mudskippers lack specialized accessory respiratory organs. Their gas exchange surface is composed of epithelia lining the buccal, pharyngeal and opercular cavities, the gills and the skin. The lack of specialized respiratory organs is not unique to mudskippers but prevalent among marine amphibious, air-breathing fish, such as blennies. The suprabranchial chamber, which was reported for Periophthalmodon schlosseri by and for another oxudercine Pseudapocryptes lanceolatus (=Pseudapocryptes elongatus, the latter is the valid scientific name of the species, by Yadav and Singh , appears to be no more than a simple expansion of the bucco-pharyngeal cavity and, in structural complexity, is nothing like the suprabranchial chambers of the freshwater air-breathing fishes.

Mudskippers retain the basic gill plan seen in purely aquatic teleosts, as in eel gobies. However, the gills of mudskippers show various degrees of development, from well-developed gills in more aquatic Scartelaos and Boleophthalmus to highly atrophied morphology, seen in Periophthalmodon and Periophthalmus species. Gill filaments of Periophthalmodon and Periophthalmus are short, sparsely spaced and twisted (Periophthalmus) or frequently bifurcated. The secondary lamellae of Periophthalmodon schlosseri are covered by thick epithelia and frequently fused together to the extent that led Wilson et al. to conclude that the gill lamellae of the fish are generally ineffective for gas exchange.

It is generally held that the gross anatomy of the central vascular system in mudskippers is identical to that seen in typical teleosts. However, there are a few papers that have reported anomalous vascular connections in mudskippers. For instance, Schöttle reported that the ventral aorta of Boleophthalmus viridis (this scientific name is no longer valid, possibly Scartelaos histophorus) extends further anteriorly after giving off the first afferent branchial arteries and bifurcates to become the hyomandibular arteries, which supply the respiratory surfaces of the buccopharyngeal and opercular cavities. Similarly, reported the presence of the bucco-opercular artery branching off from the first afferent branchial artery in Boleophthalmus boddarti, which is not shown in the Schöttle's figure for the same species. In addition, described that the first afferent branchial artery in Pseudapocryptes elongatus sends off the buccoopercular artery at about one third of its origin from the ventral aorta.

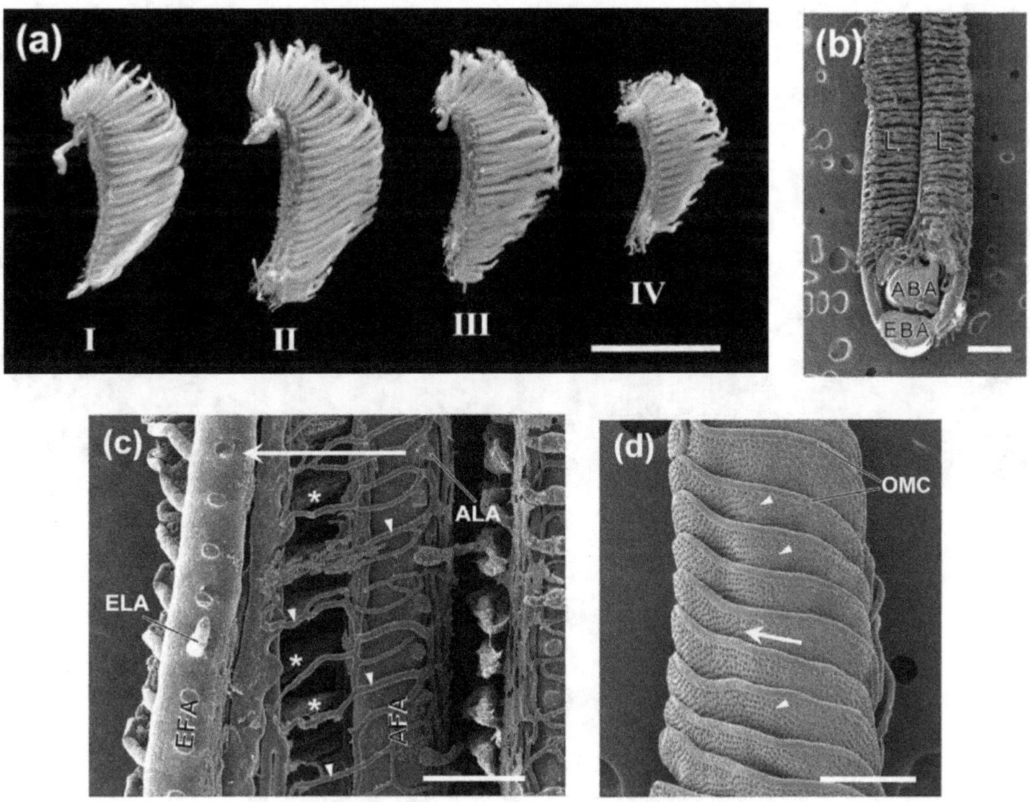

Gill vasculature of Odontamblyopus lacepedii.

Corrosion casts showing the four intact branchial arches (I–IV) of the left gills (a) and vascular details of the filaments and lamellae (b–d). A pair of filaments (holobranch) bears regularly spaced, well-developed lamellae (b). A network of vessels is apparent in the afferent but not efferent filament (c). Several lamellae have been removed from the single filament (hemibranch) to expose the intrafilamental circuit (arrowheads in c). The pillar cells (holes in the lamellae denoted by arrowheads in d) are arranged more or less in parallel with the outer marginal channel (OMC, d). Asterisks in (c) denote the discontinuous inner marginal channels (IMCs) of the lamellae. Arrows in (c and d) indicate the direction of blood flow. ABA afferent branchial artery; AFA afferent filamental artery; ALA afferent lamellar arteriole; EBA efferent branchial artery; EFA efferent filamental artery; ELA efferent lamellar areriole; L lamella. Scale bars: (a) 1 cm; (b) 500 μm; (c, d) 200 μm.

In spite of the general retention of a teleostean circulatory pattern at a gross anatomical level, more subtle modifications are evident in the mudskipper's circulatory system at microscopic levels. The most obvious is rich vascularization of the inner epithelia of the bucco-opercular cavity and the skin. The capillary density in the inner epithelia of the bucco-opercular cavity is so high in Periophthalmodon schlosseri that a surface view of its vascular casts looks as if there is no space between the neighboring capillaries. In addition, the entire length of the branchial arches and the leading, but not trailing, edge of the filaments are supplied by a dense network of capillary vessels. Figure compares the respiratory capillary density of the tongue (a), palate (b) and inner operculum (c) in a purely water-breathing goby, Acanthogobius hasta, a facultative air breathing eel goby, Odontamblyopus lacepedii and two species of mud skippers. Increasing air breathing capability is reflected by the progressively higher density of the capillaries in these epithelia, culminating in Periophthalmodon schlosseri, which has nearly the same capillary density on all these epithelia. Zhang examined the histology of the respiratory epithelia in oxudercine gobies and demonstrated rich vascularization of the inner epithelium of the operculum in Boleophthalmus boddarti, B. dussumieri, B. pectinirostris, Oxuderces dentatus, Periophthalmodon septemradiatus, P. schlosseri, Periophthalmus chrysospilos, P. modestus, Pseudapocryptes elongatus, and Scartelaos histophorus.

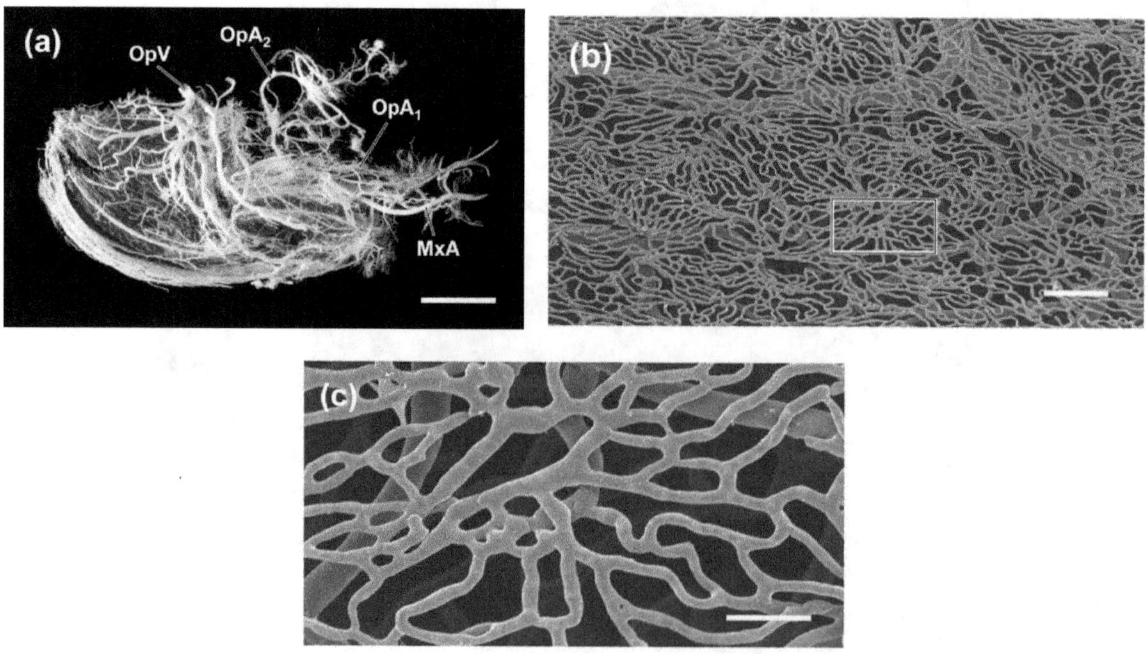

Lateral view of the corrosion cast of the right operculum of Odontamblyopus lacepedii.

Anterior to the right (a). The inner epithelial surface of the operculum is overlaid by a dense network of capillary vessels (b). The boxed area in (b) is shown in a higher magnification SEM photograph in (c). MxA maxillary artery; OpA1 opercular artery from the anterior end of the 1st efferent branchial artery; OpA2 opercular artery from the distal end of the 1st efferent branchial artery; OpV opercular vein. Scale bars: (a) 5 mm; (b) 200 μm; (c) 50 μm.

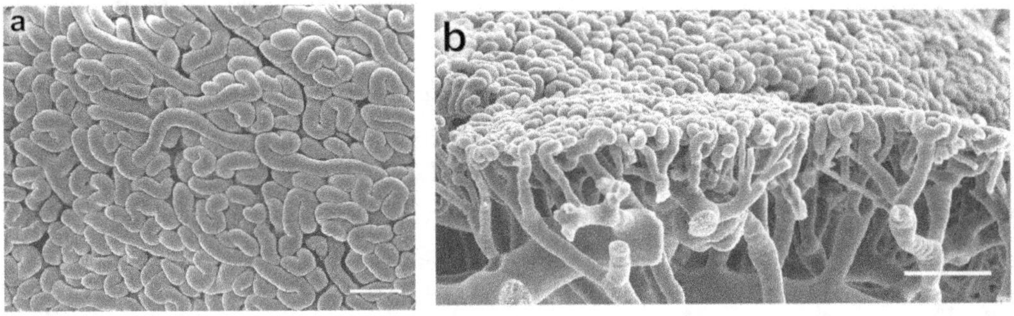

Corrosion casts showing the surface.

(a) transverse (b) views of the capillary network covering the inner operculum of Periophthalmodon schlosseri. In transverse view, the capillaries are shown connected to an underlying intermeshed network of arteries and veins. Each capillary forms a spiral coil as it emerges from the surface (b). Bars 20 μm (a), 50 μm (b).

Periophthalmodon schlosseri shows a specialized microvascular arrangement of the secondary lamellae. The vascular pathway of the secondary lamellae consists of 5–8 parallel blood vessels, rather than forming lamellar sinusoids (narrow vascular spaces delineated by pillar cells). The internal diameter of the lamellar vessels of P. schlosseri (14.1 ± 2.5 μm) is larger than the size of the red blood cells (11.1 ± 8.8 μm) and would therefore impose less vascular resistance to blood than typical lamellar sinusoids. In aquatic fishes with typical secondary lamellae, the erythrocytes are squeezed, deformed and meander during their transit through the lamellar sinusoids, possibly to enhance oxygen uptake. During transit through the gills, 20 to 40% of the ventral aortic pressure is lost in typical water-breathing fishes. The channelization of the lamellar vessels in P. schlosseri, together with interlamellar fusion and thickening of the lamellar epithelia, would also help maintain sufficient blood flow to perfuse the aerial respiratory capillaries and the systemic beds, both located downstream of the lamellae, when the fish breathes air. Similar channelization of lamellar vasculature has also been reported in a few freshwater air-breathing fishes. It is currently unknown whether other mudskippers have similar vascular specialization in their lamellae.

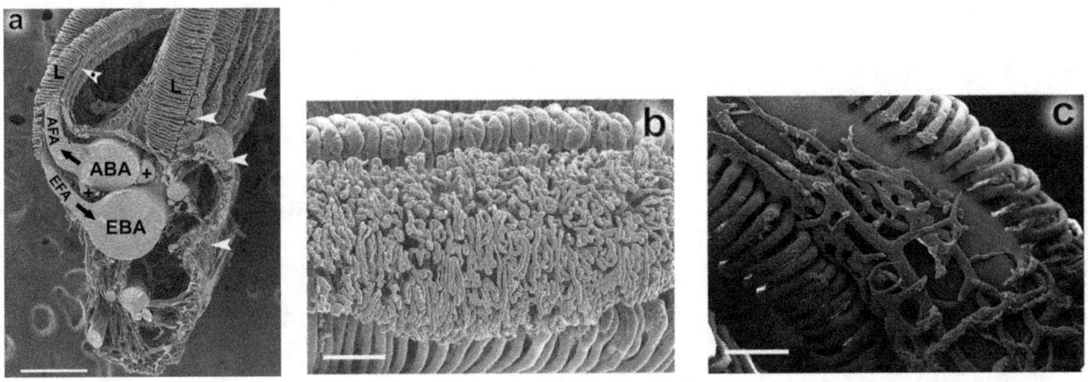

Cross-section through a corrosion cast of the right first branchial arch in Periophthalmodon schlosseri.

(a). Surface views of the leading (b) and trailing (c) edges of a filament are shown with a higher magnification. The branchial arch and leading edge of the filaments are supplied by a dense network of capillary vessels (white arrowheads in a), while no such capillary network was observed on the trailing edge (white arrowhead with black dot inside in a). Plus signs are the branchial veins. Black arrows in (a) indicate the direction of blood flow. Bars 500 μm (a), 100 μm (b, c). ABA afferent branchial artery; AFA afferent filamental artery; EBA efferent branchial artery; EFA efferent filamental artery; L lamella.

The mudskipper skin is in general richly vascularized, although capillary density in the outermost layer of the epidermis varies between species and also among body parts within a species. Skin capillaries are distributed in higher densities and with shorter diffusion distance over more exposed parts of the body than in ventral surfaces, which are in contact with the mud when the mudskippers emerge. In Boleophthalmus and to a lesser extent in Scartelaos, the entire body surface, except the ventral surface, is studded with characteristic papillary protrusions called "dermal bulges". An artery runs through the dermis into each dermal bulge at its center and sends off radially arranged capillaries. Effluent blood from the dermal bulge is drained by a vein running parallel to the artery Periophthalmodon or Periophthalmus lacks a dermal bulge over their skin.

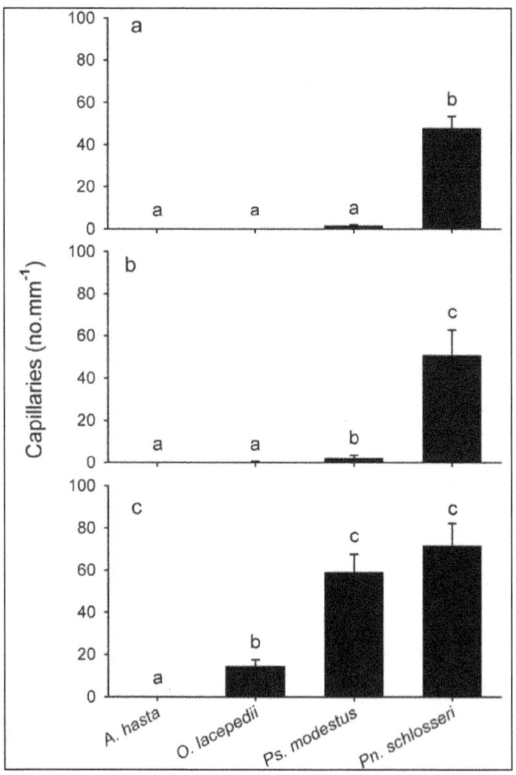

Comparison of capillary densities within the 0–10 μm diffusion distance range in the tongue (a), palate (b) and operculum (c) among four species of gobies. Acanthogobius hasta is a non-air breather, Odontamblyopus lacepedii is a facultative air breather and Periophthalmus modestus and Periophthalmodon schlosseri are amphibious mudskippers. Values are mean + s.d. for 3 individuals. Different letters above the bars in each panel indicate significant differences between species in capillary density in respective epithelia.

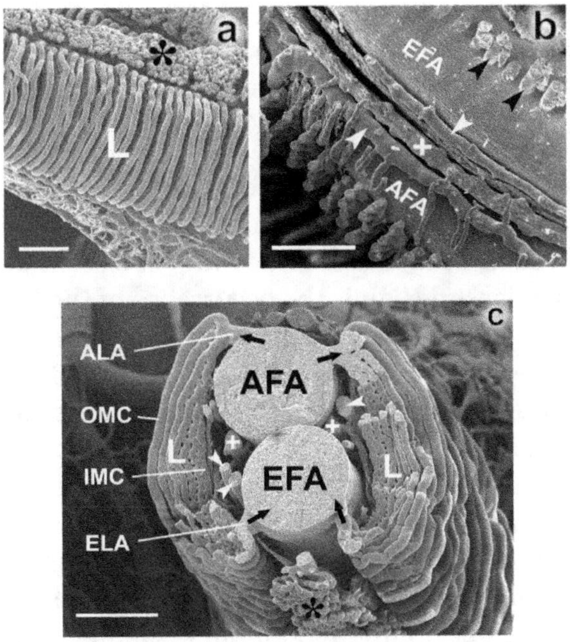

A corrosion cast of Periophthalmodon schlosseri to show a filament of the left second branchial arch showing the lamellae covering one side of the filament (a) and the interlamellar circuit when lamellae were removed (b). A cross-section through a filament shows the vascular details of the arterio-arterial and interlamellar circuits (c). Plus signs in (b) and (c) are filamental veins; white arrowheads in (b) and (c) are interlamellar vessels of unknown origin; black arrowheads in (b) are the efferent lamellar arterioles; asterisks in (a) and (c) are capillaries in the leading edge of the filament. Black arrows in (c) indicate the direction of blood flow. Bars 100 μm (a–c). AFA afferent filamental artery; ALA afferent lamellar arteriole; EFA efferent filamental artery; ELA efferent lamellar arteriole; IMC inner marginal channel; L lamella; OMC outer marginal channel.

Respiratory capillaries of the bucco-opercular cavity and the skin are believed to originate from the systemic arteries such that systemic beds and aerial respiratory beds are connected in parallel. In this design, the aerial respiratory capillaries are supplied with arterial blood having the same Po2 and oxygen content as systemic capillaries would receive, which may be considered as physiologically ineffective. Oxygenated blood from the respiratory epithelia is drained by systemic veins and therefore, likely mixed with deoxygenated systemic venous blood before returning to the heart.

Swamp Eels

The main aerial respiratory surface resides in the buccopharyngeal and opercular epithelia in Monopterus swamp eels. As is generally the case in other air-breathing fishes, the gills become atrophied with increasing reliance on air breathing. Thus, exclusively waterbreathing Macrotrema caligans has four pairs of holobranchs resembling those of purely aquatic teleosts, as does facultative air-breathing Synbranchus, whilst the filaments are arranged alternately along the length of a gill arch in Synbranchus. Species of Monopterus show different degrees of reductions in gill dimension; the most advanced stage of gill specialization is seen in Monopterus boueti, in which “the gill filament becomes a small, finger-like appendage devoid of secondary filaments”. Rosen and Greenwood also stated “Monopterus boueti has no free gill filaments on any arch; the

1st to 3rd arches each carrying a narrow flange of tissue, the flange on the 1st arch being the smallest". Monopterus cuchia has filaments or filamentallike structures only on the 2nd and 3rd gill arches, but Liem's diagram indicates that the gill filaments also occur on the 1st gill arch.

Cardiovascular anatomy has been described in detail only for Monopterus ablus and Monopterus cuchia. Rosen and Greenwood presented simplified diagrams of the central vascular system of Synbranchus marmoratus, Synbranchus madeirae, Ophisternon afrum, Monopterus albus, Monopterus cuchia and Monopterus boueti Anatomical issues). A consistent deviation of the synbranchid vasculatures from the typical teleost design is the occurrence of the hyoidean artery branching off from the elongate ventral aorta to supply the respiratory vasculature within the mouth. Thus, the respiratory capillaries of this region are perfused directly with ventral aortic blood (similar to the descriptions for Boleophthalmus viridis, Boleophthalmus boddarti and Pseudapocryptes elongatus. Even more notable is the presence of large thoroughfare (shunt) vessels in the 4th gill arch in all Monopterus species, which form direct conduits between the ventral and dorsal aortae. To my knowledge, a direct vascular connection between the two aortae has been described only for these Monopterus species among teleosts and possibly, for Electrophorus electricus . Oxygenated blood from the aerial respiratory surface is drained by the anterior cardinal vein and the inferior jugular vein as in Odontamblyopus, Periophthalmodon and Channa. demonstrated for Monopterus cuchia that afferent and efferent filamental arteries are connected by single channel capillaries or several parallel capillaries supplied by the lamellar arterioles.

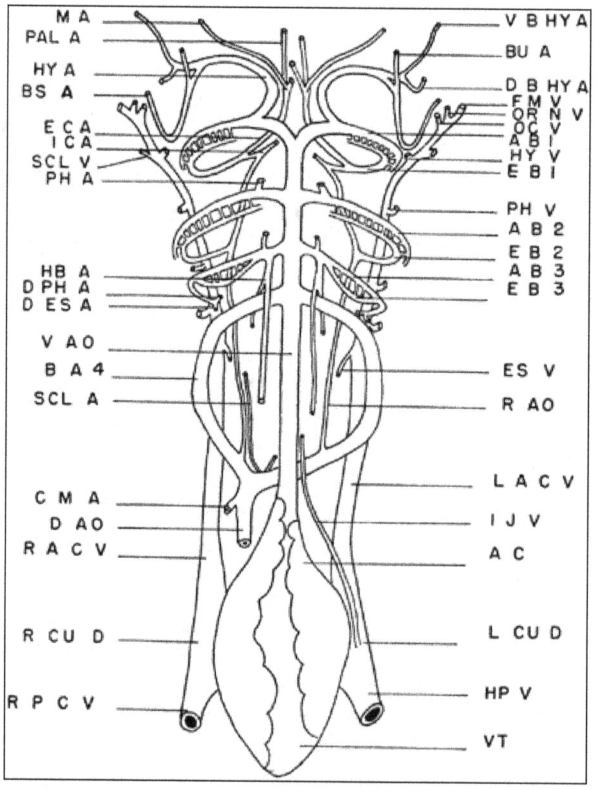

The branchial and cranial blood vessels of Fluta alba. Ventral view.

AB1-3 1st to 3rd afferent branchial arteries; AC auriculae cordis; BA4 4th branchial artery; BS A branchiostegal artery, BU A buccal artery; CMA coeliacomesenteric artery; D AO dorsal aorta; D B HY A dorsal branch of hyoidean artery; D ES A dorsal esophageal artery; D PH A dorsal pharyngeal artery; EB1-3 1st to 3rd efferent branchial arteries; E C A external carotid artery; ES V esophageal vein; F M V facialis - mandibularis vein; HB A hypobranchial artery; HP V hepatic vein; HY A hyoidean artery; HY V hyoidean vein; I C A internal carotid artery; I J V inferior jugular vein; L A C V left anterior cardinal vein; L CU D left Cuvierian duct; MA mandibular artery; OC V occipital vein; OR N V orbitonasal vein; PAL V palatine vein; PH A pharyngeal artery; PH V pharyngeal vein; RAC V right anterior cardinal vein; R AO radix aorta; R CU D right Cuvierian duct; RPCV right posterior cardinal vein; SCL A subclavian artery; SCL V subclavian vein; V AO ventral aorta; V BH A ventral branch of hyoidean artery; VT ventricle.

The cardiac anatomy of Monopterus albus was described by Liem and Lai et al. and that of Monopterus cuchia by Munshi and Mishra. The heart of all Monopterus species is situated far posterior to the usual position of teleost hearts at the level of the pectoral girdle and therefore has an unusually long ventral aorta. Liem described that the fish has three sino-atrial valves in Monopterus albus. Sinoatrial valves are absent from Monopterus cuchia. Liem stated "A rather distinct dorsal median groove seems to divide the sinus venosus partly into two halves", which was later confirmed by Munshi and Mishra. The atrium of Monopterus has abnormal morphologies located dorsal to and surrounding the bulbus arteriosus. It is unknown whether these peculiarities have any functional significance.

Snakeheads

The air-breathing organ of snakeheads forms a pair of diverticula, often called suprabranchial chambers, located in the dorsal side of the buccopharyngeal cavity. The organ is protected dorsally by the heavy, bony plates of the skull and posterolaterally communicates with the opercular cavities. Into the lumen of the suprabranchial chambers, three pairs of cartilaginous processes project, anteroposteriorly from the parasphenoid, hyomandibular and epibranchial of the 1st gill arch. Among them, the hyomandibular process is the largest and almost completely traverses the lumen of the organs. The respiratory surface of the organ is studded with small papilla-like projections, which increase in number and complexity with fish growth. In addition, the aerial respiratory surface extends to the roof of the buccal cavity anterior to the suprabranchial chambers and on the surface of the tongue. The gill filaments are shorter compared with those of the strictly water-breathing fishes. The 1st to 3rd gill arches are larger and composed of complete holobranchs, while the 4th arch is much smaller and the inner hemibranch of the 4th arch is rudimentary in Channa argus, Channa maculata, and Channa marulius. In comparison, the 4th gill arch of facultative air-breathing Channa punctata is anatomically similar to the other three arches, although smaller in size.

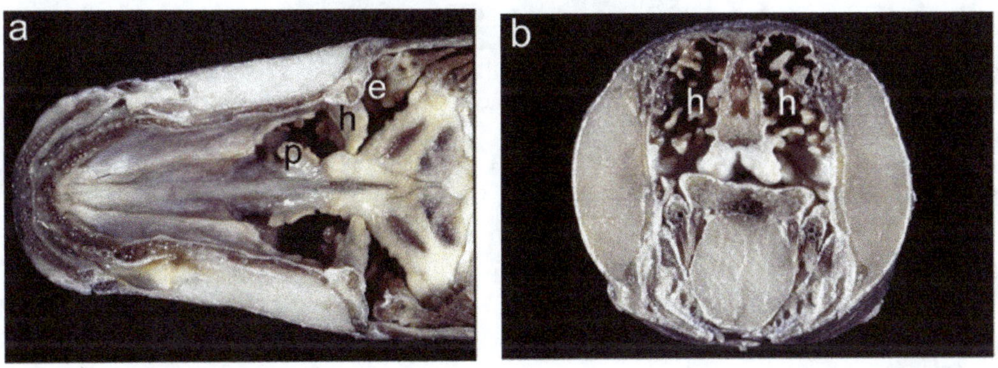

Horizontal (palatine view, a) and cross (frontal view, b) sections of the head of Channa argus to show the suprabranchial chambers. Note that cartilaginous processes from the parasphenoid (p), hyomandibular (h) and the epibranchial of the 1st gill arch (e) projecting into the chambers (a). The inner margin of the hyomandibular processes (h) abuts to the inner wall of the chambers and thereby divides the cavity when the operculum is adducted (b). The capillary network extends to the palatine region anterior to the chambers and the surface of the tongue.

The central cardiovascular system of Channa species shows several unique deviations from typical teleost design. Most notable is the presence of two (anterior and posterior) ventral aortae. Equally remarkable is the almost complete separation of the respiratory and systemic circuits on the arterial side. Thus, the vascular route originating from the anterior ventral aorta takes on the majority of the respiratory role, whereas the posterior vascular pathway is specialized almost as a conduit between the ventral and dorsal aortae in snakeheads. The anterior ventral aorta gives off the afferent branchial arteries of the 1st and 2nd arches, which exclusively perfuse the capillary bed of the air breathing organ but have no vascular connection with the systemic arteries. On the other hand, the posterior ventral aorta sends blood to the 3rd and 4th afferent branchial arteries, which after passing through gill arch vasculatures, supply the entire systemic arterial system. The description by Munshi et al. on the postero-ventral and antero-dorsal origins of the posterior and anterior ventral aortae, respectively from the bulbus arteriosus, seems to be due to misinterpretation of their casts (Ishimatsu, personal observation). Earlier descriptions also support the dorsal origin of the ventral aorta to the 3rd and 4th gill arches. Other morphological specializations of the Channa heart include absence of the sino-atrial valves, the highly trabeculate nature of the ventricle and muscular ridges running longitudinally on the wall of the bulbus arteriosus.

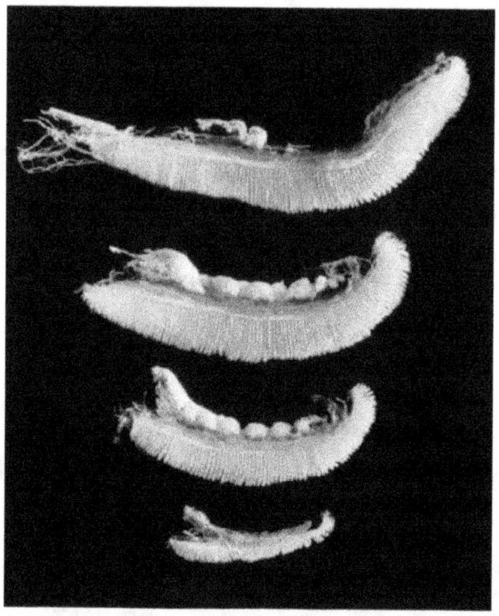

Lateral views of the corrosion casts of the left gill arches of Channa argus. From top to bottom, the 1st to 4th gill arches. Note the much smaller size of the 4th gill arch.

Respiratory vasculature of the air-breathing organ originates from the 1st and 2nd efferent branchial arteries; the 1st one irrigates the antero-dorsal and lateral surfaces of the organ, while the 2nd one supplies the postero-ventral surfaces. Each terminal respiratory capillary forms a coiled spiral with several loops parallel to the respiratory surface. Blood oxygenated in the airbreathing organ is drained by tributaries of the anterior cardinal vein, which also drains systemic vascular beds of the head. The inferior jugular vein drains the respiratory surface on the tongue.

At microscopic levels, the lamellar vasculature of the 1st to 3rd gill arches is modified into parallel channels, as in Periophthalmodon schlosseri. The filaments in the 4th gill arch of obligate air-breathing snakeheads are atrophied to the extent that the inner hemibranch bears only very short filaments with nearly no secondary lamellae and has direct vascular connections (shunts) between the afferent and efferent branchial arteries. The 4th - arch gill vasculature of facultative airbreathing C. punctata was reported to lack shunt vessels.

Lungfishes

Lungfishes are distinct from most other living airbreathing fishes in that they possess "true" lungs. The lungs of all lungfishes are positioned dorsal to the alimentary canal as a single (in Neoceratodus) or paired (in Protopterus and Lepidosiren) sac. The pneumatic duct opens in the ventral wall of the esophagus and runs dorsally round the right side of the esophagus. Internally, the lungs are subdivided into numerous compartments or alveoli by progressively dividing septa and trabeculae.

Neoceratodus has four pairs of holobranchs and a pseudobranch on each side of the head, enclosed by an operculum. Each holobranch bears well-developed gill filaments, which carry densely spaced secondary lamellae on either side. The gill microvasculature of Neoceratodus is reminiscent of that reported for elasmobranches; each afferent filamental artery delivers blood to a flattened plexus of interconnecting vessels, the corpus cavernosum, from which the afferent lamellar arterioles feed the lamellar sinusoids of the well-developed secondary lamellae. The efferent branchial arteries of all four gill arches of N. forsteri unite to form the dorsal aorta and other major arteries of the head. In addition, the pulmonary artery originates from the 4th efferent branchial artery (i.e., 6th aortic arch) of either side. Parallel to the curved course of the pneumatic duct described above, the left pulmonary artery from the 4th left efferent branchial artery loops around the esophagus before it reaches the ventral wall of the lung. The 4th right pulmonary artery branches off from the right efferent branchial artery to perfuse the dorsal wall of the lung. The two pulmonary veins drain the lung, join to form the common pulmonary vein running down on the right side of the esophagus and empty into the left side of the atrium.

The gills of Protopterus are remarkably specialized. The 1st arch bears a hemibranch, which Laurent et al. considered to be the pseudobranch. The 2nd and 3rd arches almost completely lack the gill lamellae and instead, form direct pathways between the heart and the dorsal aorta. The posterior three pairs of arches (the 4th to 6th arches) bear the gills. The gill respiratory microvasculature of Protopterus is composed of a loose network of interconnecting vessels, which is radically different from the lamellar sinusoids of Neoceratodus and most teleosts. The primary lamellae of Protopterus aethiopicus have afferent-efferent shunts, which dilate during aestivation. The efferent branchial arteries of these posterior gill arches unite and send off two major vessels, i.e., the pulmonary artery to perfuse the lungs and the ductus arteriosus leading to the dorsal aorta.

Comparison of the central circulatory system of Channa and of typical teleosts (a) and the photograph of the heart of Channa argus after a sagittal section was cut on the median plane (b). Two ventral aortae arise from the bulbus arteriosus. The anterior aorta extends anteriorly but is partly cut between the 1st and 2nd afferent branchial arteries in b. The posterior aorta runs more dorsally and soon divides into the 3rd and 4th afferent branchial arteries. Note also the highly trabeculate ventricular lumen, muscular ridges running on the inner wall of the bulbus and the absence of the sino-atrial valve. The ventral vascular wall of the buccal cavity is also highly vascular, supplied with blood from the efferent side of the 1st and 2nd gill arches (dotted vessels in a). Arrows in a indicate the direction of blood flow. A atrium; ABO air-breathing organ; acv anterior cardinal vein; ava anterior ventral aorta; BA bulbus arteriosus; ccv common cardinal vein (=ductus Cuvieri); da dorsal aorta; hv hepatic vein; ijv inferior jugular vein; pcv posterior cardinal vein; pva posterior ventral aorta; scv subclavian vein; SV sinus venosus; V ventricle; va ventral aorta; VW vascular wall of the buccal cavity.

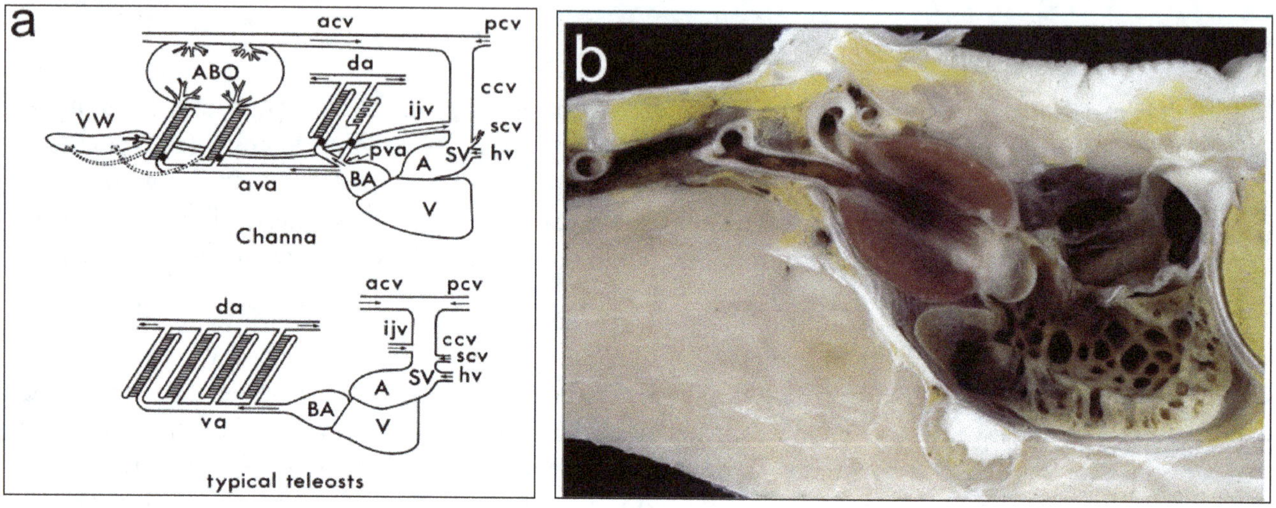

Information on the gills of Lepidosiren is more limited and confusing. reported that "In Lepidosiren, there are six branchial arches which are short and rodlike. Gills are found on the last three branchial arches only. These are small and each consists of 8–15 lobe-like filaments of varying size, some with irregular, knob-like secondary and often tertiary branches". In contrast, de Moraes et al. demonstrated that gills are present on five gill arches; "irregularly arranged, short, papillar gill lamellae are supported by the five gill arches: hyoid and branchial arches 1, 2, 3, and 4". Regarding the central vasculature of Lepidosiren, Robertson described that "four afferent vessels arise in two sets of two in close proximity on either side, from the very short ventral aorta at the anterior end of the bulbus cordis". According to Robertson, these are the aortic arches 3, 4, 5, and 6. She also described afferent and efferent vessels of the hyoidean hemibranch originating from the 3rd aortic arch. I am not aware of any recent publication on the vascular anatomy of Lepidosiren.

The atrium and ventricle of all lungfishes are incompletely divided into right and left sides by median septa. Systemic venous blood from the central veins (the vena cava, the right and left ductus Cuvieri and the hepatic veins) is drained into the sinus venosus and conveyed to the right side of the atrium and ventricle. Pulmonary venous blood is separately drained by the pulmonary vein, which fuses with the pulmonary fold within the sinus venosus, and returns to the left side of the atrium. Protopterus and Lepidosiren show a somewhat higher degree of the specialization of the cardiac anatomy than Neoceratodus, in line with the higher dependence on aerial gas exchange in the former two genera. The bulbus cordis of Protopterus and Lepidosiren takes a characteristic S-shaped bend and contains a spiral fold and an opposing shorter fold, which nearly completely divides the lumen into two channels. Most anteriorly, a horizontal septum divides the bulbar lumen to the ventral and dorsal channels; the ventral channel gives rise to the vessels to the anterior arches, while the dorsal channel sends vessels off to the posterior arches. The bulbus cordis of Neoceratodus also takes a similar S-shaped course and contains a spiral fold in its proximal portion. However, it has four rows of longitudinal valves in the distal portion and lacks the septum in its most cranial end. The disposition of the vessels emanating from the bulbus is similar to that in the other two genera, i.e., the anterior arch arteries come out ventrally and the posterior arch arteries dorsally.

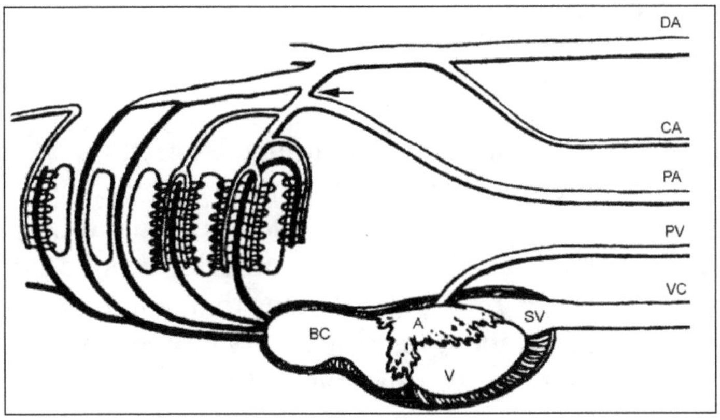

Schematic drawing of the central circulation in Protopterus aethiopicus. A atrium; BC bulbus cordis; CA coeliac artery; DA dorsal aorta; PA pulmonary artery; PV pulmonary vein; SV sinus venosus; V ventricle; VC vena cava.

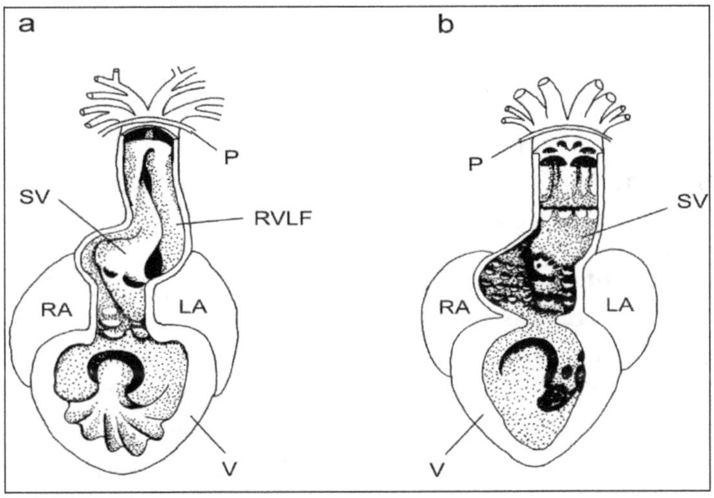

The hearts of Protopterus (a) and Neoceratodus (b).

The ventricle and bulbus cordis are cut to show the internal structure (ventral view). LA, left part of the atrium; P, pericardium; RA, right part of the atrium; RVLF, right ventrolateral folds; SV, spiral valve; V, ventricle.

Physiology of the Cardiorespiratory System in Aquatic Air-breathing Fishes

Eel Gobies

Relatively little is known about the cardiorespiratory physiology of eel gobies, except the air-breathing threshold at which Odontamblyopus lacepedii switches from aquatic to bimodal respiration, total gas volume within the air-breathing organ and the tidal volume relative to the total gas volume. Odontamblyopus lacepedii is a facultative air-breather, which satisfies its oxygen requirement entirely through aquatic gas exchange in normoxic water, and starts air breathing only when water Po2 is decreased to around 20 torr. The volume of its air breathing organ is about 5% of its body volume and the entire volume of the gas is renewed in each breathing cycle, as evidenced by near matching of the total gas volume and the tidal volume. The duration of breath holding ranges from a few seconds up to about 30 min.

Mudskippers

Mudskippers are distinct from most aquatic airbreathing fishes in their respiratory capability on land. First, mudskippers are capable of maintaining oxygen uptake at a similar or even higher rate in air than in water. Second, mudskippers are able to eliminate CO_2 at a normal rate while out of water, and thereby maintain blood pH. Third, mudskippers do not need water for aerial ventilation of their air-breathing organ. However, other than comparisons of aerial and aquatic respiratory rates, not much is known about the respiratory physiology of mudskippers. The aerial tidal volume of Periophthalmodon schlosseri and found that it amounts to 54% of the total volume of the air-breathing organ (buccopharyngeal and opercular cavities). Similar to aerial ventilatory responses to air and water Po_2/Pco_2 shown for Channa argus, aquatic hypoxia had little, if any, effect on aerial ventilation, while aerial hypoxia greatly stimulated it. Unlike the findings for Channa argus, however, both ventilatory frequency and tidal volume increased under hypoxia. Thus, Periophthalmodon schlosseri is able to improve gas exchange efficiency by renewing a larger percentage of the gas in the airbreathing organ in response to limited O2 availability. The respiratory system of Periophthalmodon schlosseri is likely better adapted to an aerial rather than aquatic environment. This has been supported by the fact that the fish can repay oxygen debts after exhaustive exercise in air but not in water. When the fish was placed in air after exhaustive exercise, oxygen uptake rate immediately increased 2.5 times, whereas no such increase occurred when the fish was confined in water after exercise.

Compared with the amount of data available on the gas exchange physiology of mudskippers, far less is known about the blood gas levels and cardiovascular physiology of these fishes. This lack of data stems mainly from the difficulty of working with mudskippers, which usually weigh from only a few grams to 50 g. Electrocardiograms have been recorded for only a few species of mudskipper. An exception is Periophthalmodon schlosseri (larger specimens weighing over 200 g). We have shown that Periophthalmodon schlosseri maintained normal heart rate and blood gas levels when they were placed on land. Arterial Po_2 was unaffected, Pco_2 slightly decreased and pH became higher by air exposure. Neither arterial blood pressure nor heart rate significantly changed during the air exposure or recovery. In contrast, forced submersion may elicit significant responses to the cardiorespiratory parameters of mudskippers. Garey was the first to demonstrate a lowering of heart rate upon submersion (water oxygen level not stated) in a mudskipper (Periophthalmus australis = Periophthalmodon freycineti. We also have shown in Periophthalmodon schlosseri that heart rate rapidly decreased in response to forced submersion in normoxic seawater but rose to a higher than pre-submersion level when the fish took the first air breath upon re-emersion. Blood pressure was depressed during submersion and did not show any overshoot upon re-emersion. Kok et al. reported that a similar decrease in heart rate occurred during submersion in Periophthalmodon schlosseri but not in Boleophthalmus boddarti. Martin and Bridges reported relative changes in heart rate, cardiac output and stroke amplitude of Periophthalmus argentilineatus under aquatic hypoxia, hyperoxia and emersion.

Swamp Eels

Swamp eels are relatively well studied for their cardiorespiratory functions because of the peculiar arrangement of the central vascular connections. In addition, the relatively large size of these fishes (up to the body length of 1.5 m,) facilitates the application of modern physiological techniques for blood flow and pressure measurements, and blood sampling. The facultative air-breathing Synbranchus marmoratus relies exclusively on water breathing in normoxic water and switches to air breathing when water Po_2 declines to 40–50 torr. In contrast, the obligate airbreathing Monopterus cuchia satisfies about 65% of its O_2 requirements by aerial gas exchange even in normoxic water with free access to air. In response to breathing air in hypoxic water, cardiac output increased by 70% as a result of a ca. 35% increase in stroke volume and heart rate in S. marmoratus. Mean central filling pressure also increased in response to air breathing, which likely resulted in a rise in stroke volume in spite of an increased heart rate.

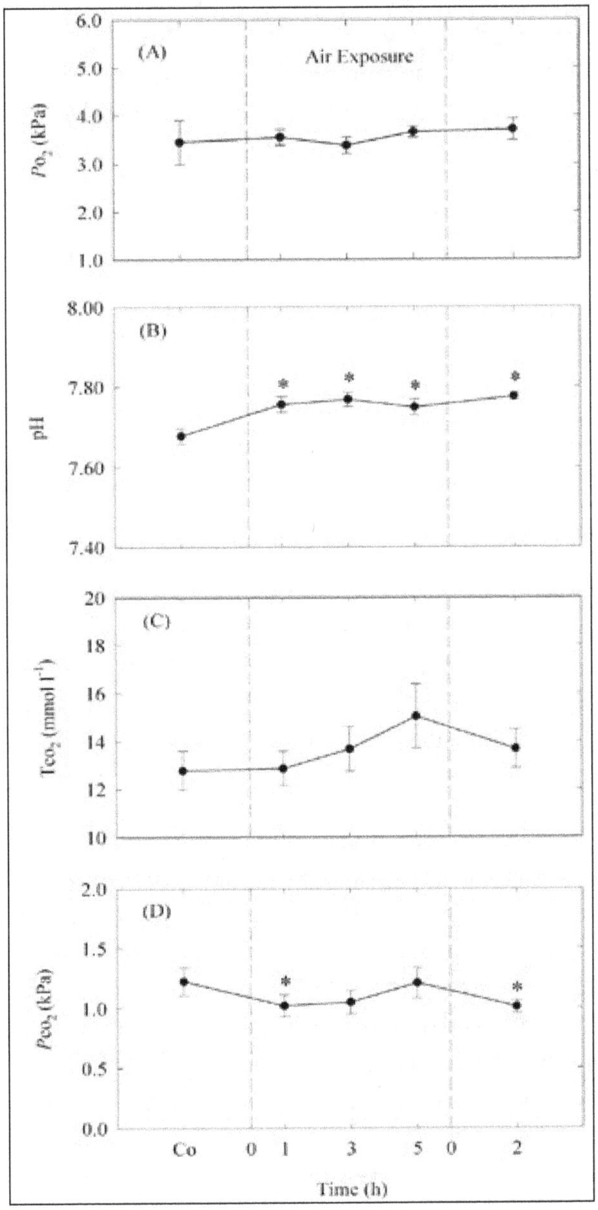

Changes in arterial Po_2 (A), pH (B), total CO_2 (TCO_2 , C) and Pco_2 (D) of Periophthalmodon schlosseri under control (Co) conditions and during 6 h of air exposure and 2 h of recovery. Before and after the air exposure, the fish were partially immersed in normoxic 50% sea water with free access to air. Values are means ± S.E.M. (N = 7). Asterisks indicate significant differences from the corresponding control value ($P < 0.05$).

Blood flow pattern during air breathing and breath holding was estimated for Monopterus cuchia by Johansen from the data of Lomholt and Johansen. The analysis was based upon the measured values of Po_2 and pH of arterial and venous blood, O_2 capacity and O_2 dissociation curve of the blood, and oxygen uptake rates, assuming complete equilibration of blood and gas across the capillary in the air-breathing organ. Even though some assumptions were not fully explained in the text, the analysis demonstrated that during air breathing when the air-breathing organ was filled with air of relatively high Po_2, a large fraction of cardiac output was directed to the airbreathing circuit, whilst during breath holding when the air was expelled from the organ, by far the greater portion of cardiac output perfused the systemic vascular bed. This led Johansen to the idea of temporal, not spatial, separation of systemic and airbreathing circuits in these fishes. Johansen assumed total mixing of systemic venous blood and O_2- rich blood from the air-breathing organ on the basis of the anatomy of the central circulatory system.

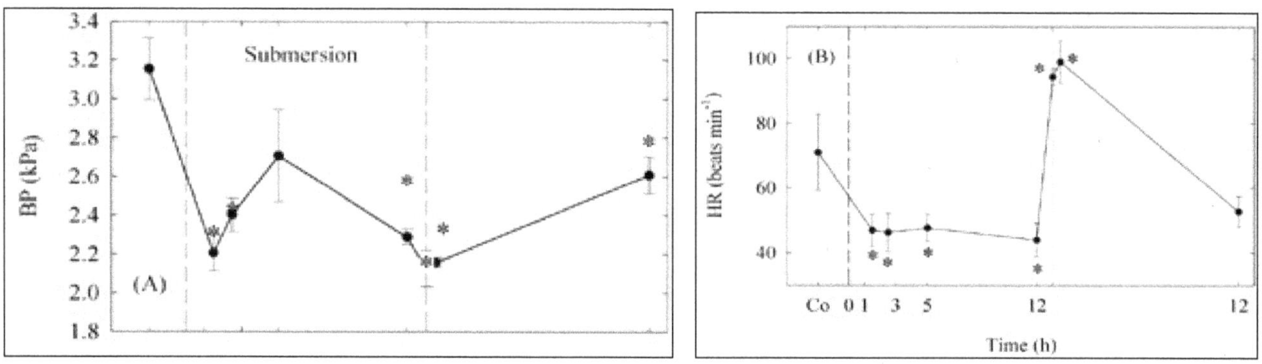

Changes in blood pressure (A) and heart rate (B) of Periophthalmodon schlosseri under control (Co) conditions and during 12 h of forced submersion in normoxic water without access to air and 12 h of recovery. Before and after the submersion period, the fish were partially immersed in normoxic 50% sea water with free access to air. Values are means ± S.E.M. (N = 6). Asterisks indicate significant differences from the corresponding control value ($P < 0.05$).

Snakeheads

Channa argus takes up 60% and 85% of its total oxygen requirement by air breathing when breathing bimodally in normoxic and hypoxic (water Po2 ~ 40 torr) water, respectively. In contrast, aerial gas exchange makes only a minor (15%) contribution to total CO_2 elimination, as in most aquatic air-breathing fishes. The breathing pattern always consists in a rapid breathing movement interspersed with variable periods of breath holding. Upon surfacing, the fish expires the gas inside the air-breathing organ before inspiring fresh air. Aerial ventilation of the air-breathing organ is effective only in water since the gas is expelled by a reversed flow of water from the gill openings, which will completely replace the gas in the organ and thereby attain a complete renewal of the gas upon inspiration. When the fish is experimentally placed out of water, the blood Pco_2 in creases significantly and the fish suffers from uncompensated respiratory acidosis in spite of greatly increased ventilatory frequency. During bimodal respiration, aerial ventilation is strongly enhanced by aerial hypoxia but not by lowered aquatic Po_2. The enhanced aerial ventilation is largely due to increased ventilatory frequency, with little changes in tidal volume. Temperature affects the aerial ventilatory responses to hypoxia such that much higher responses were evoked at 25°C than at 15°C. Aerial hypercapnia had little effect on aerial ventilation of Channa argus.

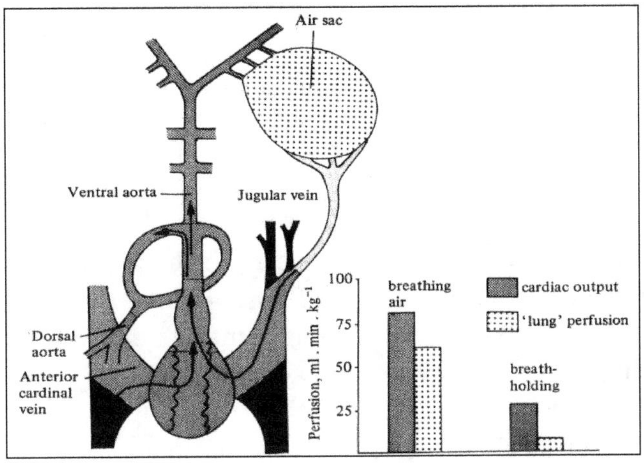

Left: Schematic drawing of the heart and its major in- and outflow vessels in Amphipnous cuchia Note the paired shunt vessels from the ventral aorta forming the dorsal aorta. The effluent blood from the air-breathing organ converges with systemic venous blood in the jugular veins. Right: Air breathing increases cardiac output and the fraction of the cardiac output perfusing the air breathing organ. During breath holding, a major fraction of the cardiac output is shunted from the ventral to the dorsal aorta..

The unusual cardiovascular function of snakeheads is that the fish has the capacity of functional separation of O_2-rich blood from the air-breathing organ and O_2-poor systemic venous blood with no morphological septation within the heart. Determinations of blood oxygen levels from chronically cannulated free-swimming fish revealed that both Po_2 and O_2 content are significantly higher for the blood sampled from the 3rd and 4th afferent branchial arteries originating the posterior ventral aorta than the blood sampled from the 1st and 2nd afferent branchial arteries derived from the anterior ventral aorta. The Po_2 and O_2 content differences between the two ventral aortic blood was 12–22 torr and 0.6–1.4 mmol l^{-1}. Thus, there must be at least partially separated blood streams within the heart of Channa argus, which is presumably brought about by the presence of laminar blood streams since the heart of C. argus lacks any anatomical septation. The morphological specializations of the heart likely help prevent total mixing of the two types of blood and also channeling the outflow from the heart.

Flow dynamics in the central circulatory system of Channa argus was examined by angiocardiography. The results demonstrated nearly complete ventricular emptying at the end of systole and high elasticity of the bulbus arteriosus, which almost completely absorbed ventricular output shortly after the completion of ventricular systole. A preference for a greater volume of blood from the air-breathing organ through the heart to the posterior ventral aorta was suggested by the larger diameter of the X-ray image of and a higher density of contrast medium within the posterior ventral aorta, even though this needs more rigorous investigation.

Regarding possible regulatory mechanisms for bloodflow partitioning between the air-breathing organ and systemic circulations in C. argus, a preliminary data by Ishimatsu et al. on perfused head preparations illustrated that both acetylcholine and adrenaline added to the perfusion fluid increased the resistance of the vascular bed downstream of the anterior ventral aorta, and redistributed perfusion flows away from the air-breathing circuit (as determined in the anterior venous flow) to the general systemic circuit (as determined in the dorsal aortic flow). Stimulation of a vagal branch innervating the air-breathing organ mimicked the effects of acetylcholine or adrenaline. This innervation is probably non-adrenergic, since no adrenergic nerve fibers could be demonstrated in the vasculature of the air-breathing organ with Falck-Hillarp histochemistry.

Lungfishes

Protopterus and Lepidosiren are obligate air-breathing fishes. These lungfishes satisfy nearly 90% of their O_2 requirement through aerial respiration even in normoxic water, although the relative importance of aerial O_2 uptake is size-dependent. Thus, Johansen et al. reported that Protopterus with body weight of 4 g took up 70% of total oxygen demand aquatically but the value diminished to 10–15% in fish larger than 50 g. In comparison, aquatic respiration is generally more important for CO_2 elimination such that 70% of CO_2 elimination is through aquatic breathing even in adult Protopterus. Somewhat in conflict with those earlier findings, Perry claimed that the lung was responsible for 91% of O_2 uptake and 76% of CO_2 elimination in P. dolloi. However, this might be due to very high pulmonary ventilation frequencies of their animals (6–48 breaths min−1 at 25°C), which would enhance aerial CO_2 excretion. Working on Lepidosiren, Amin-Naves et al. tested the effects of acute temperature changes on respiratory partitioning of aerial and aquatic gas exchange. Pulmonary ventilation increased from 0.5 to 8.1 ml BTPS (body temperature and pressure, saturated with water vapor) kg−1 min−1 when temperature was raised from 15 to 35°C, exclusively due to increased frequency of air ventilation. Concomitantly, the relative contribution of aerial respiration to CO_2 elimination increased from 13 to 74% (as estimated from their Figure). Neoceratodus satisfied almost all of its O2 demand through aquatic gas exchange when the fish rested in normoxic water, but commenced air breathing in response to decreasing water Po_2 or during exercise. Control of ventilation in lungfishes has been reported for Neoceratodus by Johansen et al. and Fritsche, for Lepidosiren by Sanchez and Glass for Protopterus by Johansen and Lenfant and Perry. Lungfishes breathe air through the mouth as in all other air-breathing fishes.

The investigation into the cardiovascular dynamics of lungfishes was pioneered by Kjell Johansen and his colleagues. Heart rate and cardiac output of Protopterus aethiopicus increased immediately after each air breath and subsequently subsided, as has been shown for swamp eels and electric eel. A radiological study of P. aethiopicus by Johansen and Hol revealed that injections of a contrast medium in the pulmonary vein preferentially filled the left cardiac chambers to be selectively dispatched to the anterior gill-less arches, which give rise to the major systemic arteries. In contrast, injections of the medium in the vena cava was conveyed into the sinus venosus and then to the right, posterior part of the atrium. Johansen estimated distribution of the systemic and pulmonary venous blood into anterior and posterior arches of Protopterus aethiopicus. Their calculations estimated that over 90% of blood flow through the anterior arches was derived from the pulmonary venous blood immediately after air breathing and then the fraction decreased down to 65% in four minutes. However, these calculations used blood Po_2 but not O_2 content or saturation and therefore, probably overestimated the fraction mainly because of low systemic venous Po_2.

In addition, their experimental animals may not have recovered sufficiently from rather invasive surgical procedures (implantation of five catheters and four electromagnetic flow probes), guessed from the very high frequency of air-breathing (once in several minutes) reported in these studies, as compared with the much lower air-breathing frequency (ca. 5 breaths h^{-1}) reported for intact Lepidosiren by Sanchez and Glass, which is allegedly more dependent on air breathing than Protopterus. Incomplete recovery may also have resulted in a relatively low arterial Po_2 (28–38 torr) reported for Lepidosiren by Johansen et al, as compared with the values (80–90 torr) reported for the same species subjected to less invasive surgery (a dorsal aortic catheter alone) by da Silva et al. estimated %R-L shunt (=Qsh / Qtot × 100, where Qsh is systemic venous blood flow recirculated into the systemic circuit and Qtot is the total systemic blood flow) of Lepidosiren to be 19%, based on the equation given by. This means that systemic blood is composed of 81% of pulmonary venous blood and 19% of systemic venous blood, which agrees with the estimate given by Johansen et al. for Protopterus in the middle of a breath-holding period. Nearly nothing is known about blood separation in Neoceratodus; blood gas analysis by Johansen et al. indicated that 84% of blood flow through the anterior arch originated from the pulmonary venous blood in hypoxic water, but the value declined to 33% in normoxic water.

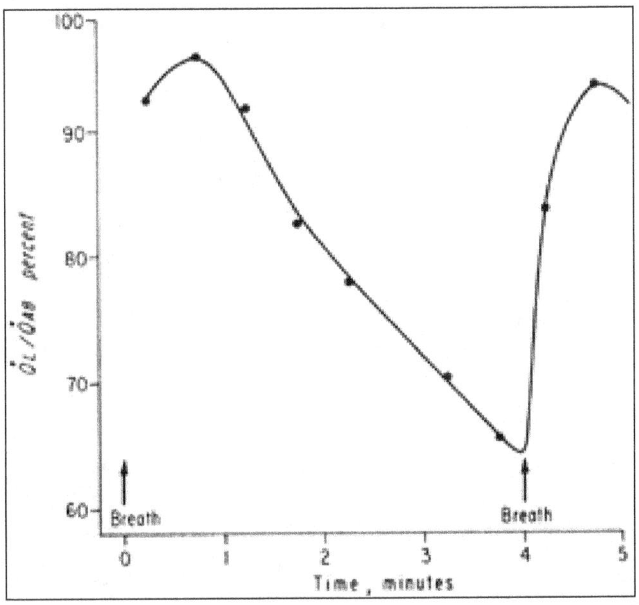

Time course of the proportion of pulmonary flow (Q_L) to total flow perfusing the anterior gill-less branchial arteries (Q_{AB}) during an interval between air breaths in Protopterus aethiopicus.

In his influential review on the evolution of double circulation in vertebrates, Foxon listed common features shared by vertebrates with double circulation. These are (1) separate return of oxygenated and deoxygenated blood into the heart, (2) at least a considerable amount of anatomical guiding of the two streams in the heart, if not complete separation and (3) separate vessels by which the oxygenated and deoxygenated bloods leave the heart.

Separate vessels to drain the air-breathing organ ocur in lungfishes and several 'primitive' air-breathing fishes but it is only in lungfishes that these vessels directly open into the heart. This condition prevails in all three genera of lungfish (note that the statement of pulmonary return only into the sinus venosus for Neoceratodus given by. Amia, Gymnarchus and Polypterus have separate air-bladder veins but these do not reach the heart. Cardiac septum is almost nonexistent in airbreathing fishes except lungfishes. A central septum was reported for the sinus venosus of Monopterus but no detailed account was given. Apart from lungfishes, separate outflow vessels from the heart to the air-breathing organ circuits have been found in Polypterus, Gymnarchus, two closely related catfishes, Clarias and Heteropneustes in the superfamily Siluroidea, Anabas and Channa. According to Allis, the truncus arteriosus of Polypterus gives off, immediately after issuing from the pericardial chamber, a large vessel on either side, which splits into the 2nd afferent branchial artery and the united trunk of the 3rd and 4th afferent branchial arteries. The latter sends off the pulmonary artery. The bulbus arteriosus of Gymnarchus sends off four and that of Heteropneustes and Clarias emits three separate vessels. Thus, the occurrence of more than a single ventral aorta is more frequently seen among airbreathing fishes than the two other conditions, even though it is still quite rare considering the large number of air-breathing fishes known. The case of Channa indicates that separate afferent conduits to the air-breathing organ are the most crucial of the three conditions proposed by Foxon to achieve partial if not complete double circulation.

Partial separation of the O_2-rich and O_2-poor blood seen in lungfishes and C. argus is likely to be of physiological significance, although the Po_2 difference may be quite small. The difference in the systemic and pulmonary arterial blood Po_2 in lungfishes ranges from 7 to 16 torr but because the blood Po_2 levels are at the steep portion of their oxygen dissociation curve, the difference in O_2 saturation level could be up to 30%. Similarly, the Po_2 difference in the anterior and posterior ventral aortic blood in Channa argus ranges from 12 to 22 torr but O_2 content in the posterior ventral aortic blood is on the average 34% higher than that in the anterior ventral aortic blood range 18% to 52%.

It is worth noting that both Protopterus and Channa show differential development of gill tissues between the arches; those gill arches carrying O_2-rich blood have become either gill-less or reduced in size, with shunt vessels connecting the afferent and efferent arteries. Neoceratodus does not show such differentiation in gill anatomy, probably reflecting the facultative air-breathing habit of the fish. The vascular anatomy in Lepidosiren is obscure in this regard. Reductions in gill surface area in airbreathing fishes have been interpreted as an adaptation to reduce oxygen loss to hypoxic water for aquatic air-breathing fishes. Gill tissues are proven to be highly plastic. Gills can change surface area by filling or exposing interlamellar space with cell mass in response to ambient conditions such as hypoxia and temperature, probably to balance physiological needs for gas exchange and unavoidable passive loss or gain of water and ions across the gill surface. The gills of obligate air-breathing Arapaima gigas change anatomy with growth. Thus, the secondary lamellae protrude from the filamental tissues as in ordinary fishes in small individuals but are embedded in filamental tissues in larger fish, which probably rely upon air breathing more heavily than smaller fish. Further, Bond demonstrated a reverse relationship between gill area and ambient oxygen levels for salamanders. It then is tempting to speculate that those gill arches receiving more oxygenated blood slow their development rates or stop development at some stage and maintain a pedomorphic state. Then, those fishes showing differential development among arches might have the capacity of separate blood streaming through the heart and then the gill arches.

The above reasoning leads to the idea that double circulation in vertebrates initially evolved as a means of reducing oxygen loss to hypoxic surrounding water, in which air breathing probably evolved. Physiological measurements must be made to clarify whether double circulation occurs or not in modern airbreathing fishes. The reasons why modern amphibious fishes can invade land to some extent without restructuring their cardiovascular system (and therefore possibly without the capacity of double circulation) might be because of the location of the gas exchange surface within the mouth and their small size. Gulping air inside the mouth and not ventilating the gills during breath holding, mudskippers are probably not subjected to the risk of oxygen loss through the gills. In addition, the small size of mudskippers should facilitate oxygen uptake through the skin, due to the large surface-to-volume rate of the body. The Devonian sarcopterygians are generally much larger. For instance, Panderichthys had a total body length of over a meter, the lower jaws alone of Tiktaalik range from 170 to 310 mm, the skulls of Ichthyostega and Acanthostega can be 250 and 200 mm long, respectively. These large animals might have needed to develop a more efficient cardiovascular system to sustain activities on land with a lesser contribution of cutaneous gas exchange.

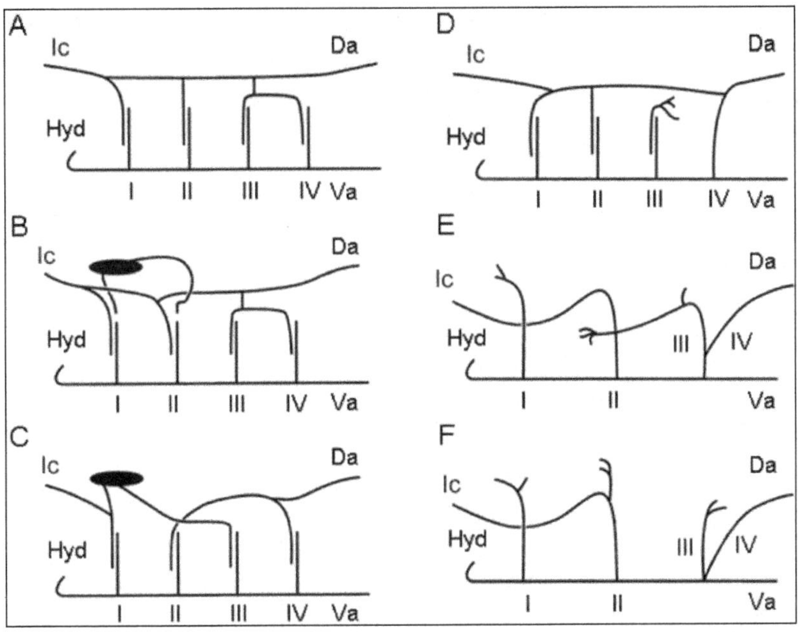

Schematics of the central circulation in synbranchids. A Synbranchus marmoratus, B Synbranchus madeirae, C Ophisternon afrum, D Monopterus albus, E Monopterus cuchia, F Monopterus boueti. Da dorsal aorta; Hyd hyoidean artery; Ic internal carotid artery; Va ventral aorta. Roman numerals indicate the number of gill arches. Black ellipsoids in B and C represent the median sinus in the buccopharyngeal epithelium.

Future Research Topics

The following lists several more important questions in the anatomical and physiological research of airbreathing fishes. It is most important to analyze blood gas levels of the different outflow vessels from the heart for those fish in which double circulation is a possibility.

Anatomical Issues

Early descriptions of the cardiovascular anatomy of air-breathing fishes often lack sufficient accuracy for the interpretation of function and therefore, need to be re-examined with modern techniques such as corrosion vascular casting. The more important species for the study include Lepidosiren, Electrophorus, Gymnarchus and synbranchids.

The cardiovascular anatomy of Lepidosiren is obscure. Laurent illustrated the schematics of the arrangement of gill arches in different fish groups including Lepidosiren but did not cite original sources. According to Laurent, the central vasculature of Lepidosiren is different from that of Protopterus in that the efferent arteries of the posterior arches do not bifurcate and that there is no afferent or efferent vessel on the posterior wall of the 6th gill cleft.

Electrophorus is the only teleost, together with Monopterus, that might have direct vascular connections between the ventral and dorsal aortae. The diagram in Carter illustrates that all four branchial arteries connect the ventral aorta directly to the dorsal aorta. Gymnarchus is the only fish that is said to have partial separation of the heart. A description of the peculiar anatomy drawn by Graham, i.e., partly divided ductus Cuvieri, direct drainage of venous blood from the air-breathing organ (gas bladder) into the left side of the atrium and almost separate origins of four afferent branchial arteries from the heart, requires careful investigation.

Figure demonstrates diagrams of synbranchid central vessels reported by Rosen and Greenwood. The diagrams depict only the course of major vessels. Nonetheless, it is apparent that gradual loss of gills and re-modeling of vascular connections occurred presumably with an increasing dependency on air breathing. The two species of Synbranchus and one of Opisternon possess gills on all four arches but species of Monopterus are radically different from them in gill vascular anatomy.

Physiological Issues

If differential development of gill size/vasculature is a feature to indicate the presence of separate blood streams as hypothesized above, then fishes with different sizes of gill arches may be of higher priority for research. These include Anabas and other anabatoids and synbranchids. The climbing perch Anabas testudineus shows a central vascular anatomy almost identical with that in Channa. Gill vascular specialization is even further advanced in A. testudineus such that shunt vessels between the afferent and efferent gill vessels occur not only in the 4th but also in the 3rd gill arch. Other anabantoids likely share the central vascular pattern known for A. testudineus, including two ventral aortae and thereby meet the supposedly basic anatomical requirement for double circulation. Synbranchids also show a variable degree of development between gill arches but the presence of an unusually long ventral aorta would promote blood mixing. Although Polypterus, Amia and Gymnarchus show different courses of pulmonary veins, their gills are more or less equally developed between arches. Still, these are important species in the consideration of cardiovascular evolution in fishes and therefore, should be tested for their capacity of double circula tion. Farrell discussed the cardiovascular anatomy and physiology of polypterids, gars, Amia and lungfishes.

Full analysis of blood partitioning within the central vasculature of Channa is challenging. Assuming that total mixing of blood from the air-breathing organ and systemic beds in the common cardinal vein and no O_2 uptake through the gills (!), Tazawa and Johansen arrived at equations for shunt fractions (percent of common cardinal vein blood directed to the anterior ventral aorta and percent of hepatic venous blood directed to the posterior ventral aorta) within snakehead's heart. Their analysis still requires blood oxygen data from five different sites in addition to aerial O_2 uptake. Even if one assumes the complete equilibrium of the blood draining the air-breathing organ and the gas within the organ, the fish must be implanted with four catheters (the two ventral aortae or afferent branchial arteries from them, a systemic vein and the common cardinal vein). I have attempted such surgery but found it to be technically highly difficult. There is also a possibility that different types of blood within the anterior and common cardinal veins are not totally mixed but conveyed as separate streams. Implantable micro-Po_2 sensors might prove useful. Measurements of blood pressure on a freely swimming C. argus demonstrated nearly simultaneous upstroke of blood pressures in the two artae (Ishimatsu, unpublished). Resolving how blood streams can be separated within the central vascular system of Channa needs novel techniques of blood flow and pressure determination and tracers that are applicable to free-ranging animals.

The hemodynamic basis for the blood separation in lungfishes is almost totally unknown. In this regard, knowledge is more advanced for amphibians and reptiles. Partial mixing of blood likely occurs within the ventricle and the conus arteriosus in amphibians, whereas the volume of blood residing in the cavum venosum is thought to be an important factor for blood mixing in reptiles. Within the lungfish heart, neither intra-atrial nor intra-ventricular septa divide the respective lumens completely and therefore, volume outputs into the two sides of the bulbus cordis are likely determined by the vascular resistance ratios of the pulmonary and systemic circuits, which are related to diastolic pressures in the two circuits. In this consideration, the position of the orifice of the bulbus cordis in relation to the position of the inter-ventricular septum is also important. Szidon et al. demonstrated in Protopterus aethiopicus that the orifice is far to the left of the inter-ventricular septum. If this is the case, the difference in diastolic pressure may be an important determinant for blood distribution between the pulmonary and systemic circuits in this fish, as proposed for the turtle by Hicks. In contrast, Icardo et al. stated that the orifice of the bulbus cordis is positioned only slightly to the right of the ventricular septum in Protopterus dolloi.

Digestion in Fishes

Jaws allow fish to eat a wide variety of food, including plants and other organisms. Fish ingest food through the mouth and break it down in the esophagus. In the stomach, food is further digested and, in many fish, processed in finger-shaped pouches called pyloric caeca, which secrete digestive enzymes and absorb nutrients. Organs such as the liver and pancreas add enzymes and various chemicals as the food moves through the digestive tract. The intestine completes the process of digestion and nutrient absorption.

In most vertebrates, digestion is a four-stage process involving the main structures of the digestive tract, starting with ingestion, placing food into the mouth, and concluding with the excretion of undigested material through the anus. From the mouth, the food moves to the stomach, where as bolus it is broken down chemically. It then moves to the intestine, where the process of breaking the food down into simple molecules continues and the results are absorbed as nutrients into the circulatory and lymphatic system.

Although the precise shape and size of the stomach varies widely among different vertebrates, the relative positions of the oesophageal and duodenal openings remain relatively constant. As a result the organ always curves somewhat to the left before curving back to meet the pyloric sphincter. However, lampreys, hagfishes, chimaeras, lungfishes, and some teleost fish have no stomach at all, with the oesophagus opening directly into the intestine. These animals all consume diets that either require little storage of food, or no pre-digestion with gastric juices, or both.

The small intestine is the part of the digestive tract following the stomach and followed by the large intestine, and is where much of the digestion and absorption of food takes place. In fish, the divisions of the small intestine are not clear, and the terms *anterior* or *proximal* intestine may be used instead of duodenum. The small intestine is found in all teleosts, although its form and length vary enormously between species. In teleosts, it is relatively short, typically around one and a half times the length of the fish's body. It commonly has a number of *pyloric caeca*, small pouch-like structures along its length that help to increase the overall surface area of the organ for digesting food. There is no ileocaecal valve in teleosts, with the boundary between the small intestine and the rectum being marked only by the end of the digestive epithelium.

There is no small intestine as such in non-teleost fish, such as sharks, sturgeons, and lungfish. Instead, the digestive part of the gut forms a spiral intestine, connecting the stomach to the rectum. In this type of gut, the intestine itself is relatively straight, but has a long fold running along the inner surface in a spiral fashion, sometimes for dozens of turns. This valve greatly increases both the surface area and the effective length of the intestine. The lining of the spiral intestine is similar to that of the small intestine in teleosts and non-mammalian tetrapods. In lampreys, the spiral valve is extremely small, possibly because their diet requires little digestion. Hagfish have no spiral valve at all, with digestion occurring for almost the entire length of the intestine, which is not subdivided into different regions.

The large intestine is the last part of the digestive system normally found in vertebrate animals. Its function is to absorb water from the remaining indigestible food matter, and then to pass useless waste material from the body. In fish, there is no true large intestine, but simply a short rectum connecting the end of the digestive part of the gut to the cloaca. In sharks, this includes a *rectal gland* that secretes salt to help the animal maintain osmotic balance with the seawater. The gland somewhat resembles a caecum in structure, but is not a homologous structure.

As with many aquatic animals, most fish release their nitrogenous wastes as ammonia. Some of the wastes diffuse through the gills. Blood wastes are filtered by the kidneys.

Saltwater fish tend to lose water because of osmosis. Their kidneys return water to the body. The reverse happens in freshwater fish: they tend to gain water osmotically. Their kidneys produce dilute urine for excretion. Some fish have specially adapted kidneys that vary in function, allowing them to move from freshwater to saltwater.

In sharks, digestion can take a long time. The food moves from the mouth to a J-shaped stomach, where it is stored and initial digestion occurs. Unwanted items may never get past the stomach, and instead the shark either vomits or turns its stomachs inside out and ejects unwanted items from its mouth. One of the biggest differences between the digestive systems of sharks and mammals is that sharks have much shorter intestines. This short length is achieved by the spiral valve with multiple turns within a single short section instead of a long tube-like intestine. The valve provides a long surface area, requiring food to circulate inside the short gut until fully digested, when remaining waste products pass into the cloaca.

Nervous System of Fishes

Central Nervous System

Fish typically have quite small brains relative to body size compared with other vertebrates, typically one-fifteenth the brain mass of a similarly sized bird or mammal.However, some fish have relatively large brains, most notably mormyrids and sharks, which have brains about as massive relative to body weight as birds and marsupials.

Fish brains are divided into several regions. At the front are the olfactory lobes, a pair of structures that receive and process signals from the nostrils via the two olfactory nerves. The olfactory lobes are very large in fish that hunt primarily by smell, such as hagfish, sharks, and catfish. Behind the olfactory lobes is the two-lobed telencephalon, the structural equivalent to the cerebrum in higher vertebrates. In fish the telencephalon is concerned mostly with olfaction. Together these structures form the forebrain.

Connecting the forebrain to the midbrain is the diencephalon (in the diagram, this structure is below the optic lobes and consequently not visible). The diencephalon performs functions associated with hormones and homeostasis. The pineal body lies just above the diencephalon. This structure detects light, maintains circadian rhythms, and controls color changes.

The midbrain (or mesencephalon) contains the two optic lobes. These are very large in species that hunt by sight, such as rainbow trout and cichlids.

The hindbrain (or metencephalon) is particularly involved in swimming and balance. The cerebellum is a single-lobed structure that is typically the biggest part of the brain.Hagfish and lampreys have relatively small cerebellae, while the mormyrid cerebellum is massive and apparently involved in their electrical sense.

The brain stem (or myelencephalon) is the brain's posterior. As well as controlling some muscles and body organs, in bony fish at least, the brain stem governs respiration and osmoregulation.

Sense Organs

Most fish possess highly developed sense organs. Nearly all daylight fish have color vision that is at least as good as a human's. Many fish also have chemoreceptors that are responsible for extraordinary senses of taste and smell. Although they have ears, many fish may not hear very well. Most fish have sensitive receptors that form the lateral line system, which detects gentle currents and vibrations, and senses the motion of nearby fish and prey.Some fish, such as catfish and sharks, have the Ampullae of Lorenzini, organs that detect weak electric currents on the order of millivolt. Other fish, like the South American electric fishes Gymnotiformes, can produce weak electric currents, which they use in navigation and social communication.

Fish orient themselves using landmarks and may use mental maps based on multiple landmarks or symbols. Fish behavior in mazes reveals that they possess spatial memory and visual discrimination.

Vision

Vision is an important sensory system for most species of fish. Fish eyes are similar to those of terrestrial vertebrates like birds and mammals, but have a more spherical lens. Their retinas generally have both rods and cones (for scotopic and photopic vision), and most species have colour vision. Some fish can see ultraviolet and some can see polarized light. Amongst jawless fish, the lamprey has well-developed eyes, while the hagfish has only primitive eyespots. Fish vision shows adaptation to their visual environment, for example deep sea fishes have eyes suited to the dark environment.

Hearing

Hearing is an important sensory system for most species of fish. Fish sense sound using their lateral lines and their ears.

Cognition

New research has expanded preconceptions about the cognitive capacities of fish. For example, manta rays have exhibited behavior linked to self-awareness in mirror test cases. Placed in front of a mirror, individual rays engaged in contingency testing, that is, repetitive behavior aiming to check whether their reflection's behavior mimics their body movement.

Wrasses have also passed the mirror test in a 2018 scientific study.

Cases of tool use have also been noticed, notably in the Choerodon family, in archerfish and Atlantic cod.

Capacity for Pain

Experiments done by William Tavolga provide evidence that fish have pain and fear responses. For instance, in Tavolga's experiments, toadfish grunted when electrically shocked and over time they came to grunt at the mere sight of an electrode.

In 2003, Scottish scientists at the University of Edinburgh and the Roslin Institute concluded that rainbow trout exhibit behaviors often associated with pain in other animals. Bee venom and acetic acid injected into the lips resulted in fish rocking their bodies and rubbing their lips along the sides and floors of their tanks, which the researchers concluded were attempts to relieve pain, similar to what mammals would do. Neurons fired in a pattern resembling human neuronal patterns.

Professor James D. Rose of the University of Wyoming claimed the study was flawed since it did not provide proof that fish possess "conscious awareness, particularly a kind of awareness that is meaningfully like ours". Rose argues that since fish brains are so different from human brains, fish are probably not conscious in the manner humans are, so that reactions similar to human reactions to pain instead have other causes. Rose had published a study a year earlier arguing that fish cannot feel pain because their brains lack a neocortex. However, animal behaviorist Temple Grandin argues that fish could still have consciousness without a neocortex because "different species can use different brain structures and systems to handle the same functions".

Animal welfare advocates raise concerns about the possible suffering of fish caused by angling. Some countries, such as Germany have banned specific types of fishing, and the British RSPCA now formally prosecutes individuals who are cruel to fish.

ANATOMY OF REPTILES

Respiration in Reptiles

The trachea of snakes and lizards supported by cartilaginous rings on the dorsal surface of the accession to the thin tracheal smooth muscle. The inner surface of the trachea is lined with multi-row ciliated columnar epithelium, comprising a plurality of goblet cells secrete mucus. Depending on the type of secreting mucins or serous acini secret can be presented to a greater or lesser extent, are more caudally in the trachea and main bronchi department. The submucosal layer in this area are usually small units scattered lymphoid cells. The bronchi retain the histological structure of the trachea to the level of major bronchioles. Bronchiolar and respiratory epithelium from columnar flattened to nearly flat with a decrease in airway diameter.

The primary function of the respiratory system is to supply the blood with oxygen in order for the blood to deliver oxygen to all parts of the body. The respiratory system does this through breathing. When we breathe, we inhale oxygen and exhale carbon dioxide. This exchange of gases is the respiratory system's means of getting oxygen to the blood. Reptiles and amphibians have lungs and exchange gases in the capillaries like mammals, but there are some differences in how they ventilate their respiratory systems. Reptiles don't typically breathe the same way as mammals, since many reptiles lack a diaphragm. Reptiles use their axial muscles, the ones attached to their ribs, to expand their ribcage for breathing. During periods of intense activity, reptiles might be forced to hold their breath, as they use those muscles for running away. Apart from their capillaries, amphibians perform exchange directly through their skin. This works for them because their skin has lots of blood vessels very close to the permeable skin surface. Diffusion can take place right through the skin. In fact, some salamanders have no lungs at all, and they get all of their oxygen through their skin.

Part land animals, part water creatures, reptiles have developed unique respiratory systems. Reptile respiration differs among these classes of reptiles:

- The crocodilians.
- The squamates (snakes and lizards).
- The testudines (turtles).

  Snakes and lizards don't have diaphragm muscles for lung ventilation: muscles used for locomotion are the same used for their respiratory systems. Contracting and flexing body muscles moves their ribs and fill their lungs. Most squamate reptiles must hold their breath when they engage in bursts of intense physical activity. Some lizards complement lung ventilation with buccal pumping. This involves using throat muscles to help fill their lungs. Buccal pumping is when an animal uses the muscles of the mouth and throat to pull air into tJhe lungs. Muscles pull air through the mouth or nose into a buccal cavity. Throat muscles then pump and move the floor of the mouth up in a way that's visible from the outside. This forces air out of the mouth and into the lungs. This is what amphibians do, by puffing up their chinny-chin-chins to get the air in.

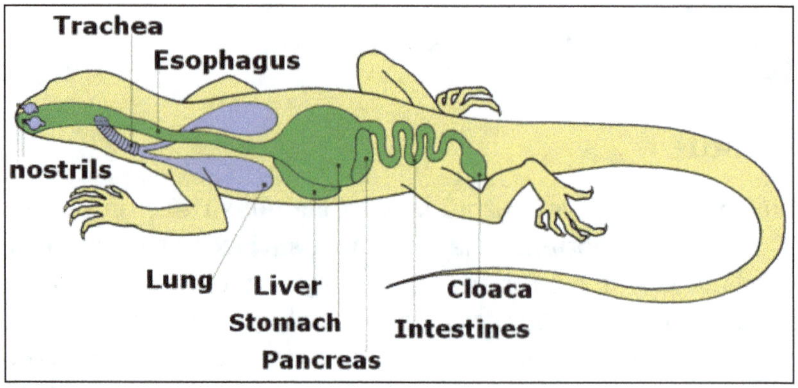

Crocodilians use a diaphragm muscle for lung ventilation, but their respiratory system differs from mammals. They use a hepatic piston, a process involving muscular movement of the pelvis and liver rather than the rib cage. This organ movement allows the lungs to expand.

The hard shell that turtles and tortoises have makes lung expansion a bit more difficult. Most must use their limbs for lung ventilation. Some turtle lungs are covered with a sheet of muscle. Pulling their limbs in expels air from the lungs and pushing their limbs out of their shells expands the lungs. The lungs of many turtles attach to both the shell and the internal organs. By contracting and flexing a series of muscles, these turtles can push these organs up and down, contracting and expanding the lungs. Movement of their limbs aids this respiratory process.

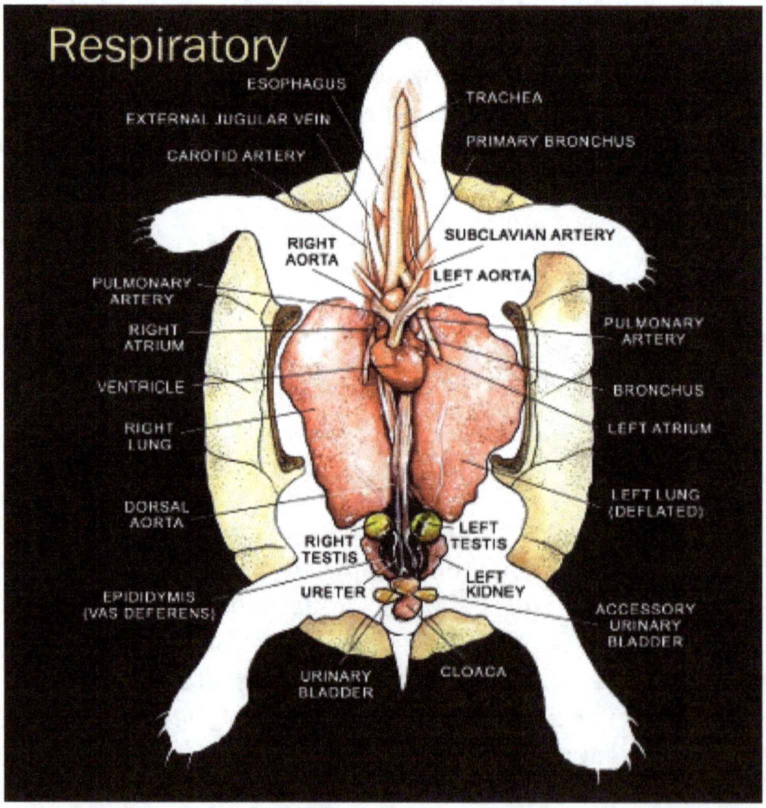

Most reptiles must hold their breath when they swallow but crocodilians and skinks have a second palate made of bone that allows them to breathe under water. Snakes extend their windpipes so that they don't suffocate during the swallowing process. Turtle respiration while in motion varies among species. Green sea turtles don't breathe when they're moving along nesting beaches, but the North American box turtle breathes continuously while moving, and this respiration doesn't coincide with their limb movement. Red-eared sliders take smaller breaths while moving than at rest, and box turtles can breathe without extending their limbs.

Blood Circulation in Reptiles

The location of the heart within the body cavity varies according to species. In most chelonians the heart lies on midline just caudal to the thoracic girdle, ventral to the lungs. The heart of some chelonians such as soft-shelled turtles is pushed to the side of the body cavity in order to accommodate the retracted neck. The heart of most lizards lies within the thoracic girdle, with the exception of some species such as monitors and tegus (as well as crocodilians) in which the heart lies farther back in the coelomic cavity. Cardiac location varies in snakes according to species, but usually is found at the junction of the first and second quarter of the animal's body length. Typically arboreal snakes' hearts are found more cranially in the body than in terrestrial animals. Snake's hearts are fairly mobile within the coelomic cavity helping to facilitate the ingestion of large prey items.

The cardiac structure of reptiles is significantly different from that of mammals. Please note that the following descriptions are very general, and that significant variation exists between species. Most reptiles have three chambered hearts with two atria and one common ventricle. The right atrium receives blood returning from the systemic circulation via the sinus venosus, which is formed by the confluence of the right and left precaval veins and the single postcaval vein. The walls of the sinus venosus contain cardiac muscle and the pacemaker of the heart. The left atrium receives oxygenated blood from the lungs via the pulmonary vein(s). The atrioventricular valves are bicuspid, membranous structures. Under normal conditions the three chambered heart functions much like a four chambered structure, therefore relatively little mixing of oxygenated and de-oxygenated blood occurs. Three cavities exist within the ventricle and can be functionally separate; the cavum venosum, cavum arteriosum and the cavum pulmonale. These cavities are partially separated by two muscular ridges found within the ventricle. These ridges vary in prominence in different species, but are generally well-developed in chelonians. The muscular ridge divides the cavum pulmonale and the cavum venosum. The vertical ridge divides the cavum venosum and cavum arteriosum. The cavum pulmonale receives blood from the right atrium through the cavum venosum and directs flow into the pulmonary circulation. The cavum arteriosum receives blood from the pulmonary veins and then directs oxygenated blood to the cavum venosum. The paired aortic arches arise from the cavum venosum and lead to the systemic circulation. The right and left aortic arches come together to form a single aorta at variable distances caudal to the heart. Differential blood flow and separation of oxygenated and de-oxygenated blood is maintained by pressure differences of the outflow tracts and the muscular ridges that partially divide the ventricle. In most non-crocodilian reptiles the ventricle function as a single pump, meaning that the same pressures are generated by both the cavum pulmonale and cavum venosum. This is not the case in monitor lizards and at least one species of python in which significantly higher pressures are generated in the cavum venosum and thus the systemic arches.

Due the unique anatomy, both right to left and left to right shunts are possible in the reptilian heart. Shunting does occur during apnea, though all the details regarding the exact purpose of shunts are unclear. Multiple theories exist regarding the purpose of right to left cardiac shunting in reptiles including the conservation of cardiac energy, facilitation of warming, reduction of plasma filtration into the lungs, reduction of carbon dioxide flux into the lungs and the metering of oxygen stores from the lung(s) during apnea. Theories to explain the purpose of left to right shunting include facilitation of carbon dioxide elimination from the lung(s), minimization of ventilation/perfusion mismatches and improvement of systemic oxygen transport. In times of oxygen deprivation (diving in some reptiles, consumption of large prey in snakes), reptiles can shunt blood away from the lungs. Right to left cardiac shunting in the non-crocodilian heart can be facilitated by an increase in pulmonary vascular resistance and action of the muscular and vertical ridge. Resumption of breathing results in a decrease in pressures within the pulmonary vasculature and restoration of pulmonary blood flow.

Crocodilians are the only reptiles which possess four chambered hearts comparable to mammals. Even so, crocodilian cardiac anatomy is quite different from what is seen in birds and mammals. Crocodilians possess two aortas; the right arising from the left ventricle and the left from the right ventricle. Both aortas route blood to the systemic circulation. The right and left aortas are connected near the base of the heart by the foramen of Panizza. The foramen allows blood from the right ventricle to bypass the pulmonary circulation when necessary. A valve exists at the opening of the pulmonary artery which has interdigitating muscular projections, hence the commonly used name “cog-wheel valve”. When the animal holds its breath, the cog-wheel valve closes and blood that would have normally entered the pulmonary circulation is diverted into the left aorta. It should be noted that most veterinary texts incorrectly report that the location of the foramen of Panizza is in the ventricular septum or atrial septum.

Heart rate of reptiles depends on species, size, temperature and activity/level of metabolic function. An equation employing metabolic scaling for determination of the “appropriate” heart rate in reptiles has been proposed: Heart rate = 33.4(Weight in kilograms$^{-0.25}$). This equation assumes that the reptile is within its preferred optimal temperature zone.

Digestion in Reptiles

The digestive system of modern reptiles is similar in general plan to that of all higher vertebrates. It includes the mouth and its salivary glands, the esophagus, the stomach, and the intestine and ends in a cloaca. Of the few specializations of the reptilian digestive system, the evolution of one pair of salivary glands into poison glands in the venomous snakes is the most remarkable.

During development the embryos of higher vertebrates (reptiles, birds, and mammals) consecutively develop three separate sets of kidneys; these are arranged in longitudinal sequence in the body cavity. The first set, the pronephroi, are vestigial organs left over from the evolutionary past that soon degenerate and disappear without having had any function. The second set, the mesonephroi, are the functional kidneys of adult amphibians, but their only contribution to the lives of reptiles is in providing the duct (the Wolffian duct) that forms a connection between the testes and the cloaca. The operational kidneys of reptiles, birds, and mammals are the last set, the metanephroi, which have separate ducts to the cloaca. The principal functions of the kidney are the removal of nitrogenous wastes resulting from the oxidation of proteins and the regulation of water loss. Vertebrates eliminate three kinds of nitrogenous wastes: ammonia, urea, and uric acid. Ammonia and urea are highly soluble in water, but uric acid is not. Ammonia is highly poisonous, urea is slightly poisonous, and uric acid is not poisonous at all.

Among reptiles the form taken by the nitrogenous wastes is closely related to the habits and habitat of the animal. Aquatic reptiles tend to excrete a large proportion of these wastes as ammonia in aqueous solution. This method uses large amounts of water and is no problem for a freshwater resident, such as an alligator, which eliminates between 40 and 75 percent of its nitrogenous wastes as ammonia. Terrestrial reptiles, such as most snakes and lizards, must conserve body water, and they convert their nitrogenous wastes to insoluble, harmless uric acid, which forms a more or less solid mass in the cloaca. In snakes and lizards, these wastes are eliminated from the cloaca together with wastes from the digestive system.

Prior to the evolution of the metanephric kidney, the products of the male gonad, the testis, traveled through the same duct with the nitrogenous wastes from the kidney. But with the appearance of the metanephros, the two systems became separated. The female reproductive system never shared a common tube with the kidney. Oviducts in all female vertebrates arise as separate tubes with openings usually near, but not connected to, the ovaries. The oviducts, like the Wolffian ducts of the testes, open to the cloaca. Both ovaries and testes lie in the body cavity near the kidneys.

With the evolution of the reptilian egg, internal fertilization became necessary. The males of all modern reptiles, with the exception of tuatara, have functional copulatory organs. The structures vary from group to group, but all include erectile tissue as an important element of the operating mechanism, and all are protruded through the male's cloaca into that of the female during copulation. Unlike the penis of turtles and crocodiles, the copulatory organ of lizards and snakes is paired, with each unit being called a hemipenis. The hemipenes of lizards and snakes are elongated tubular structures lying in the tail. The penis of a crocodile or turtle is protruded through the cloacal opening wholly by means of a filling of blood space (sinuses) in the penis; protrusion of a lizard's or snake's hemipenis, however, is begun by a pair of propulsor muscles. Completion of the erection is brought about by blood filling the sinuses in the erectile tissue. Only one hemipenis is inserted into a female, but which one is a matter of chance. Unlike the penis of mammals, the copulatory organs of reptiles do not transport sperm through a tube. The ducts from the testes, as already mentioned, empty into the cloaca, and the sperm flow along a groove on the surface of the penis or hemipenis.

Nervous System of Reptiles

The diversity of reptiles and their evolutionary relationship to mammals make reptilian brains great models to explore questions related to the structural and functional evolution of vertebrate neural circuits. To this end, comparative studies seek to identify homologies — structural or molecular similarities that are due to common ancestry — at a variety of levels, for example, brain regions, circuits or cell types. Homologies can be inferred from extant species by using a comparative approach within a phylogenetic framework. Vertebrate brains have been classically compared in terms of morphology, connectivity, and neurochemistry; however, adult neuroanatomy may not be sufficient to determine homologies without ambiguity.

Vertebrates share among themselves the highest morphological resemblance not as adults, but at their so-called 'phylotypic stage' of embryonic development, a resemblance mirrored by similarity across transcriptomes. Identification of conserved brain subdivisions, established by conserved signaling centers and uniquely defined by the combinatorial expression of transcription factors during development, demonstrates that all of the general brain regions found in mammals, including the cerebral cortex, have homologies in reptiles. The cerebral cortex is part of the pallium, a developmental subdivision of the telencephalon delineated by the expression of transcription factors such as Pax6, Emx1 and Tbr1, which is conserved in all vertebrates. Gene expression data show that the same fundamental subdivisions of the pallium — lateral, ventral, medial and dorsal, the latter giving rise to the neocortex in mammals — occur in developing vertebrates, despite the divergent morphologies of pallial structures in adults. The cerebral cortex is thus not a mammalian invention, but rather an ancient pallial structure that predates the split between sauropsids and theropsids, the mammals' precursors.

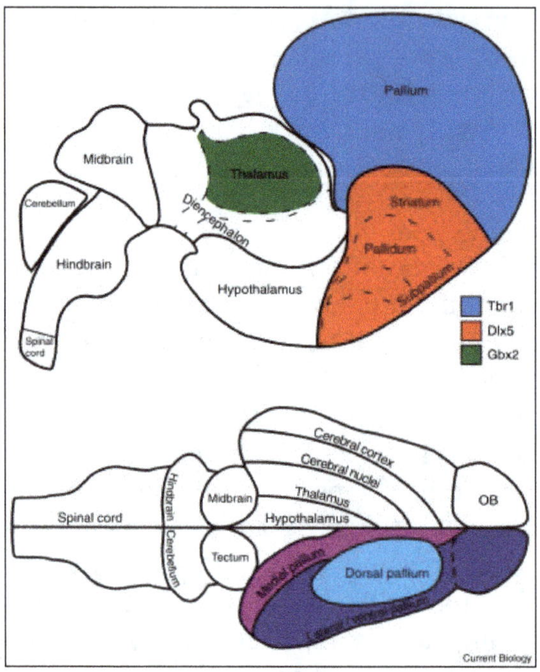

Developmental and adult bauplan of the vertebrate brain.

Upper panel: simplified developmental scheme of vertebrate brain regions. Combinatorial expression of transcription factors such as Tbr1, Dlx5, and Gbx2 defines brain regions during development and sets the stage for further differentiation in adults. Note that Tbr1 and Dlx5 delineate further regions not shown here. Lower panel: schematic illustration of the adult turtle brain showing major subdivisions present in all vertebrates. Lateral and medial pallium express different sets of molecular markers during development. Which regions of the adult brain correspond to lateral and medial pallium is an active topic of research and thus we combine both pallial subdivisions into a single region.

Comparative studies of brain structure and development have revealed a general *Bauplan*, an outline of the fundamental large-scale architecture of the vertebrate brain, reflecting its basic functional organization. The telencephalon integrates and stores multimodal information, and is also the higher center of action selection and motor control (basal ganglia). The hypothalamus is a conserved area controlling homeostasis and behaviors essential for survival, such as feeding and reproduction.

In all vertebrates, behavioral states are controlled by common brainstem neuromodulatory circuits, such as the serotoninergic system. Finally, vertebrates harbor a diverse set of sense organs, and their brains share pathways for processing incoming sensory inputs. For example, in all vertebrates, visual information from the retina is relayed and processed to the pallium through the tectum and the thalamus, while olfactory input from the nose first reaches the olfactory bulb and then the pallium.

Although pallial structures exist in amphibians and fish, reptiles and mammals are the only vertebrates to have a cerebral cortex with a clear, though simple, three-layered structure, similar to that of mammalian allocortex. The reptilian ventral pallium also gives rise to the dorsal ventricular ridge, a structure that dominates the bird pallium and contributes to the complex cognitive abilities of birds, but whose mammalian equivalent is still the subject of debate among comparative anatomists. The reptilian cortex contains far fewer subdivisions than that of rodents, carnivores, or primates: it is subdivided into a medial cortex, often called hippocampus by anatomists; a lateral cortex, equivalent to the mammalian piriform cortex; and a dorsal cortex in between, which receives multimodal inputs (for example, visual inputs in turtles). There is little evidence for motor and somatosensory areas in the reptilian cortex, but pallial motor control may have evolved early in vertebrate evolution. Owing to this simplicity, the reptilian brain facilitates the study of primordial cortical function as a whole, and points to the origins of cortex as fulfilling general associative functions.

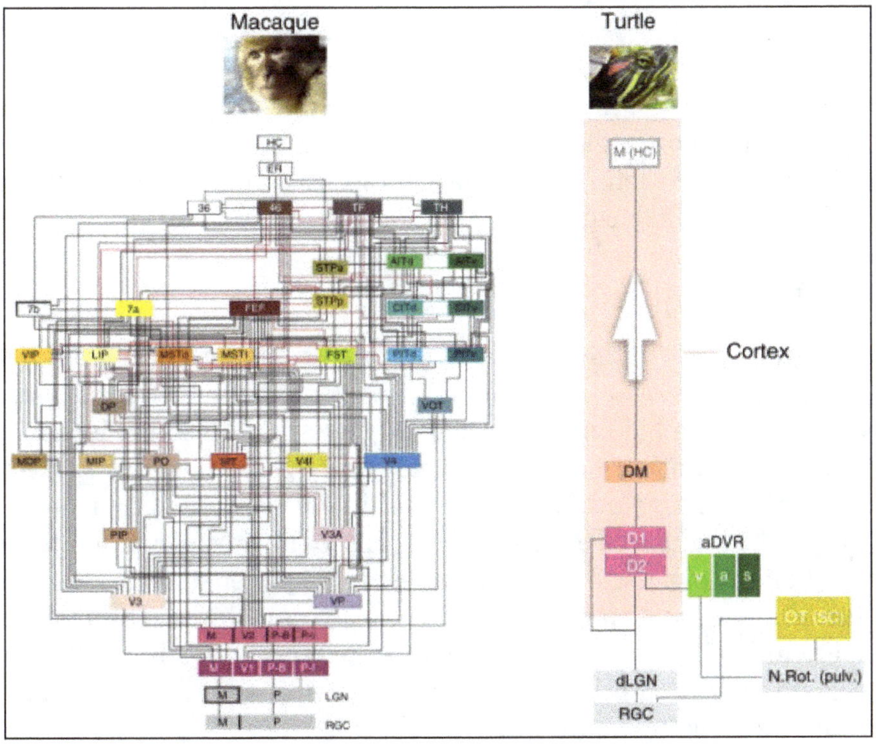

Visual circuits in monkey and turtle.

Comparison of known visual cortical circuits between macaque (left) and freshwater turtle (right), emphasizing the great simplicity of reptilian cortex. Arrow indicates direction of information flow from sensory periphery. aDVR, anterior dorsal ventricular ridge (a: auditory, s: somatosensory, v: visual); dLGN, dorsal lateral geniculate nucleus; D1 and D2, dorsal cortex areas D1 and D2; DM, dorsomedial cortex; M (HC), medial cortex ('hippocampus'); N.Rot., nucleus rotundus ('pulvinar'); OT, optic tectum ('superior colliculus'); RGC, retinal ganglion cells.

Besides sharing pallial modules, mammals and reptiles also share a complement of cortical cell types, suggesting that some structural elements of cortical circuits arose early in amniote evolution. Like the mammalian cortex, the reptilian cortex contains excitatory, glutamatergic neurons and inhibitory (GABAergic) interneurons. In both mammals and reptiles, these neurons have a common developmental origin: excitatory neurons are generated by multipotent cortical progenitors, whereas inhibitory neurons are born in the subpallium before migrating to the cortex.

Classical studies suggest that the reptilian main cortical cell layer (layer 2, L2) corresponds to the deep, output layers of mammalian neocortex, whereas its layer 1 (L1) is equivalent to mammalian layer I. According to this view, mammalian cortical evolution would have included the incorporation of new, intermediate cell layers acting as input stations and internal circuitry. Morphologically, L2 pyramidal neurons of the reptilian dorsal cortex are most similar to mammalian hippocampal excitatory neurons. Indeed, reptilian pyramidal neurons have, depending on the area, little to no basal dendritic field, and several densely spine-studded apical dendrites, quite different from the single, long, apical dendrite of neocortical pyramidal neurons. Consistent with this correspondence between layers, reptilian subpallial cells transplanted into mammalian embryos generate GABAergic neurons that can populate only the deeper cortical layers.

Challenging this view, however, recent molecular studies have found that turtle and lizard cortical neuroblasts generate neurons that express upper layer molecular markers, in a developmental sequence similar to that observed in mammals. Although the molecular characterization of neuronal types in the reptilian cortex is still in its infancy, it is possible that the reptilian cortex represents an ancestral blueprint for the more elaborate mammalian cortical circuits. For example, reptilian cortical neurons, or subsets of them, might share molecular (and functional) features with both upper and lower layer mammalian cells.

Cortical Function

Cortical circuitry is where most of the comparative work on the reptilian brain has been done. In the three-layered reptilian cortex, afferent inputs travel medially through superficial L1, where they fan out in a non-topographic manner. Their targets are GABAergic interneurons (with cell bodies in L1 and L3) and glutamatergic pyramidal neurons, with cell bodies packed in L2, and dendrites in L1 and L3. Interneurons project locally within cortex, providing feedforward and feedback inhibition to specific regions of the pyramidal cells' dendrites. Several interneuron subtypes can be defined based on the expression of a subset of common genetic markers used in mammals. For example, in turtle cortex, some interneuron types express calbindin, others express neuropeptide-Y, while parvalbumin (PV)-positive interneurons appear to be absent. Pyramidal cells make reciprocal connections with each other and with interneurons locally, with subcortical afferent structures, and with other cortical areas.

The lateral cortex receives olfactory input from the olfactory bulb, and projects to the medial cortex (hippocampus). The dorsal cortex receives input from the thalamus; in many species this input is visual, originating in the thalamic lateral geniculate nucleus (LGN) and also eventually reaches the hippocampus. Pyramidal cells of the hippocampus project back to dorsal and lateral cortices, forming an internal cortical loop. Different mammalian species can exhibit vast elaborations in the number and connectivity of cortical subregions. However, by examining homologous structures, shared circuit motifs can be recognized.

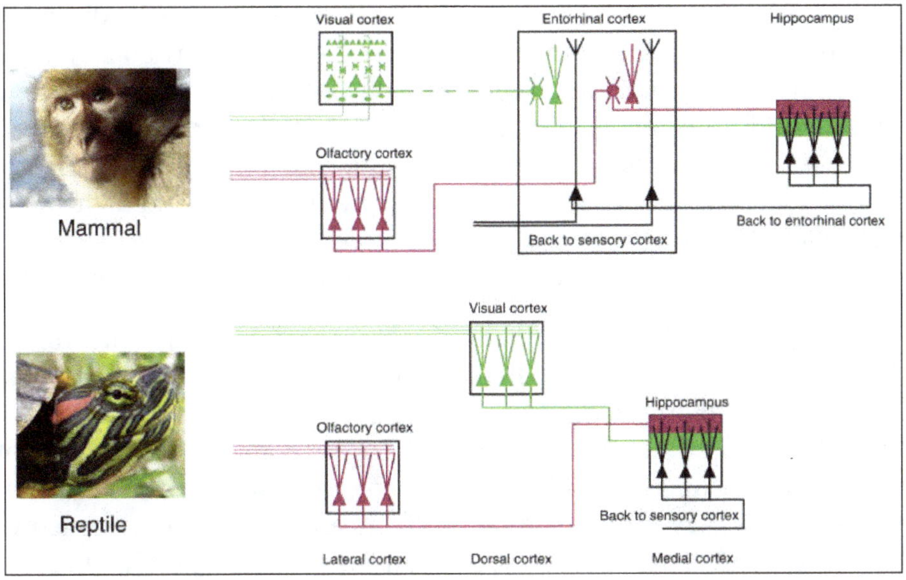

Shared circuit motifs between reptilian and mammalian cortex.

Whereas processing steps are fewer in reptiles, they reach the same target as in mammalian cortex. Cortical architecture is more similar across regions in reptiles, suggesting similar and possibly general computations are performed on different sensory inputs.

In turtles, visual stimulation triggers propagating waves of neural activity that travel across the cortex. These waves are slower and simpler than those observed in mammalian neocortex. They are accompanied by relatively slow oscillations, which are most prominent in the 20 Hz frequency band. Whereas the so-called gamma oscillations in mammalian cortex are typically around and above 40 Hz, recent results in mice indicate that the 20 Hz band dominates when PV-interneuron development is artificially arrested, consistent with the above observation that turtle cortex lacks PV interneurons. The computational role, if any, of such dynamics is unknown at present. Progress will require new experimental approaches that allow the simultaneous sampling of large neuronal populations. Specific and data-driven theories of computation in reptilian cortex thus await further study. To the extent that modern reptilian cortex resembles that in the common ancestor of reptiles and mammals, understanding reptilian cortex function may reveal some of the fundamental associative computations that early cortical circuits evolved to carry out.

Physiological Adaptations of the Reptilian Brain

Reptiles are ectotherms and must behaviorally or physiologically adapt to varying temperatures, making reptile brains notable for their adaptation to extreme conditions. The turtle brain, for example, has evolved remarkable adaptations to hypoxic conditions, which have long been known to experimentalists interested in hearing, olfaction, motor control or cerebellar physiology. Semi-aquatic freshwater turtles, such as *Chrysemys picta,* are remarkable for their ability to survive hibernation in hypoxic waters during cold northern winters, and have been reported to survive anoxic periods of over four months at 3°C. This resilience points to several physiological adaptations to survive cold, anoxic conditions during hibernation and subsequent re-oxygenation in the spring. The mechanisms underlying such tolerance are interesting — and also possibly applicable to the treatment of brain injury due to ischemia.

Anoxia tolerance critically depends on the ability to reduce energy expenditure while elevating anaerobic ATP production. ATP consumption in neural tissues, particularly those involved in sensory functions, requires high ATP turnover. Free glucose and glycogen can serve as metabolic substrates in anoxic conditions, and stores of glycogen are released from the liver and skeletal muscles to facilitate ATP production during hibernation. But, by shifting to anaerobic glycolysis, hibernating turtles must buffer the resulting metabolic product lactic acid to prevent fatal acidosis. In turtles, extracellular buffering of lactic acid is achieved by having large volumes of pericardial and peritoneal fluids, and through sequestration in the mineralized shell and skeleton.

During hibernation, turtles reduce ATP consumption in neural tissue by reducing membrane permeability and thus the need for pump-mediated restoration of ionic balance. This phenomenon has been observed in turtle neurons for voltage-gated Na^+ channels, the oxygen-sensitive calcium-activated K^+ channels, as well as AMPA and NMDA receptors. Neuronal depolarization and NMDA receptor over-activation are a hallmark of anoxia in mammalian neural tissue and lead to excitotoxic cell death via increased intracellular Ca^{2+} concentrations. In turtle neurons, NMDA receptor opening times are reduced by 65% after 60 minutes of anoxia, and NMDA receptor currents are actually silenced by anoxia-triggered cytosolic Ca^{2+} release from mitochondrial stores. Despite reduced channel activity and consequent reduced electrical activity during anoxia, turtles appear to remain vigilant and responsive to visual stimuli during hibernation (but less so to tactile stimuli). This maintenance of visual responses in the absence of tactile responses during hibernation suggests a differential down-regulation of sensory modalities. Because changes in daily illumination contribute to signaling the end of hibernation, it may be a selective advantage for energy-conserving mechanisms to be applied less to the visual system than to other brain areas.

When oxygen again becomes available, the re-oxygenation of tissue, such as that observed after ischemia-reperfusion events caused by myocardial infarction or stroke, is problematic in mammals because it is associated with the excessive production of reactive oxygen species (ROS) that damage cells through protein oxidation and lipid peroxidation. Turtles have evolved effective physiological mechanisms that allow for repeated cycles of anoxia and re-oxygenation without the free-radical damage induced by ischemia-reperfusion of tissues observed in mammals. These mechanisms involve an evolutionarily old up-regulation of nitric oxide and its metabolites, which limit ROS generation and tissue damage in cardiac and brain tissue. Thus, the turtle brain, with these physiological adaptations, is a very useful model for the study of neural circuit function. Neural tissue remains alive and viable for days *ex vivo*, and tissue can be routinely stored overnight in a refrigerator at 3–5°C.

Brain and Behavior

Reptiles express a number of complex behaviors normally attributed to mammals. They can, for example, learn to navigate mazes as well as birds or mammals do and likely use a hippocampal structure to do so. Remarkably, sea turtles navigate across the ocean to their home beach, possibly using magnetic, olfactory, and visual cues. Reptiles also express interesting social behaviors: pythons and crocodiles, for example, incubate their eggs after hatching, and mothers guard the nest and provide infant care. Some lizard species exhibit pair bonding, parental care, and form families. Others construct networks of tunnels and may live in social communities like those of naked-mole rats.

Reptiles inhabit a variety of ecological niches, such as terrestrial deserts, temperate forests, tropical jungles, the sea, and fresh waters. They are found in all oceans and on all continents except Antarctica. Semi-aquatic turtles, which live in primarily aquatic habitats but make terrestrial excursions to lay eggs, offer an interesting model animal in which to study navigation. Many turtles are known for their impressive navigational skills and high nest-site fidelity, or the tendency for individuals to return to the same geographic location or microhabitat to nest. The mechanisms underlying these behaviors are not entirely understood, but navigational ability in the fresh water turtle *Chrysemys picta* seems to depend on experience gained during a critical period (when younger than four years of age). Translocated resident adults, for example, can quickly and consistently find new aquatic habitats within 33 hours, often using precise and predictable routes, whereas translocated non-resident adults appear unable to find new aquatic habitats even after 21 days.

The hippocampus is critical for map-like or relational memory representations of allocentric space, where objects are represented in relation to other objects. In birds and mammals, lesions to the hippocampal formation produce selective impairments in spatial tasks that require the encoding of relationships among multiple environmental features (place learning) but not in tasks that require approaching a single cue or simple non-spatial discriminations. Whereas extensive comparative research supports the idea that the reptilian medial cortex is homologous to the hippocampal formation of mammals and birds, only a few studies have examined the neural function of this brain structure or its role in place learning. In one such study, Rodriguez evaluated the effects of lesions to the hippocampus of turtles in place and cue-maze tasks. Hippocampal lesioned (and sham-lesioned) animals performed cue-discrimination tasks correctly, but failed at the place learning that relied on allocentric space learning. These results indicate that lesions to the hippocampus of turtles selectively impair map-like memory representations of the environmental space, mirroring the effect of hippocampal lesions in mammals and birds. Thus, reptilian hippocampus may also share a central role in navigation.

ANATOMY OF AMPHIBIA

Respiration in Amphibians

The living amphibians (frogs, toads, salamanders, and caecilians) depend on aquatic respiration to a degree that varies with species, stage of development, temperature, and season. With the exception of a few frog species that lay eggs on land, all amphibians begin life as completely aquatic larvae. Respiratory gas exchange is conducted through the thin, gas-permeable skin and the gills. In addition to these structures, frog tadpoles use their large tail fins for respiration; the tail fins contain blood vessels and are important respiratory structures because of their large surface area. As amphibian larvae develop, the gills (and in frogs, the tail fin) degenerate, paired lungs develop, and the metamorphosing larvae begin making excursions to the water surface to take air breaths.

The lungs of amphibians are simple saclike structures that internally lack the complex spongy appearance of the lungs of birds and mammals. The lungs of most amphibians receive a large proportion of the total blood flow from the heart. Even though the amphibian ventricle is undivided, there is surprisingly little mixture of blood from the left and right atrial chambers within the single ventricle. As a consequence, the lungs are perfused primarily with deoxygenated blood from the systemic tissues.

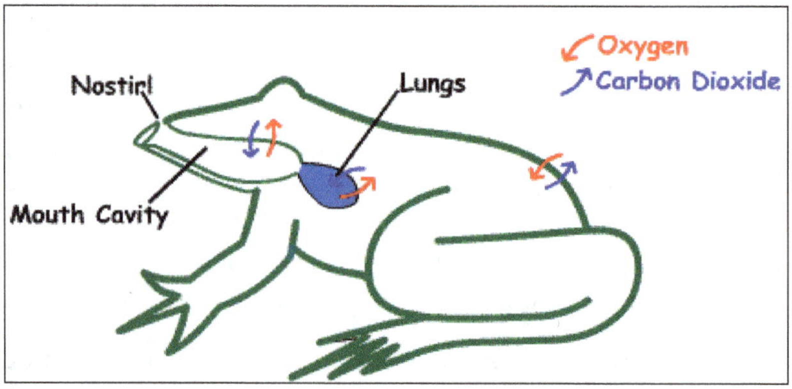

By the time the larva has reached adult form, the lungs have assumed the respiratory function of the larval gills. A few species of salamanders (for example, the axolotl) never metamorphose to the adult stage, and although they may develop lungs for air breathing, they retain external gills throughout life. Another exception to the usual pattern of respiratory development is seen in the Plethodontidae family of salamanders, which lose their gills upon metamorphosis but never develop lungs as adults; instead, gas exchange is conducted entirely across the skin. In almost all amphibian species, the skin in adults continues to play an important role in gas exchange.

The relative contributions of lungs and skin, and even local areas of skin, to gas exchange differ in different species and in the same species may change seasonally. In frogs, the skin of the back and thighs (the areas exposed to air) contains a richer capillary network than the skin of the underparts and therefore contributes more to gas exchange. The aquatic newt *Triton* utilizes both lung and skin respiration, the skin containing about 75 percent of the respiratory capillaries. At the other extreme, the tree frog *Hyla arborea* is much less aquatic, and its lungs contain over 75 percent of the respiratory capillary surface area. Similar differences are found even in closely related forms: In the relatively more terrestrial frog *Rana temporaria*, uptake of oxygen across the lung is about three times greater than across the skin; in *R. esculenta*, which is more restricted to water, the lungs and skin function about equally in the uptake of oxygen. Carbon dioxide is eliminated mainly through the skin in both these species; in fact, the skin appears to be a major avenue for carbon dioxide exchange in amphibians generally.

In temperate climates, as winter approaches, the colder environmental temperature (and thus lower body temperature) induces a marked lowering of the metabolic rate in amphibians. Terrestrial forms (*e.g.*, toads and some salamanders) may burrow into the ground to overwinter. Aquatic species burrow into the mud at the bottom of lakes or ponds. Because their metabolic rate is much lower during winter, adequate gas exchange can be provided entirely by the skin in either terrestrial or aquatic habitats.

The mechanism of lung inflation in amphibians is the buccal cavity (mouth-throat) pumping mechanism that also functions in air-breathing fishes. To produce inspiration, the floor of the mouth is depressed, causing air to be drawn into the buccal cavity through the nostrils. The nostrils are then closed, and the floor of the mouth is elevated. This creates a positive pressure in the mouth cavity and drives air into the lungs through the open glottis. Expiration is produced by contraction of the muscles of the body wall and the elastic recoil of the lungs, both acting to drive gas out of the lungs through the open glottis. In aquatic amphibians the pressure of water on the body wall can also assist expiration. Many amphibians show rhythmic oscillations of the floor of the mouth between periods of lung inflation; these oscillations are thought to be involved in olfaction by producing a flow of gas over the olfactory epithelial surfaces.

Circulation of Blood in Amphibians

Modern amphibians are characterized by the flexibility of their gaseous exchange mechanisms. Amphibian skin is moistened by mucous secretions and is well supplied with blood vessels. It is used for respiration to varying degrees. When lungs are present, carbon dioxide may pass out of the body across the skin, but in some salamanders there are no lungs and all respiratory exchanges occur via the skin. Even in such animals as frogs, it seems that oxygen can be taken up at times by the skin, under water for example. Therefore, regulation of respiration occurs within a single species, and the relative contribution of skin and lungs varies during the life of the animal.

The amphibian heart is generally of a tripartite structure, with a divided atrium but a single ventricle. The lungless salamanders, however, have no atrial septum, and one small and unfamiliar group, the caecilians, has signs of a septum in the ventricle. It is not known whether the original amphibians had septa in both atrium and ventricle. They may have, and the absence of septa in many modern forms may simply be a sign of a flexible approach to the use of skin or lung, or both, as the site of oxygen exchange. In addition, the ventricle is subdivided by muscular columns into many compartments that tend to prevent the free mixing of blood.

The conus arteriosus is muscular and contains a spiral valve. Again, as in lungfishes, this has an important role in directing blood into the correct arterial arches. In the frog, Rana, venous blood is driven into the right atrium of the heart by contraction of the sinus venosus, and it flows into the left atrium from the lungs. A wave of contraction then spreads over the whole atrium and drives blood into the ventricle, where blood from the two sources tends to remain separate. Separation is maintained in the spiral valve, and the result is similar to the situation in lungfishes. Blood from the body, entering the right atrium, tends to pass to the lungs and skin for oxygenation; that from the lungs, entering the left atrium, tends to go to the head. Some mixing does occur, and this blood tends to be directed by the spiral valve into the arterial arch leading to the body.

Blood returning from the skin does not enter the circulation at the same point as blood from the lungs. Thus, oxygenated blood arrives at the heart from two different directions—from the sinus venosus, to which the cutaneous (skin) vein connects, and from the pulmonary vein. Both right and left atria receive oxygenated blood, which must be directed primarily to the carotid arteries supplying the head and brain. It is likely that variable shunting of blood in the ventricle is important in ensuring this. A ventricular septum would inhibit shunting; it is at least possible that its absence in amphibians is not a primitive feature but a secondary adaptation to variable gas-exchange mechanisms.

The amphibian venous system shows various features that are characteristic of land vertebrates. The posterior cardinal veins are replaced by a posterior vena cava, but they are still visible in salamanders. There is a renal portal system, and an alternative route back to the heart from the legs is provided by an anterior abdominal vein that enters the hepatic portal vein to the liver.

Amphibian larvae and the adults of some species have gills. There are four arterial arches in salamanders (urodeles) and three in frogs (anurans). These are three through six of the original series, the fifth disappearing in adult frogs. There is no ventral aorta, and the arterial arches arise directly from the conus—an important feature, given that the conus and its spiral valve control the composition of blood reaching each arterial arch. The names given to the three arterial arches of frogs are those used in all land vertebrates, including mammals. They are the carotid (the third), systemic (the fourth), and pulmonary (the sixth) arches. Blood to the lungs (and skin in frogs) is always carried by the sixth arterial arch, which loses its connection to the dorsal aorta. All land vertebrates supply their lungs with deoxygenated blood from this source.

Digestion in Amphibians

Many amphibians catch their prey by flicking out an elongated tongue with a sticky tip and drawing it back into the mouth before seizing the item with their jaws. Some use inertial feeding to help them swallow the prey, repeatedly thrusting their head forward sharply causing the food to move backwards in their mouth by inertia. Most amphibians swallow their prey whole without much chewing so they possess voluminous stomachs. The short oesophagus is lined with cilia that help to move the food to the stomach and mucus produced by glands in the mouth and pharynx eases its passage. The enzyme chitinase produced in the stomach helps digest the chitinous cuticle of arthropod prey.

Amphibians possess a pancreas, liver and gall bladder. The liver is usually large with two lobes. Its size is determined by its function as a glycogen and fat storage unit, and may change with the seasons as these reserves are built or used up. Adipose tissue is another important means of storing energy and this occurs in the abdomen (in internal structures called fat bodies), under the skin and, in some salamanders, in the tail.

Dissected frog: 1 Right atrium, 2 Liver, 3 Aorta, 4 Egg mass, 5 Colon, 6 Left atrium, 7 Ventricle, 8 Stomach, 9 Left lung, 10 Spleen, 11 Small intestine, 12 Cloaca.

There are two kidneys located dorsally, near the roof of the body cavity. Their job is to filter the blood of metabolic waste and transport the urine via ureters to the urinary bladder where it is stored before being passed out periodically through the cloacal vent. Larvae and most aquatic adult amphibians excrete the nitrogen as ammonia in large quantities of dilute urine, while terrestrial species, with a greater need to conserve water, excrete the less toxic product urea. Some tree frogs with limited access to water excrete most of their metabolic waste as uric acid.

Nervous System of Amphibians

The nervous system is basically the same as in other vertebrates, with a central brain, a spinal cord, and nerves throughout the body. The amphibian brain is less well developed than that of reptiles, birds and mammals but is similar in morphology and function to that of a fish. It is believed amphibians are capable of perceiving pain. The brain consists of equal parts, cerebrum, midbrain and cerebellum. Various parts of the cerebrum process sensory input, such as smell in the olfactory lobe and sight in the optic lobe, and it is additionally the centre of behaviour and learning. The cerebellum is the center of muscular coordination and the medulla oblongata controls some organ functions including heartbeat and respiration. The brain sends signals through the spinal cord and nerves to regulate activity in the rest of the body. The pineal body, known to regulate sleep patterns in humans, is thought to produce the hormones involved in hibernation and aestivation in amphibians.

Tadpoles retain the lateral line system of their ancestral fishes, but this is lost in terrestrial adult amphibians. Some caecilians possess electroreceptors that allow them to locate objects around them when submerged in water. The ears are well developed in frogs. There is no external ear, but the large circular eardrum lies on the surface of the head just behind the eye. This vibrates and sound is transmitted through a single bone, the stapes, to the inner ear. Only high-frequency sounds like mating calls are heard in this way, but low-frequency noises can be detected through another mechanism. There is a patch of specialized haircells, called *papilla amphibiorum*, in the inner ear capable of detecting deeper sounds. Another feature, unique to frogs and salamanders, is the columella-operculum complex adjoining the auditory capsule which is involved in the transmission of both airborne and seismic signals. The ears of salamanders and caecilians are less highly developed than those of frogs as they do not normally communicate with each other through the medium of sound.

The eyes of tadpoles lack lids, but at metamorphosis, the cornea becomes more dome-shaped, the lens becomes flatter, and eyelids and associated glands and ducts develop. The adult eyes are an improvement on invertebrate eyes and were a first step in the development of more advanced vertebrate eyes. They allow colour vision and depth of focus. In the retinas are green rods, which are receptive to a wide range of wavelengths.

ANATOMY OF AVES

Respiration in Aves

Due to the high metabolic rate required for flight, birds have a high oxygen demand. Their highly effective respiratory system helps them meet that demand.

Although birds have lungs, theirs are fairly rigid structures that do not expand and contract as they do in mammals, reptiles and many amphibians. Instead, the structures that act as the bellows that ventilate the lungs are the air sacs, which are distributed throughout much of the birds' bodies. The airsacs move air unidirectionally through the parabronchi of the rigid lungs. Although bird lungs are smaller than those of mammals of comparable size, the air sacs account for 15% of the total body volume, whereas in mammals, the alveoli, which act as the bellows, constitute only 7% of the total body volume. The walls of the air sacs do not have a good blood supply and so do not play a direct role in gas exchange.

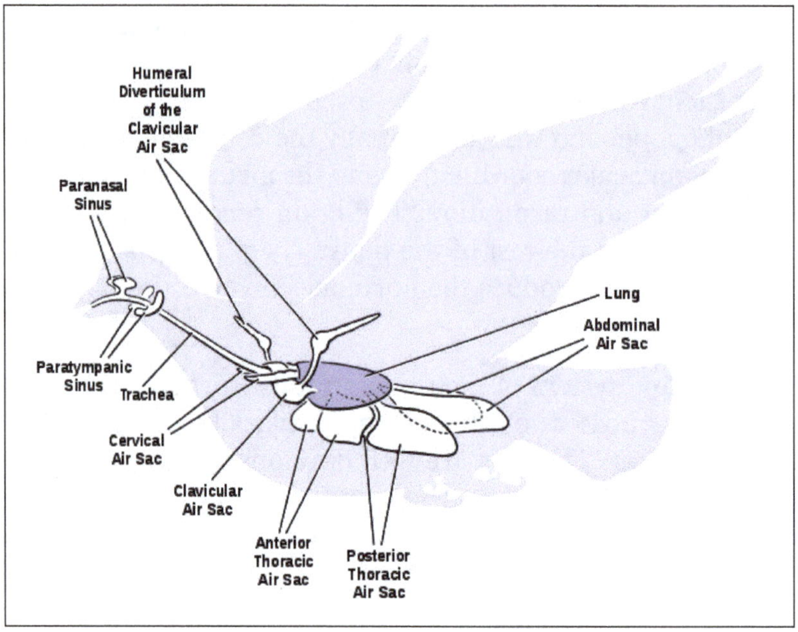

The arrangement of the air sacs, and lungs in birds.

Birds lack a diaphragm, and therefore use their intercostal and abdominal muscles to expand and contract their entire thoraco-abdominal cavities, thus rhythmically changing the volumes of all their air sacs in unison. The active phase of respiration in birds is exhalation, requiring contraction of their muscles of respiration. Relaxation of these muscles causes inhalation.

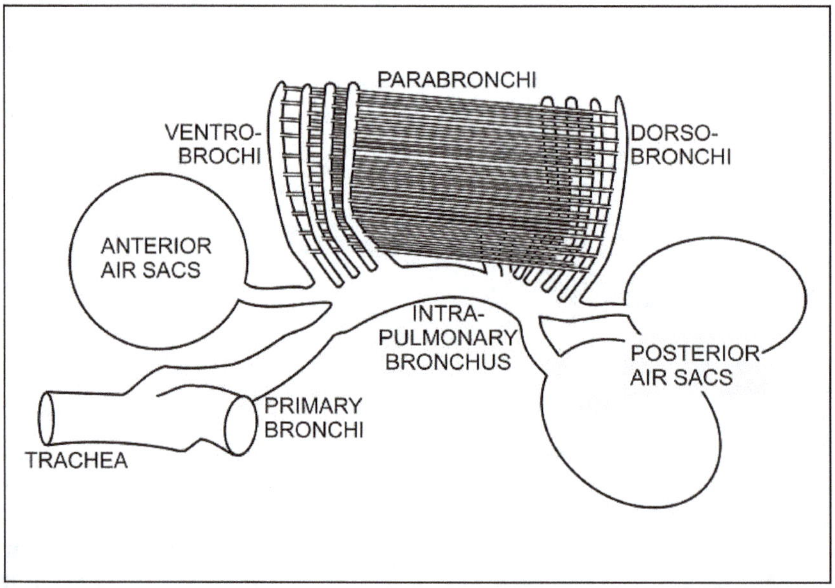

The anatomy of bird's respiratory system, showing the relationships of the trachea, primary and intra-pulmonary bronchi, the dorso- and ventro-bronchi, with the parabronchi running between the two. The posterior and anterior air sacs are also indicated, but not to scale.

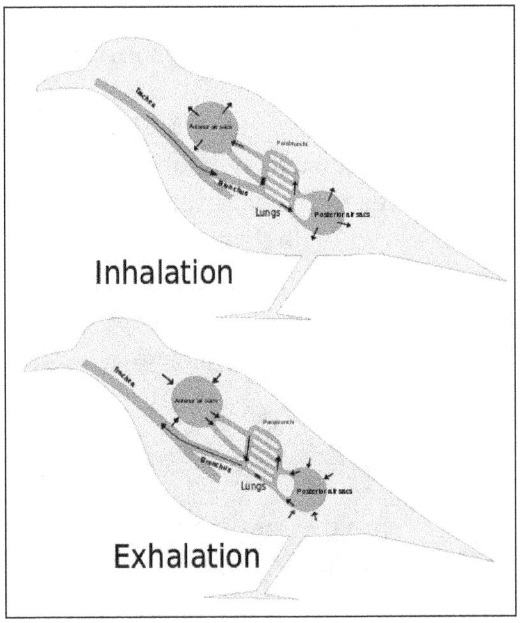

Inhalation–exhalation cycle in birds.

Three distinct sets of organs perform respiration — the anterior air sacs (interclavicular, cervicals, and anterior thoracics), the lungs, and the posterior air sacs (posterior thoracics and abdominals). Typically there are nine air sacs within the system; however, that number can range between seven and twelve, depending on the species of bird. Passerines possess seven air sacs, as the clavicular air sacs may interconnect or be fused with the anterior thoracic sacs.

During inhalation, environmental air initially enters the bird through the nostrils from where it is heated, humidified, and filtered in the nasal passages and upper parts of the trachea. From there, the air enters the lower trachea and continues to just beyond the syrinx, at which point the trachea branches into two primary bronchi, going to the two lungs. The primary bronchi enter the lungs to become the intrapulmonary bronchi, which give off a set of parallel branches called ventrobronchi and, a little further on, an equivalent set of dorsobronchi. The ends of the intrapulmonary bronchi discharge air into the posterior air sacs at the caudal end of the bird. Each pair of dorso-ventrobronchi is connected by a large number of parallel microscopic air capillaries (or parabronchi) where gas exchange occurs. As the bird inhales, tracheal air flows through the intrapulmonary bronchi into the posterior air sacs, as well as into the *dorso*bronchi (but not into the ventrobronchi whose openings into the intrapulmonary bronchi were previously believed to be tightly closed during inhalation. However, more recent studies have shown that the aerodynamics of the bronchial architecture directs the inhaled air away from the openings of the ventrobronchi, into the continuation of the intrapulmonary bronchus towards the dorsobronchi and posterior air sacs). From the dorsobronchi the air flows through the parabronchi (and therefore the gas exchanger) to the ventrobronchi from where the air can only escape into the expanding anterior air sacs. So, during inhalation, both the posterior and anterior air sacs expand, the posterior air sacs filling with fresh inhaled air, while the anterior air sacs fill with "spent" (oxygen-poor) air that has just passed through the lungs.

During exhalation the intrapulmonary bronchi were believed to be tightly constricted between the region where the ventrobronchi branch off and the region where the dorsobronchi branch off. But it is now believed that more intricate aerodynamic features have the same effect. The contracting posterior air sacs can therefore only empty into the dorsobronchi. From there the fresh air from the posterior air sacs flows through the parabronchi (in the same direction as occurred during inhalation) into ventrobronchi. The air passages connecting the ventrobronchi and anterior air sacs to the intrapulmonary bronchi open up during exhalation, thus allowing oxygen-poor air from these two organs to escape via the trachea to the exterior. Oxygenated air therefore flows constantly (during the entire breathing cycle) in a single direction through the parabronchi.

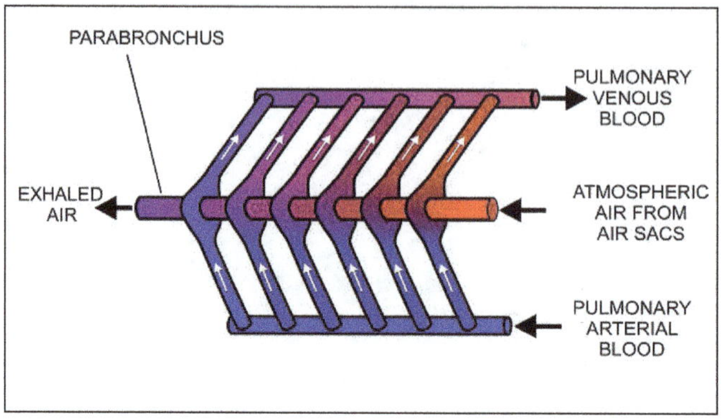

The cross-current respiratory gas exchanger in the lungs of birds.

Air is forced from the air sacs unidirectionally (from right to left in the diagram) through the parabronchi. The pulmonary capillaries surround the parabronchi in the manner shown (blood flowing from below the parabronchus to above it in the diagram). Blood or air with a high oxygen content is shown in red; oxygen-poor air or blood is shown in various shades of purple-blue.

The blood flow through the bird lung is at right angles to the flow of air through the parabronchi, forming a cross-current flow exchange system. The partial pressure of oxygen in the parabronchi declines along their lengths as O_2 diffuses into the blood. The blood capillaries leaving the exchanger near the entrance of airflow take up more O_2 than do the capillaries leaving near the exit end of the parabronchi. When the contents of all capillaries mix, the final partial pressure of oxygen of the mixed pulmonary venous blood is higher than that of the exhaled air, but is nevertheless less than half that of the inhaled air, thus achieving roughly the same systemic arterial blood partial pressure of oxygen as mammals do with their bellows-type lungs.

The trachea is an area of dead space: the oxygen-poor air it contains at the end of exhalation is the first air to re-enter the posterior air sacs and lungs. In comparison to the mammalian respiratory tract, the dead space volume in a bird is, on average, 4.5 times greater than it is in mammals of the same size. Birds with long necks will inevitably have long tracheae, and must therefore take deeper breaths than mammals do to make allowances for their greater dead space volumes. In some birds (e.g. the whooper swan, *Cygnus cygnus*, the white spoonbill, *Platalea leucorodia*, the whooping crane, *Grus americana*, and the helmeted curassow, *Pauxi pauxi*) the trachea, which some cranes can be 1.5 m long, is coiled back and forth within the body, drastically increasing the dead space ventilation. The purpose of this extraordinary feature is unknown.

Air passes unidirectionally through the lungs during both exhalation and inspiration, causing, except for the oxygen-poor dead space air left in the trachea after exhalation and breathed in at the beginning of inhalation, little to no mixing of new oxygen-rich air with spent oxygen-poor air (as occurs in mammalian lungs), changing only (from oxygen-rich to oxygen-poor) as it moves (unidirectionally) through the parabronchi.

Avian lungs do not have alveoli as mammalian lungs do. Instead they contain millions of narrow passages known as parabronchi, connecting the dorsobronchi to the ventrobronchi at either ends of the lungs. Air flows anteriorly (caudal to cranial) through the parallel parabronchi. These parabronchi have honeycombed walls. The cells of the honeycomb are dead-end air vesicles, called *atria*, which project radially from the parabronchi. The *atria* are the site of gas exchange by simple diffusion. The blood flow around the parabronchi (and their atria), forms a cross-current gas exchanger.

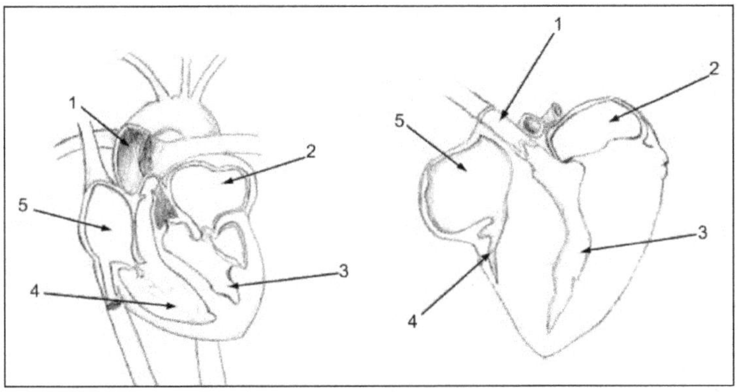

The human heart (left) and chicken heart (right) share many similar characteristics. Avian hearts pump faster than mammalian hearts. Due to the faster heart rate, the muscles surrounding the ventricles of the chicken heart are thicker.

Both hearts are labeled with the following parts: 1. Ascending Aorta 2. Left Atrium 3. Left Ventricle 4. Right Ventricle 5. Right Atrium. In chickens and others birds, the superior cava is double.

All species of birds with the exception of the penguin, have a small region of their lungs devoted to "neopulmonic parabronchi". This unorganized network of microscopic tubes branches off from the posterior air sacs, and open haphazardly into both the dorso- and ventrobronchi, as well as directly into the intrapulmonary bronchi. Unlike the parabronchi, in which the air moves unidirectionally, the air flow in the neopulmonic parabronchi is bidirectional. The neopulmonic parabronchi never make up more than 25% of the total gas exchange surface of birds.

The syrinx is the sound-producing vocal organ of birds, located at the base of a bird's trachea. As with the mammalian larynx, sound is produced by the vibration of air flowing across the organ. The syrinx enables some species of birds to produce extremely complex vocalizations, even mimicking human speech. In some songbirds, the syrinx can produce more than one sound at a time.

Circulation in Aves

The circulatory system is responsible for the transport of the various essential compounds and other factors around the body, as well as the removal of the metabolic wastes that accumulate in the tissues from body activities, to the appropriate places. The compounds and other factors transported around the body are blood, nutrients, medications and antibodies to fight infection, the residue of worn out cells and the wastes of metabolism. There are times when undesirable compounds and factors are found in the system as well, such as poisons or toxins and disease causing organisms.

The circulatory system consists of a number of organs and an associated transport system. These include the heart, the blood vessels, the spleen, the bone marrow, the blood and the lymph vessels. The blood and vascular system develops very early in the life of the embryo as the nutrition of the rapidly developing embryo is urgent and a transport system is required to transport the nutrients to where they are needed. Evidence of the system can be seen within about an hour and the system is clearly defined and operating within two days. By day three, it is possible to see the beat of the embryonic heart with the naked eye.

Fluids that act as a carrier of the various compounds and factors form part of the system. Each of these body fluids has its own function. There are two types of body fluids:

- Circulatory fluids (these move around):
    - Blood.
    - Lymph.
- Static body fluids (remain stationary):
    - Intracellular fluid or cell sap.
    - Intercellular, interstitial or tissue fluid.
    - Cerebrospinal fluid of the central nervous system.
    - Synovial fluid in the joints of the skeleton.
    - Aqueous and vitreous humor in the eye (fluid).
    - Endolymph and perilymph in the ear (cochlea).

The circulatory system consists of a number of organs that produce the blood cells and the organs that pump it around the body (the heart), and the blood vessels such as the arteries, veins, capillaries and lymph vessels that carry it.

After the arteries leave the heart they divide to ever smaller and smaller arteries to provide the many branches required to service the many systems, organs, tissues and cells. When they reach the tissues they divide into capillaries that are very small and very thin walled (one cell thick). This allows the nutrients and factors being delivered to move out of the capillary into the tissues and the material to be collected from the cell to enter. When they leave the tissue the capillaries rejoin to form veins until, when near the heart, the blood is moving through either the vena cava or the pulmonary veins.

The lymphatic system is a special fluid collecting system that collects the fluid left behind in the tissues by the capillaries and transports it to a region of the heart where it enters the vena cava.

Heart

In broad terms the heart acts as the pump that pumps in two directions:

- To the lungs where the carbon dioxide in the blood is removed and the oxygen replaced.

- To the rest of the body to deliver the nutrients and oxygen to the cells and to collect wastes and carbon dioxide.

The blood leaves the heart via arteries called the aorta (to the body) and the pulmonary artery (to the lungs). The blood always enters the heart via the vena cava vein (from the body) and the pulmonary vein (from the lungs).

Location and Shape

The heart is located in the thoracic cavity between the two lobes of the liver and mainly in front of that organ. It is relatively large and is enclosed in a thin membrane called the pericardium. The avian heart has two atria and two ventricles (four chambers), as is found in mammals. In general shape, it is typically conical with its apex, or pointed end, directed to the rear and slightly left of middle. A shallow groove, visible externally, indicates the division between the relatively small atria in the cranial end and the quite large ventricles making up the largest part of the organ in the caudal end.

Atria and Ventricles and Fibres

The walls of the atria are thin while those of the ventricles are quite thick. This is because of the atria only have to move the blood from the atria to the ventricle while the ventricles are responsible for pumping blood around the body. Microscopically, the muscle of the heart is similar to that of mammals as the fibres are nucleated, striated and form a syncytium (a mass of cytoplasm with numerous nuclei). Numerous fibres called Purkinje fibres are found in the heart under the endocardium and in the myocardium adjacent to the larger arteries.

The large veins draining the body (the two cranial vena cava and the caudal vena cava) enter the right atrium. The two pulmonary veins from the lungs join before entering the left atrium. The opening from the right atrium into the right ventricle is crescent-shaped. A strong muscular plate replaces the membranous tricuspid valve found in mammalian hearts. The opening between the left atrium and left ventricle is circular in shape with membranous flaps that correspond to the bicuspid valves of mammals. The septum (wall) between the two ventricles is thin with an even thinner area, called the fossa ovalis, located near the centre.

Ventricles

The external wall of the right ventricle is much thinner than that of the left. At its base is the opening for the pulmonary artery with three pocket-like semi-lunar valves. The openings of these pockets face away from the heart. The wall of the left ventricle is very thick except at the apex (pointed end). This is to provide the structural strength required for the heavy workload of that segment as it pumps blood around the body through the aorta. This leaves the heart from the left ventricle through an opening with semi-lunar valves similar that of the pulmonary artery. Tendinous cords called the chordae tendonae prevent too much movement of the valve segments from the force of the pumping heart (prevent the valves turning inside out).

Heart Blood Supply

The heart has its own blood supply via the two coronary arteries which divide from the aorta shortly after that artery leaves the left ventricle. The autonomous nervous system controls the pumping action of the heart.

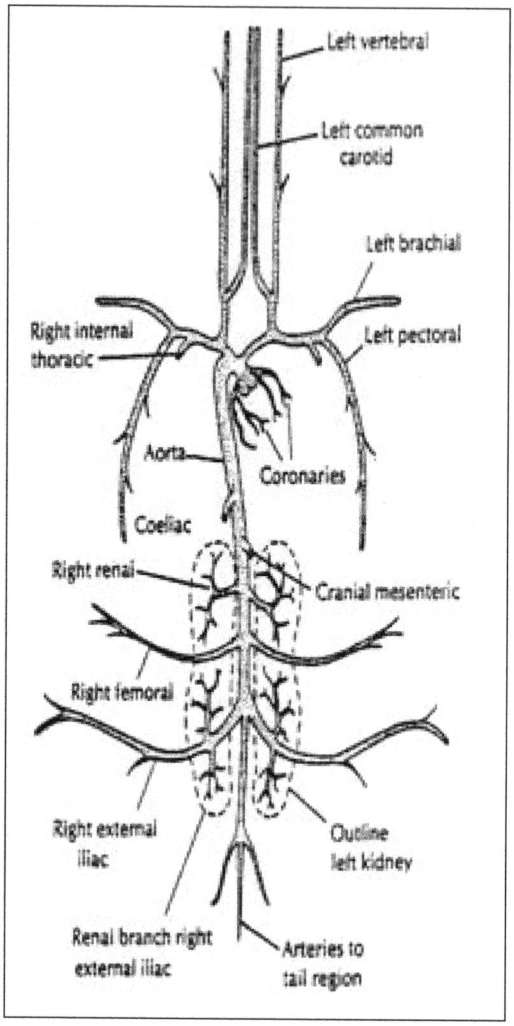

Arterial system.

Arteries

The pulmonary system starts as with the single pulmonary artery that leaves the right ventricle and shortly after divides into the right and left branches. Each of these branches enter the lungs and divide to ultimately be involved in the gaseous exchange that takes place in the lungs. The aorta leaves the heart via the left ventricle and very shortly after divides to provide the two coronary arteries that supply blood to the heart itself. Shortly after that the first of many divisions occur. Blood from the aorta is supplied to the head, the thoracic region including the wings, the abdominal region including the legs and tail.

New or oxygenated blood is bright red while old or deoxygenated blood is a very dark colour. In the pulmonary system, the arteries carry old blood while the veins carry new blood. In the main circulatory system, the arteries carry new blood and the veins carry old (arteries carry blood away from the heart and veins carry blood to the heart). As the arteries divide they become smaller and smaller until they form very small arteries which are often referred to as arterioles. When they enter the target organ they divide to become very thin walled capillaries and it is from these that the oxygen, nutrients, wastes and other factors either leave or enter the system.

Veins

The capillaries in the various tissues and organs rejoin to form the veins that carry the blood back to the heart. Those draining the body, other than the lungs, enter the right atrium as three veins called:

- Caudal vena cava.
- Right cranial vena cava.
- Left cranial vena cava.

The caudal vena cava drains the abdominal region while the two cranial vena cava drain the head via the juggler veins and the upper body including the wings and thoracic region excluding the lungs.

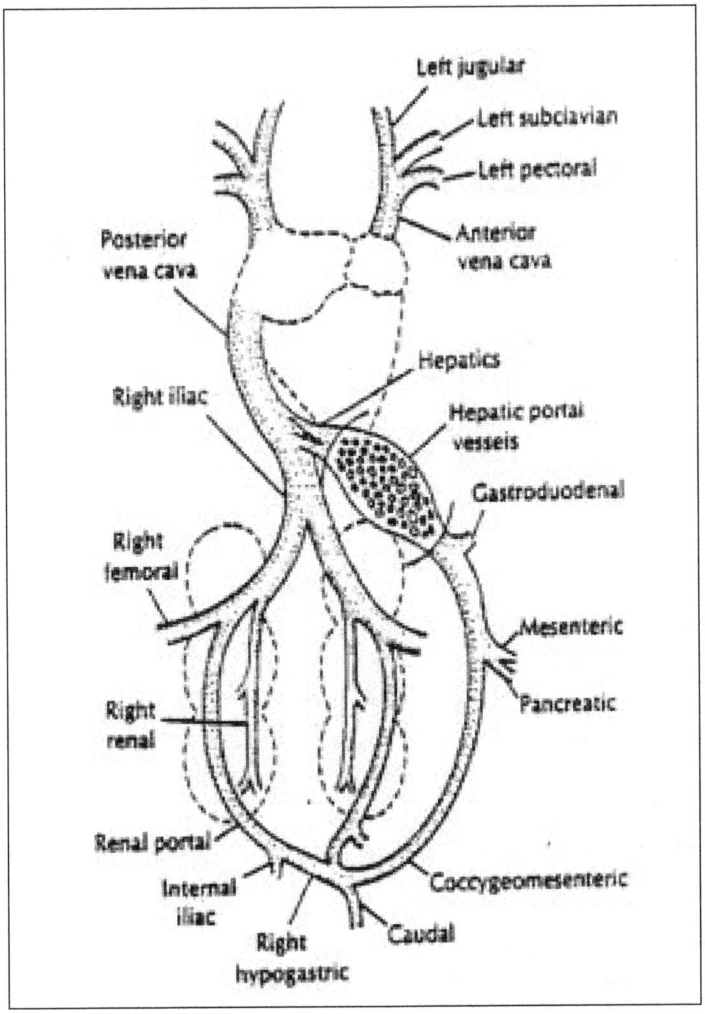

Venous system.

Kidney, Small Intestine, Liver and Lungs

The kidney system is very complex because of its function. The kidneys remove excess water and the wastes of metabolism from the blood stream. In addition they need their own supply for the maintenance of their own tissues and cells. As in mammals, blood containing nutrients absorbed from the small intestine, is carried to the liver by the portal vein. This is separate to the hepatic system that is responsible for the maintenance of the liver as an organ. The portal and hepatic systems join together in the liver to form the hepatic veins that enter the caudal vena cava near the heart. The capillaries in the lungs formed from the pulmonary arteries rejoin to form the pulmonary veins that enter the heart via the left atrium for pumping to the rest of the body.

Blood

Blood is the transport vehicle for the body. It is a very complex compound and consists of:

- Liquid intercellular substance or plasma.
- Suspended formed elements.

The formed elements include the red corpuscles or erythrocytes, the white corpuscles or leukocytes and the thrombocytes that correspond to the mammalian platelets. Blood also carries a great number and variety of other substances and factors. These include:

- Products from metabolism (wastes).
- Hormones.
- Enzymes.
- Antibodies.
- Effete products of tissues and organs (effete = worn out).
- Many other organic and inorganic compounds.

Erythrocytes

These are large, oval, flat, nucleated cells (mammalian erythrocytes are non-nucleated). Occasionally erytheroid structures without a nucleus are observed but these are normal avian erythrocytes from which the nucleus has been lost. The nucleus is centrally located and tends to be elongated in shape with a length of approximately 12 microns and a width of approximately 7 microns. The red colouring is caused by the presence of haemoglobin which is an iron compound that carries oxygen. The function of the erythrocytes is to transport the oxygen from the lungs to the tissues and the carbon dioxide from the tissues to the lungs. The erythrocytes are formed in the red bone marrow.

In the adult fowl there are 2.5 million erythrocytes in each millilitre of blood. The cell volume may be measured by taking a sample of the blood and centrifuging it at 3000 rpm for 15-20 minutes. The cells and the plasma separate and the volume of the cellular portion can be measured. This volume is called the haematocrit and, while it is mainly erythrocytes, it contains the leucocytes as well as a small quantity of plasma that has been trapped. The error as a result of the trapped plasma is approximately 5%.

Leucocytes

These are nucleated, amoeboid cells with a colourless cytoplasm. Some have fine granules in their cytoplasm, others carry coarse granules and others are non-granular. Because of the presence of granules the leucocytes are often called granulocytes. Most are phagocytic. The shape is therefore variable, but under normal conditions they are spherical. They are mono-nuclear although at times they appear to be multi-nuclear because of the shape of the nucleus as it is often multi-lobed.

The leucocytes are formed in the spleen, lymphoid tissue and in special cells in the bone marrow. The average count is approximately 30,000 per millilitre in adult birds and 10,000 per millilitre in day old chickens. Leucocytes may be classified by the character of their cytoplasm and nucleus. Protein granules in the cytoplasm may be acidic, basic or have neutral pH and this determines, to some extent, the type of stain they take up. Microscopic bodies such as cells and bacteria are usually stained in the laboratory to aid in their identification. Without staining, individually the different leucocytes are mainly colourless and it is very difficult to see their characteristics clearly even with a powerful microscope. Those identified in this way are:

- Acidic granules that stain with a basic (alkaline) stain and are called basophils.
- Basic (alkaline) granules that stain with an acidic stain and are called acidophils, oxyphils or eosinophils.
- Neutral granules may be stained with a neutral stain and are called neutrophils.

Using this system there are five main groups of leucocytes or granulocytes:

- Polymorphonuclear heterophilic granulocytes.
- Polymorphonuclear eosinophilic granulocytes.
- Polymorphonuclear basophilic granulocytes.
- Lymphocytes.
- Monocytes.

Polymorphonuclear Heterophilic Granulocytes (Heterophils)

These are known as heterophils. They are spherical cells with a diameter of 10-15 nanometres (nm). The nucleus is distinctly lobulated, with the lobes joined by thick strands of nucleoplasm (nucleus material). These cells play an important part in the body's defence against bacterial invasion as they have a bactericidal function and ingest protein.

Polymorphonuclear Eosinophilic Granulocytes (Eosinophils)

These are of similar size to the heterophils (10-15 nm), and the nucleus is frequently bi-lobed.

The eosinophilic granules in the nucleus are round and are not so numerous as the granules in the heterophils. The function of the eosinophils is not definitely known but it is believed that they increase in number in cases of internal parasite infestations. It has been suggested also that they have a detoxifying function.

Polymorphonuclear Basophilic Granulocytes (Basophils)

These are also about the same size as the heterophils and eosinophils. The nucleus is generally not lobulated but appears to be oval in shape. There tends to be fewer of these cells and their function is unknown, however it has been suggested that they are a stage in the degeneration of eosinophils.

Lymphocytes

These are the most numerous of all of the leukocytes. Their size varies significantly, the older cells being somewhat smaller than the newer cells because of the loss of cytoplasm from the older cells. The nucleus size is the same in all cells but may be kidney shaped in some cases. The mobility of these cells is greatly reduced and their phagocytic action is probably nil. Their function is thought to be that of regulating toxic materials. They also release enzymes that assist in the synthesis of nucleoplasm. In their degeneration they release special compounds called gamma globulins that have antibody properties.

Monocytes

Monocytes are the largest of all of the leukocytes. They are difficult to separate from the large lymphocytes because of a strong similarity in appearance. The nucleus may be round, oval or kidney shaped. They are very mobile and very phagocytic in their activity. Their function is to remove, by ingestion, bacteria, protozoan and tissue debris.

Thrombocytes

The thrombocytes are similar to the blood platelets of mammals, however, they are less involved with blood clotting. They are quite numerous (approximately 25000 per millilitre). They are the smallest of all of the blood cells and group together as small clumps. Thrombocytes are formed in the bone marrow in a way similar to the red blood cells.

Plasma

The plasma is the liquid or non-cellular portion of the blood. It contains a number of different compounds including:

- Glucose or blood sugar.
- Non protein nitrogen substances.
- Plasma proteins.
- Plasma liquids.

- Metallic elements.
- Plasma enzymes.
- Water.

Blood sugar

The blood sugar of birds is in the form of D-Glucose as in mammals, but usually at higher, though variable, levels. Certain hormones such as insulin, glucagon, corticoids and glucocorticoid influence the actual level. The plane of nutrition is also an influence- Whether the bird is full fed or is starved (plane of nutrition) also influences blood sugar. The amount of sugar in the blood is approximately 260 mg/100 ml, although levels as low as 110 and as high as 350 mg/100 ml have been reported.

Non-protein Nitrogen

After the removal of the plasma proteins, the remaining fluid contains a number of nitrogenous substances. The amount being in the approximate range of 15.5 mg/100 ml in the layer to 19.5 mg/100 ml in the immature bird. These include:

- Uric acid.
- Urea.
- Creatine.
- Free ammonia.
- Free amino acids.

Uric Acid and Urea

Uric acid is the end product of protein metabolism and represents the bulk of the waste nitrogen eliminated by the birds. The level in the blood varies considerably and is influenced by the sex and reproductive state. As a result, non-layers have a higher level. Diet is another influence as birds on a high protein diet have a higher level of uric acid in their blood. Urea is another nitrogen containing compound and is eliminated by the kidney. Relatively small amounts may be found in the blood (2.2-2.5 mg/100 ml), and the level is influenced by age, sex and production.

Creatine, Ammonia and Free Amino Acids

Creatine is an amino acid present in animal tissues. It is equally distributed between the plasma and the cellular portion. It combines with phosphate to form phosphocreatine, an important compound in the anaerobic (without oxygen) phase of muscle contraction. The levels found range from 0.5-1.6 mg /100 ml with an average of approximately 1.0 mg/100 ml. Ammonia, or more specifically the ammonium ion, is found in such small quantities that further consideration is unwarranted. Free amino acids are found in very small quantities. About 22 different amino acids have been detected in the blood of fowls, but not all at the same time. The total amount has been estimated at approximately 8 mg/100 ml of blood.

Plasma Proteins

Avian plasma contains a number of protein compounds which may be grouped into the following:

- Albumins: Transport the fatty acids, calcium and other minerals, and hormones.
- Immunoglobulins: Produced by lymphocytes as a result of antigenic demand.
- Transferrin: Transports iron (Fe) across the intestine wall. Forms part of the secondary defence mechanism against bacteria and viruses by making iron unavailable to them.
- Ceruloplasmin: Copper transport.
- Phosvitin: Carries calcium, iron and phosphate to the yolk and calcium to the shell gland.
- Lipoproteins: Carry fat-soluble vitamins, hormones and some minerals in the blood.
- Fibrinogen: Is involved in blood clotting and is also present in the blood.

Lymph system

The body of the fowl is richly served with lymph vessels. Lymph is derived from the body fluid found within the interstitial space (the space between the important parts of the body). The lymphatic system has the function of draining the body systems of fluid that is left behind by the blood vessels, although the lymph fluid does ultimately return to the main circulatory system when the lymph vessel enters the vena cava near the heart. There are no lymph nodes in fowls. Lymph plexuses (an intertwining of the very small lymph vessels) are found instead of mammalian lymph nodes.

Spleen

The spleen is located immediately to the right of the junction of the glandular stomach (proventriculus) and the gizzard. The main function of the spleen is to filter out the unwanted particles from the blood. Another function is its involvement in the formation of lymphocytes. It is reddish brown in colour and generally round in shape. A thin, fibrous capsule with few muscle fibres surrounds it.

Digestion in Aves

The digestive system in the domestic fowl is very simple but efficient when compared to many other species, such as cattle. In the process of evolution, those avian species that developed simple but effective digestive systems were more able to fly and hence survive, as the simple digestive system would be lighter in weight. It is necessary that the diet provided to fowls be of high quality and easily digestible due to the simplicity in the structure and function of their digestive system. This is especially important if the birds are to attain the expected productive performance. The metabolism of carbohydrate, fat and protein is closely linked to the digestive system.

The digestive system consists of the alimentary canal along which the food passes after eating to where the residual wastes are eliminated from the body, together with the liver and the pancreas. The digestive system is responsible for the ingestion of food, its breakdown into its constituent nutrients and their absorption into the blood stream, and the elimination of wastes from this process.

The liver produces bile and is associated with the metabolism of nutrients together with a number of other functions. The main function of the pancreas is the production of digestive enzymes and special compounds called hormones.

The Alimentary Canal

The alimentary canal is a long tube-like organ that starts at the beak and ends with the vent or cloaca in the abdominal region. Generally the alimentary canal has layers of muscle that run lengthwise and around it and is lined with mucous membranes. Glands that produce important digestive juices are found in different locations of the canal. The nutrients from the food, after digestion, are absorbed through the wall of the alimentary canal into the circulatory system for transport to the liver or other parts of the body. The waste remaining is eliminated from the body via the cloaca or vent.

Mouth Structure

Fowls don't have lips and cheeks, they instead have a beak which is an area of dense and horny skin lying over the mandible and incisive bones that serve as the bony foundation. There are no teeth. The so called egg tooth found on the end of the beak of newly hatched chickens is an aid to their escape from the egg at hatching and disappears after a day or two. The hard palate that forms the roof of the mouth, presents a long, narrow median (median – along the middle) slit that communicates with the nasal cavity. The hard palate has five transverse rows of backwardly pointing, hard, conical papillae. Numerous ducts of the salivary glands pierce the hard palate to release their secretions into the mouth cavity.

Salivary Glands

A thick layer of stratified squamous epithelium covers the free surface. The salivary glands run the whole length of the hard palate, the groups of glands merging to form one mass of glandular tissue under the epithelium. Lymphoid tissue is found in most glands.

The salivary glands are:

- Maxillary: In the roof of the mouth.
- Palatine: On either side of the nasal opening in the roof of the mouth.
- Apheno-pteryoid glands: In the roof of the pharynx on each side of the common opening for the eustachian tubes (the eustachian tubes connect the middle ear to the mouth and their function is to equalise the air pressure on each side of the tympanic membrane in the ear).
- Anterior sub-mandible glands: In the angle formed by the union of the upper and lower beaks or mandibles.
- Posterior sub-mandibular glands.
- Lingual glands: In the tongue.

- Crico-arytenoid glands: Around the glottis.
- A small gland in the angle of the mouth.

Pharynx and Tongue

The pharynx is continuous with, and follows, the mouth. The combined cavity of the mouth and the pharynx is often referred to as the oropharynx. The common opening for the two eustachian tubes is located in the middle of its dorsal wall (roof). The tongue is long and pointed and conforms to the shape of the beak in which it operates. The epithelium of the tongue is thick and horny, especially towards the tip. A transverse row of simple, large and horny papillae with their tips directed towards the rear of the mouth cavity are located on the posterior end. The hyoid bone provides the framework to support the tongue. The entoglossal bone extends longitudinally in the median plane. Small patches of lymphatic tissue are located throughout the corium. Mucous glands are located in the tongue with short ducts directed towards the rear. Some believe that there are taste buds located on the tongue, but this belief is not universally held. In any case, the sense of taste appears to be very weak if at all present.

The mouth has two major functions:

- To pick up the food particles.
- To direct the food into the oesophagus – as part of the bird's eating behaviour.

Oesophagus, Crop and Proventriculus

The oesophagus is wide and is capable of being significantly stretched. It connects the mouth region to the crop in close association with the trachea. The crop is a large dilation of the oesophagus located just prior to where the oesophagus enters the thoracic cavity. The crop provides the capacity to hold food for some time before further digestion commences. This capacity enables the bird to take its food as "meals" at time intervals but permits continuous digestion. Inside the thoracic cavity, the oesophagus enters or becomes the proventriculus which is a very glandular part of the digestive tract (often called the glandular stomach).

The wall of the oesophagus is composed of four layers of tissue, the innermost being mucous membrane. The mucous membrane is an important barrier to the entry of microbes and the mucous it produces is a lubricant that aids the passage of the food along the alimentary canal. The structure below the crop is similar to that above except there is less lymphoid tissue below the crop. The crop structure is similar to that of the oesophagus except there are no glands present in fowls. Ducks and geese have glands in the crop mucous membranes. In pigeons the surface cells of the crop slough off during brooding to form pigeon's milk which is used to feed the baby pigeons in the nest.

Proventriculus

The glandular stomach, or proventriculus, is relatively small and tubular. The wall is very thick and is composed of five layers:

- Outer serous membrane.
- Muscle layer composed of three separate layers:

  - Two thin longitudinal layers.
  - Thick circular layer.
- Layer of areolar tissue containing blood and lymph vessels.
- Thick layer composed mainly of glandular tissue.
- Mucous membrane.

The glands form the greater part of the thickness of the organ. Simple single glands group to form lobules each of which converges into a common cavity near the surface. The cavities converge to form a common duct that leads to the surface through the apex of a small papilla. These glands produce a number of juices or enzymes that are used in the digestion or breaking down of food into its constituent nutrients. The mucous membrane is raised into folds and between these folds are numerous simple tubular glands that produce hydrochloric acid as well as lymphoid tissue.

Gizzard

The muscular stomach or gizzard is located immediately after the proventriculus, partly between the lobes and partly behind the left lobe of the liver. It has a flattened, rounded shape somewhat like a convex lens, with one side slightly larger than the other. Each surface is covered by a glistening layer of tendinous tissue which is thicker at the centre and becoming thinner towards the edges. Under this outer layer there are located very powerful masses of red muscle. The inner surface is lined with a creamy-coloured, thick, horny tissue raised in ridges. The gizzard almost always contains quantities of hard objects such as gravel or other grit that aids in the disintegration of food, which is the primary function of the gizzard.

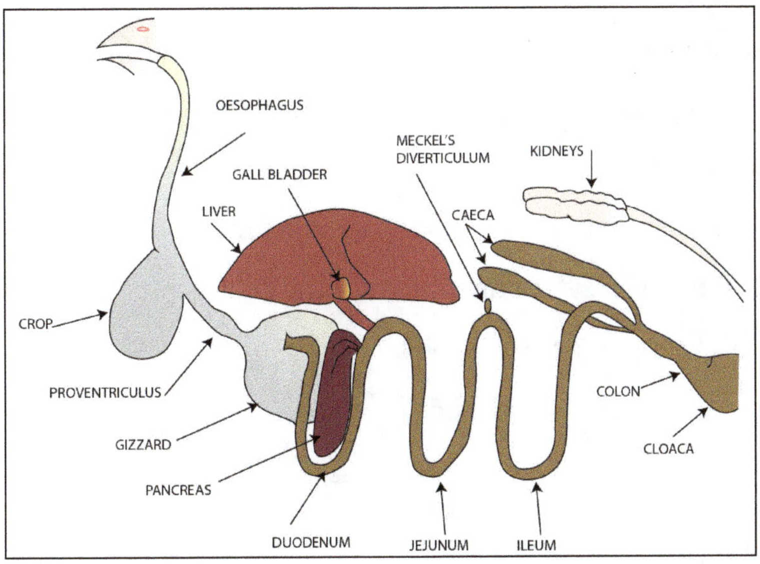

The entrance from the proventriculus and the exit to the duodenum are close together and dorsal in location. The gizzard consists of a number of layers of tissues, some of which contain straight tubular glands. The innermost layer is a strong, flexible skin that is able to withstand the potentially damaging effects of the muscular action grinding the food often in the presence of stones or other insoluble material. The glands of the gizzard produce a liquid which is a keratinised material that passes to the surface of the horny lining where it hardens to replace tissue worn away by the grinding action of the organ.

The Small Intestine

The small intestine begins at the exit from the gizzard and ends at the junction of the small intestine, caeca and colon. It is relatively long and has a constant diameter. Of the three parts of the mammalian small intestine, the duodenum, jejunum and ileum, only the duodenum can be easily distinguished in the fowl. There is no clear demarcation between the jejunum and ileum and the small intestine appears as one long tube. Much of the digestion of the food and all of the absorption of the nutrients takes place in the small intestine and hence its structure is quite important. The structure is as follows:

- Serosa: A serous membrane on the outside of the intestine.
- A layer of longitudinal muscle: Fibres run along the length of the intestine.
- A layer of circular muscle: Three times as thick as the longitudinal muscle. Located between the two muscle layers are:
    - Blood vessels.
    - Lymph vessels.
    - A network of nerve fibres.
- An ill-defined sub-mucosa – the areolar of the oesophagus.
- Mucous membrane consisting of:
    - A thick muscularis mucosae of longitudinal and circular muscle.
    - Corium – many glands, lymphoid tissue, muscle fibres and a variety of free cells.
    - Epithelium or surface.

The small intestine has a number of very important functions:

- Produces a number of enzymes involved in the digestion process.
- Site of much of the digestion of the food.
- Site of much of the absorption of food.

Villi

When a piece of the small intestine is immersed in water it takes on a very velvety appearance because of the presence of villi – long flattened, fingerlike projections that extend into the lumen (inside) of the intestine like flexible fingers. The villi are very actively involved in the absorption process. A single layer of columnar epithelium together with goblet cells covers the lining. The goblet cells secrete mucous. Permanent folds in the mucous membrane called the "valves of kerkring" are located at the proximal end (closest to the front) of the duodenum.

A lacteal (lymph vessels), capillaries, bundles of plain muscle fibres, nerves and other tissues and cells occupy the core of the villus. The villi have the function of providing a vastly increased surface area for the more efficient absorption of the nutrients. The efficiency of the absorption is influenced by the surface area available for the nutrients to move through i.e. the more villi the better the absorption. They also provide a means of concentrating the nutrients collection ability once they have moved through the intestine wall.

Duodenum

After the duodenum the small intestine forms a coil and is suspended from the dorsal wall of the abdominal wall by a thin membrane called the mesentery. This membrane carries the blood vessels associated with the intestine. The duodenum starts at the gizzard and forms an elongated loop that is approximately 20 centimetres long. The pancreas lies between the arms of the loop and is attached to, and actually holds together, each arm of the duodenum.

Lymphoid tissue in the duodenum is very plentiful and is usually located in the corium. The lymphoid tissue collects the lymph and the lymph vessels transport fluid, other than blood, that is found in the spaces between cells and tissues until it passes into the blood system. Bile ducts from the gall bladder that are attached to the liver and two to three pancreatic ducts enter the small intestine by a common papilla at the caudal end (closest to the rear) of the duodenum. The pancreas is a very important organ in the process of digesting food and it is attached to each side of the duodenal loop and lies between the two arms.

Jejunum and the Ileum

The jejunum and the ileum, together about 120 cm long, commence at the caudal end of the duodenum where the bile and the pancreatic duct papilla are located and terminates at the ileo-caecal-colic junction. This junction is where the small intestine, the two caeca and the colon all meet. This portion of the small intestine is similar in structure to the duodenum except that:

- It is suspended in the mesentery.
- The villi are shorter.
- There is less lymphoid tissue.

Meckel's Diverticulum is a constant feature about half way along the small intestine and appears as a small projection on the outer surface of the small intestine. This projection is where the yolk sac was attached during the development of the embryo.

Large Intestine

The large intestine is very short and does not differ to any extent from the calibre of the small intestine. It runs in nearly a straight line below the vertebrae and ends at the cloaca. Sometimes this section is referred to as the colon and the rectum (the rectum being the terminal section). The bursa of fabricius is located immediately above the cloaca of young birds but disappears when the birds have reached approximately one year old.

Caeca

The two caeca or blind pouches are about 16-18 centimetres long in the adult. They extend along the line of the small intestine towards the liver and are closely attached to the small intestine along their length by the mesentery. Each caecum has three main parts:

- A narrow base with thick walls arising at the ileo-colic-caecal junction.
- Middle part with thin walls.
- The wide blind apex with fairly thick walls.

The structure of the caeca is as follows:

- Serous membrane.
- Outer longitudinal muscle.
- Circular muscle.
- Inner longitudinal muscle forming the muscularis mucosae of the mucous membrane.

Cloaca

The large intestine terminates in the front part of the cloaca. The cloaca is a tubular cavity opening to the exterior of the body and is common to the digestive and urogenital tract. The structure of the cloaca is very similar to that of the intestine except that the muscularis mucosa disappears near the vent. It divides into three chambers, each separated by a constriction not readily defined:

- The copradaeum: A continuation of the colon-rectum.
- The urodaeum: Middle part into which the ureters and genital ducts open.
- The proctodaeum: Opens to the exterior of the vent. Birds less than one year old have a dorsal opening leading into the blind, rounded sac – the bursa of fabricius.

Liver

The liver is a bi-lobed organ that lies ventrally (below) and posterior (in rear of) to the heart and is closely associated with the proventriculus and the spleen. The right side lobe is larger. The liver is dark brown or chocolate in colour except for the first 10-14 days when it may be quite pale due to the absorption of lipids (fats) from the yolk as an embryo. It weighs approximately 50 grams in adult birds. The capsule, or glissosis, is the membrane that covers the liver and is thinner than that of mammals.

The gall bladder lies on the right lobe beneath the spleen. Two bile ducts emerge from the right lobe and one of these originates from the gall bladder and the second provides a direct connection from the liver to the small intestine. A system of ducts connects the right and left lobes.

There are a number of functions that are performed by the liver:

- Bile formation: Consisting of bile, various pigments and bile salts. Bile is involved in the digestion of fats to fatty acids and glycerol.
- The metabolism of:
    - Carbohydrate.
    - Lipids.
    - Protein.
- Production and destruction of blood cells.
- Synthesis of plasma proteins and fibrinogen (associated with blood clotting).
- Storage of glycogen, fat and fat-soluble vitamins e.g. vitamin A.
- Detoxification of certain substances (detoxify – destroy the poisonous effect).

The liver cells have a high rate of destruction and a good regenerative capacity (re-growth ability). Notwithstanding this, in the normal animal, much of the organ is in reserve and can be removed or destroyed without causing undue stress.

Blood Supply and Drainage

There are two blood supply systems. One originates from the coelic artery for normal maintenance of the liver as an organ and the second, called the hepatic portal system, transports the nutrients from the small intestine after absorption to the liver. This latter system enters the liver via two veins (one for each lobe). The two blood supply systems join together inside the organ.

The liver is drained via the hepatic veins into the posterior vena cava (hepatic – relating to the liver; vena cava – one of the main veins that enters the heart). The liver has a network of sinusoids (empty holes in the tissues as in a sponge). The hepatic portal system, the capillaries of the arterial blood supply and the hepatic veins are in close association with each other in these sinusoids.

Bile

The liver consists of a series of tissue sheets that are two cells thick, with a sinusoid on either side of the sheet. Bile is made by the cells. The blood vessels, when they enter these sinusoids, become closely associated with them to provide for the easy transfer of material from one system to another. Minute canals called canaliculi that have the task of collecting and transporting the bile are associated with the cells in the tissue sheets. These canals eventually join together to form the bile ducts with one going directly to the intestine and one to the gall bladder before it connects to the small intestine.

Pancreas

This organ has three lobes that occupy the space between the two arms of the duodenal loop. Two or three ducts pass the secretions of this organ into the distal end of the duodenum via papillae common with the ducts from the gall bladder and the liver. The structure is similar to that of the pancreas of mammals and consists of special secreting tissue for pancreatic juice as well as other groups of cells called the "islets of Langerhans". These are mainly associated with the production of hormones. In poultry the cells of the islets of Langerhans are less defined than those in mammals.

The functions of the pancreas are:

- Produce pancreatic juice which is a mixture of digestive enzymes.
- Produce the hormones insulin and glucagon that are involved in the metabolism of carbohydrate.

Chicken Digestion

The pattern of food intake and its passage through the digestive system are the main factors that influence secretory and hence digestive activity. Probably because of the high metabolic rate of the fowl, a more or less continuous supply of food is required by the digestive system. This is provided for by the crop that acts as a reservoir for the storage of food prior to its digestion and consequently permits the fowl to eat its food as periodic meals. There is quite wide variability between birds in relation to eating behaviour, even between those in the same flock. Some eat small amounts at short intervals while others eat larger amounts at wider intervals.

Food Storage

The food is delivered into the crop for storage after the first few boli have passed into the proventriculus. The crop is quite distensible and will hold a large amount of undigested food that is then moved on as required by the proventriculus. This function of the crop is less important when there is a plentiful supply of food available. Due to the crop's ability to hold a supply of food, when applying a food control (restriction) program, it is necessary to compensate by providing a long period of food deprivation to achieve the required degree of control. There is no relationship between the length of time of food deprivation and the amount of food consumed.

Food Intake

While there is a wide variation between the eating habits of different birds in the flock, fowls do tend to eat meals on about 15-minute intervals through the daylight hours and, to some extent, during darkness. They tend to eat larger portions at first light and in the late evening.

Factors that affect food intake include:

- Age and live weight.
- Environmental temperature.
- Energy content of the food.
- Food texture.

- Level of other key nutrients.
- Egg production.
- Water quality and temperature.
- The health status of the flock.

Similar factors affect the rate of movement of the food through the digestive system with a meal of normal food taking approximately 4 hours to pass through in the case of young stock, 8 hours in the case of laying hens and 12 hours for broody hens. Intact, hard grains take longer to digest than the cracked grain and, quite often some whole grain will pass through unchanged.

Enzyme Action

After ingestion, the food is mixed with saliva and mucous from the mouth and oesophagus and these secretions thoroughly moisten the food. The enzyme amylase, which is produced by the salivary and oesophageal glands and found in the saliva and mucous, can now commence to break-down the complex carbohydrates. However, the amount of enzyme action at this stage is minimal and the first major enzyme activity takes place in the proventriculus and in the gizzard.

Hydrochloric Acid, Pepsin and Gastrin

The secretions of the proventriculus, or glandular stomach as it is often called, include hydrochloric acid to lower the pH of the system and the food mixture, the enzyme pepsin that acts on protein, and the hormone gastrin that stimulates the production and release of gastric juice in the proventriculus and pancreatic juice from the pancreas.

Breaking the Food Particles

The gizzard is a very powerful organ which physically breaks the food particles into smaller sizes to make the work of the enzymes easier. At the same time, the enzymes previously released into the food with the saliva and by the proventriculus are thoroughly mixed into the food which improves their opportunity to carry out their work. This breaking and mixing function of the gizzard is enhanced by the presence of insoluble grit such as stones.

The food material enters the duodenum from the gizzard. Enzyme activity in this region is, in the main, a continuation of the breakdown of proteins started in the gizzard. Pancreatic juice and bile from the liver enters via ducts located at the distal end of the duodenum at about the junction of the duodenum and the jejunum if it were differentiated. However, because of back flow of pancreatic juice and bile towards the gizzard, the actions of these secretions start earlier in the digestive process than would be expected by their entry point to the small intestine. One effect is an increase in the pH of the intestinal contents of the latter half of the duodenum from strongly to weakly acid.

- In addition to enzymes, the pancreas produces insulin and sodium bicarbonate. The insulin is involved in the maintenance of blood sugar levels while the sodium bicarbonate, which is strongly alkaline, will increase the pH of the intestinal contents.

Reducing Complex Food Compounds

The small intestine also produces enzymes that plays part in the digestive process of reducing the complex food compounds eaten to the simple compounds or building blocks that can be absorbed across the intestinal wall for transport to the organ or location where either they will be further processed, stored or used. Food materials that escape enzyme action along this tract are subjected to bacterial breakdown in the caeca which provides a system of at least partial recovery of some nutrients.

Waste

The remainder of the material consists of waste and undigested food and are mixed with the urine in the cloaca and eliminated from the body as faeces. The appearance of the faeces varies considerably, but typically is a rounded, brown to grey mass topped with a cap of white uric acid from the kidneys. The contents of the caeca are also discharged periodically as discrete masses of brown, glutinous material.

- The average daily production of faeces from laying hens is between 100 and 150 grams. These fresh droppings are approximately 75% water and will air dry under favourable conditions to approximately 30% water.

Nervous System of Aves

The nervous system and the important sensory organs play a key role in the day-to-day functioning of an animal. The nervous system integrates and controls the various functions of the body, while the sensory organs detect the various stimuli in the bird's environment that it reacts to. Functions are actively (voluntarily) or automatically (involuntarily) controlled.

Voluntary control occurs where the animal, in its response to some activity or stimulus, has a choice in what actions (if any) it may take. It chooses to respond in one way or another, such as to move a particular part(s) of its anatomy. The nerves that make up this voluntary part of the nervous system are called voluntary, or somatic, nerves.

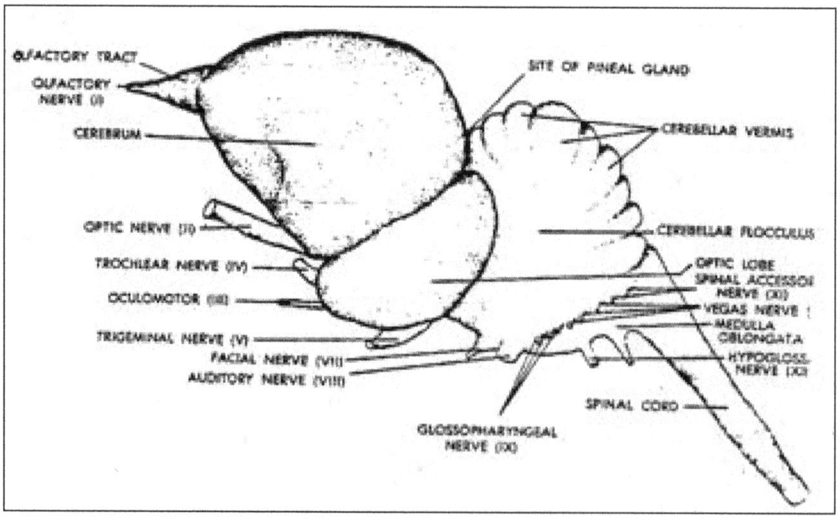

Lateral view of the chicken brain.

Involuntary control exists where the animal has no choice and the response to an activity or event occurs without the animal having any conscious control. Examples of the functioning of this part of the nervous system are the regulation of heart beat and circulation, digestion, and respiration. It is obvious that the animal cannot afford to have to remember to keep its heart beating, its digestive system functioning or even just to breath. The part of the nervous system that regulates such important functions is called the autonomic, or involuntary, nervous system.

The Brain

The brain is located in the head and is well protected by the bones of the cranium. The brain consists of a number of parts, which in turn consist of various special cells that have the ability to detect, recognise, remember and direct. Thus the brain is the control centre for the many functions and activities of the many systems, organs and tissues that make up the bird's body.

The parts and major regions that make up the avian brain are as follows:

- The forebrain which consists mainly of the cerebral hemispheres and the olfactory lobes. The hypothalamus and pituitary gland are located on the lower side of the forebrain.
- The midbrain which mainly consists of the optic lobes.
- The hindbrain which consists mainly of the cerebellum and the medulla oblongata.

The olfactory lobes of the forebrain are the receptor areas for the olfactory nerves and are the centre for smell, while the optic lobes are the receptor areas for the optic nerve and are the centre for sight. The optic lobes of poultry are very large when compared with those of other species in relation to total brain size. This indicates that sight plays a major role in the normal behaviour of fowls.

The very small gland called the pituitary gland, or hypophysis, is associated with the hypothalamus. This gland is of major importance as an endocrine gland, so much so that it is often called the "master" gland. Many of its secretions regulate the operation of many other glands as well as the functions of the systems, organs and tissues of the bird.

Spinal Cord

The spinal cord, as the name suggests, is a cord of nerve tissue that extends from the medulla oblongata of the brain along almost the full extremity of the vertebral column through the canal provided for that purpose. The spinal cord and the brain constitute the Central Nervous System (often referred to as the CNS). Like the brain, the spinal cord is well protected, firstly by its spinal fluid and the sheath that encloses it, all of which is fully enclosed within the bones of the vertebral column. The various nerves that provide for the control of the various systems, organs and tissues of the body leave the spinal cord through appropriate openings located in the joints between the different vertebrae.

If the spinal cord should be broken, the connection between the brain and the parts that it controls will be broken and hence control of the effected parts will be lost. For example, a break of the spinal cord in the lower back will result in the loss of control (paralysis) of the legs and other functions that take place below the break.

Nerves

The nerve cell or neuron consists of a cell body with one or more elongated projections extending from it. The cell body contains the nucleus. The projections are part of the cytoplasm and are called axons if they are long and singular, or dendrites if they are short and branched. Nerve endings or receptors at one end carry the sensors that respond to the stimuli, while on the opposite side, the stimulus is transferred ultimately to the brain. The nerve endings are the means by which stimuli are perceived or control exercised. The remainder of the nerve cell acts as a message carrier to the brain in much the same way that telephone line carries a message between two telephones. These messages are in the form of very weak electrical currents.

There are a number of different types of nerve endings on nerve cells depending on the task they have to carry out, such as smell, hear, see or perceive touch. The bird may then react to these stimuli and the response is directed via the nerves of a particular system, organ or tissue. Those that send messages to the muscles to respond are often called "motor nerves" or "motor neurons". The types of nerve endings associated with touch include those specifically for pain, touch, cold, heat and pressure.

The Sensory Organs

The sensory organs receive the various stimuli from the bird's environment. Depending on the mode of action of the stimuli, special endings on the nerves will perceive that stimulus. The sensory organs include:

- Eye(s): For sight.
- Ear(s): For hearing and balance.
- Olfactory organ: For smell.
- Taste buds: For taste.

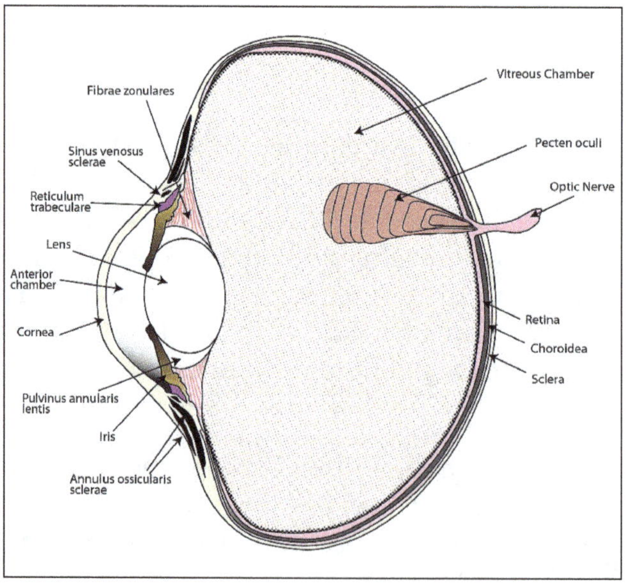

Sagittal cross section of the chicken eye.

Eyes

The eye of the fowl is fairly large in relation to the size of the bird. There is one eye on each side of the face and only a 26° angle directly to the bird's front has binocular vision (Binocular vision occurs when both eyes see an object. This is an important aid to distance perception). There are two main eyelids and a third very thin membrane called the nictitating membrane, which is located in the front corner of each eye. While the eye is open, this membranous "eyelid" is only partially visible. It is capable of covering the eyeball with a very fast movement to provide protection to that organ.

The fowl's eye is very similar to the mammalian eye, except that fowls have a higher threshold of light intensity at which they see than do some other birds and mammals. This makes them active during the day (diurnal) rather than at night (nocturnal). At night in nature, or in a free ranging situation, chickens seek the protection of trees to sleep.

Perception of the Environment

The eyes play an important part in the bird's perception of the environment and it is believed they see colours in much the same way as do humans. One exception in the structure of the eye is the presence of pectin which is believed to have the function of retaining acuity, or sharpness of vision, even after long periods of fixed staring.

The eyes function by projecting the image of the "seen" object through the lens of the eye onto the retina. Located on the retina are the receptive rod and cone cells, and the image is "seen" by these cells. It is then transmitted by the optic nerve as a "signal" to the optic lobes of the brain where the signal is recognised for what it is – the image of the object being "seen". Special muscles associated with the lens change the shape of the lens to control the focus of the eye onto objects at different distances.

The fowl's eye is well supplied with tears that drain from the eye into a collecting sac and then into the nasal cavity via a special duct. A very fine, sensitive, transparent membrane called the conjunctiva covers the surface of the eye ball and the inside of the eye socket. This membrane can become inflamed and/or infected if the bird contract conjunctivitis.

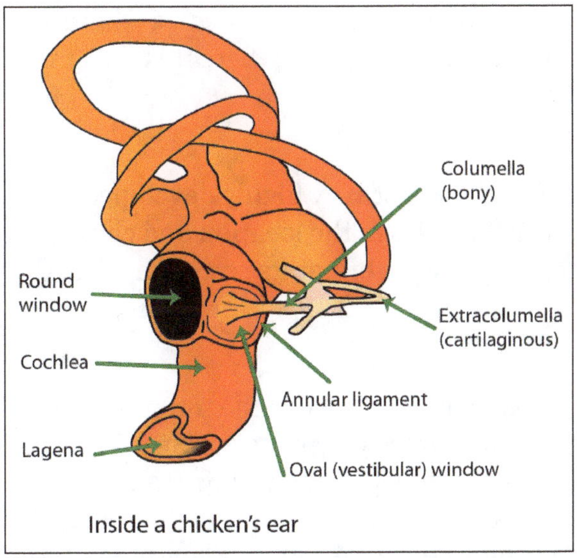

Right ear of a chicken including semicircular canals.

Ear

The ears of the fowl are located on each side of the face behind the eye. While the ear is very similar to that of mammals, there are some differences. The fowl's ear does not have a pinna, ear flap or lobe and the three bones of the mammalian middle ear have been replaced by a single structure of bone and cartilage.

The fowl ear consists of three main segments:

- The outer ear.
- The middle ear separated from the outer by the tympanic membrane (ear drum).
- The inner ear which consists of the cochlea and the three semicircular ducts.

The outer ear consists of the ear canal. Unlike the mammalian ear there is no pinna to direct the sound waves into the ear canal but it is covered by a tuft of feathers to protect it from dust and other potentially harmful materials.

The middle ear is separated from the outer ear by the tympanic membrane which is stretched across the inner end of the ear canal in a similar way to that of a percussion drum top. A rod of cartilage replaces the three bones of the mammalian ear connecting the tympanic membrane to the inner ear and bone called the columella.

The inner ear consists of the cochlea in which special nerve endings are located that receive the sound waves for transmission to the area of the brain associated with hearing. Also located in the inner ear are the semicircular ducts that are associated with the maintenance of balance. A special duct connects the middle ear with the roof of the mouth with the ducts from each ear joining before entering the mouth. The function of this duct is to regulate the air pressure in the middle ear to that of the outer ear (the environment) to prevent air pressure injury to the tympanic membrane.

The fowl hears by sound waves that enter the outer ear canal and apply pressure in waves on the tympanic membrane. The wave-like nature of sound causes the tympanic membrane to vibrate. This vibration is transmitted by the columella to the cochlea where special nerve endings receive it and transmit it by the auditory nerve to the brain where it is recognised as sound.

Olfactory Organ

This is the organ of smell. In fowls it is not very highly developed and it is believed that smell is of secondary importance to the other senses in the location and choice of food. The loss of a chicken's ability to smell does not seem to influence either its selection of food or other behaviour.

Taste Buds

The taste buds in the fowl are located at the base of the tongue and on the floor of the mouth. They contain nerve endings from the glossopharyngeal nerve. Small quantities of the chemicals of taste are recognised by the taste buds and this information is transferred to the appropriate receptors of the brain. Fowls have shown that the taste of materials in the dry state is weaker than when in the liquid state and they are more likely to refuse something on the basis of taste when it is in the water than when it is in their food. This has important application when using strong tasting medications to treat disease.

An important feature is the sensitivity of the fowl to the temperature of its drinking water. They are likely to reject water with a temperature above approximately 33°C and in any case the provision of cool water in hot weather is an important strategy to help reduce the effects of heat stress. They will drink water that has a very low temperature even just above freezing point.

ANATOMY OF MAMMALS

Respiratory System of Echidna

Mammalian lungs, including those of the Monotremata, differ from those of birds and reptiles in many ways, but principally in the great development of a branching tree-like system of intrapulmonary bronchi. The bronchi divide and subdivide forming bronchioli which lead to alveolar ducts ending blindly in alveoli. The bronchial tree is usually asymmetrical and this is the case with echidnas where two lobes of lung are present on the right side and only one on the left. Narath described the bronchial tree in Tachyglossus in detail.

Monotremes exhibit that characteristic mammalian feature, a diaphragm, which is very broad, muscular towards the periphery, but tendinous at the centre immediately under the heart.

Parameters of respiration have been determined by Bentley, Herreid & Schmidt-Nielsen and by Parer & Hodson. The former found end tidal (alveolar) CO_2 concentration was 5.3% while the concentration of O2 was 14.5%; similar concentrations of CO_2 and O_2 in end-tidal air are found in man (5.95% CO_2, 14.0% O_2) seals (6.3%, 11.6%) and dogs (5.68%, 13.6%). Mean tidal volume was found to be 26.9 ml per breath and mean minute volume 137 ml/kg body weight. Parer & Hodson recorded an average minute volume of only 76 ml/kg. found, however, that minute volume could be altered. The changes related to changes in respiratory rate: a decrease in respiratory rate is accompanied by a decrease in minute volume.

While echidnas, forced to inspire high concentrations of CO_2, increased their respiratory minute volume, the change was brought about by change in tidal volume, not by increase in respiratory rate. The tidal and minute volume levels observed are similar to those of eutherians, as is the extraction of O_2 from inspired air, although O_2 consumption at thermoneutrality is only 0.22 ml/g/h, about half that of eutherians of comparable size. Such a comparison is dependent on the eutherian selected; sloths, for example, averaging the same weight as adult Echidnas, have a very low O_2 consumption. and Dawson, Grant & Fanning confirm that tachyglossids at thermoneutrality have a low O_2 consumption reporting levels of 0.15 ml/g/h and 0.132 ml/g/h, respectively.

Echidnas are tolerant of very high levels of CO_2 in inspired air , and voluntarily remain in situations where CO_2 concentrations range as high as 6.9%. At least one other mammal, the rabbit Oryctolagus cuniculus, voluntarily withstands equally high CO_2 concentrations: the air in nesting chambers containing litters exhibits concentrations of 7–14%.

The respiratory and cardiac responses of echidnas buried in earth and breathing asphyxial air created by their own respiration. Following burial with about 200 mm of earth over the dorsal surface, bradycardia or slowing of the heart occurred; little change in respiration rate and a sustained rise in CO_2 concentration (hypercapnia) in the air around the snout were observed. When levels of 10–12% CO_2 were reached, the surface of the soil was seen to heave and crack open. After a few seconds this activity stopped, but the 'bellows' action was repeated at regular intervals leading to maintenance of about 10% CO_2 in the inspired air for the duration of the experiments (4 hours). The Short-beaked Echidna is well adapted physiologically for burrowing. Apart from CO_2 insensitivity, the blood has a high oxygen affinity and a well-defined Bohr effect of 0.49. This ensures that the circulating blood of buried animals has a degree of O_2 saturation assisting it to cope with hypoxic inspired air. Measurement of plasma in individuals buried in soil reveals that pCO_2 increases to quite high values with a concurrent drop in plasma pH. During hyperventilation or 'bellows' episodes, bradycardia was replaced by tachycardia for a short period. No major adaptations of enzyme content, enzyme organisation, myoglobin content or tissue and blood buffering capacity were found to be required to deal with the two accumulations other than the tachycardia and coupled hyperventilation. These were found to be adequate to maintain metabolic homeostasis and to sustain tachyglossids submerged in soil for hours at a time.

Respiratory System of Platypus

The lungs of the Platypus consist of two lobes on the right side and one on the left. There is a well-developed diaphragm at the base of the rib cage.

Corresponding to its haematological characteristics, the blood of the Platypus has been found to have a high oxygen carrying capacity (about 23 ml O_2/100 ml blood). Oxygen dissociation curves for blood are shown in Figure. These workers found a marked Bohr effect and bradycardia in response to forced diving which they related, along with the haematological values, to the diving physiology of the species. The Platypus does not dive for long periods or to great depths, usually spending less than 2 minutes below the water surface at one time during its foraging activities. Adult animals spend as much as 50% of their day in the burrow. Nestlings are confined to the breeding burrows for between 3–4 months, so it is possible that the physiological mechanisms are as much related to coping with hypoxia and hypercapnia within the burrow as to diving. The rapid reduction in the oxygen content of the blood and increased CO_2 content, followed by a dramatic recovery over two minutes after a forced dive in their experimental animals, does fit well, however, with the normal pattern of diving observed in free-living animals.

The Platypus can adequately regulate its body temperature in air between temperatures of 0–30°C, but does not cope well with air temperatures above 30°C. It is also a competent homeotherm in water.

The normal body temperature of the Platypus is 32.0 ± 0.4°C. This low, but constant, temperature is maintained by the elevation of metabolic rate seen in all endothermic species. The standard metabolic rate at the thermoneutral temperature of 25°C in air is 2.2 W kg-0.75 and can be elevated to 1.8 times this at an air temperature of 5°C and to 3.2 times in water of the same temperature. Radiotelemetric studies on free-ranging animals in the wild have shown that body temperatures are maintained close to 32°C during foraging periods of many hours in water as cold as 0°C.

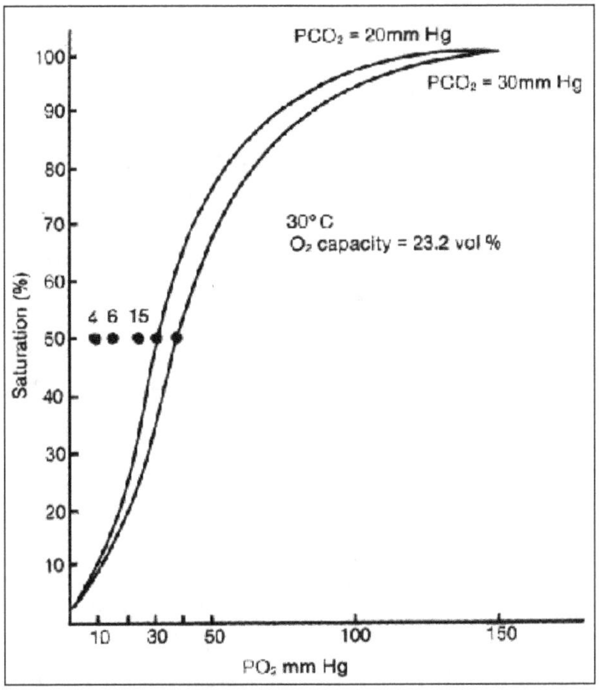

Oxygen dissociation curves of Platypus blood. Additional values of P_{50} at 4, 6 and 15 mm Hg PCO_2 are shown, indicating the Bohr Effect.

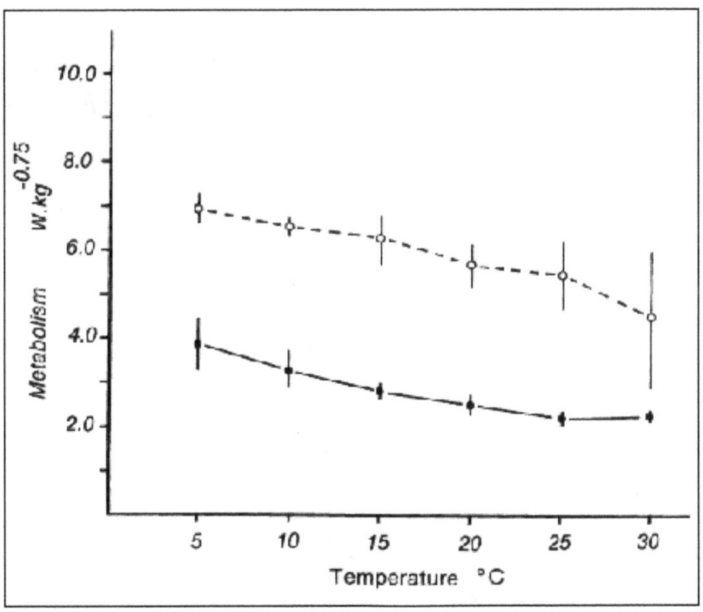

Metabolism of the Platypus in air and water of different temperatures. Means ± S.E. are shown.

Homeothermy is achieved not only by elevation of metabolic rate, but also by virtue of the high thermal insulation which results from both reduced tissue conductance at low ambient temperatures and the high insulative value of the fur and its associated layer of air (0.33 W-1 m2 °C in air). Although the fur loses 60–70% of its insulative value in water and tissue insulation becomes more important, the fur of the Platypus still contributes more significantly to total body insulation in water than the fur of many other amphibious mammals. High tissue insulation presumably is achieved by the restriction of heat loss to the environment through the unfurred extremities. Skin temperatures at low ambient air and water temperatures are kept at only 1–2°C above ambient. Occupation of a burrow, which buffers environmental temperature changes in both summer and winter conditions, also contributes to thermoregulation in the species.

There have been various reports of hibernation or torpor based on captive specimens. This has not been observed in the wild, but it is possible that the species exhibits torpor in response to cold and food shortage as does Tachyglossus aculeatus. Studies on the activities of the enzymes lactate and malate dehydrogenase from the Platypus and on oxygen consumption of various tissues in vitro over a range of temperatures have produced equivocal results in trying to relate these to the lability of body temperature in either the Platypus or the Echidna. In both studies, the responses obtained appear to be most like those expected for a homeothermic species than for a heterotherm.

Respiration in Placantal Mammals

Respiratory System of Humans

The respiratory system consists of all the organs involved in breathing. These include the nose, pharynx, larynx, trachea, bronchi and lungs. The respiratory system does two very important things: it brings oxygen into our bodies, which we need for our cells to live and function properly; and it helps us get rid of carbon dioxide, which is a waste product of cellular function. The nose, pharynx, larynx, trachea and bronchi all work like a system of pipes through which the air is funnelled down into our lungs. There, in very small air sacs called alveoli, oxygen is brought into the bloodstream and carbon dioxide is pushed from the blood out into the air. When something goes wrong with part of the respiratory system, such as an infection like pneumonia, it makes it harder for us to get the oxygen we need and to get rid of the waste product carbon dioxide. Common respiratory symptoms include breathlessness, cough, and chest pain.

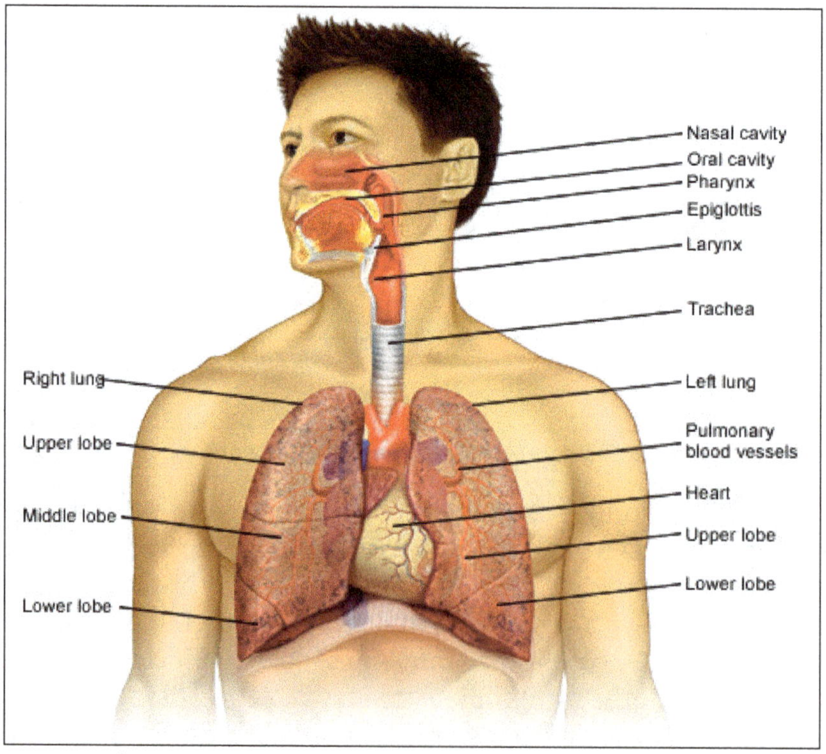

The Upper Airway and Trachea

When you breathe in, air enters your body through your nose or mouth. From there, it travels down your throat through the larynx (or voicebox) and into the trachea (or windpipe) before entering your lungs. All these structures act to funnel fresh air down from the outside world into your body. The upper airway is important because it must always stay open for you to be able to breathe. It also helps to moisten and warm the air before it reaches your lungs.

The Lungs

Structure The lungs are paired, cone-shaped organs which take up most of the space in our chests, along with the heart. Their role is to take oxygen into the body, which we need for our cells to live and function properly, and to help us get rid of carbon dioxide, which is a waste product. We each have two lungs, a left lung and a right lung. These are divided up into 'lobes', or big sections of tissue separated by 'fissures' or dividers. The right lung has three lobes but the left lung has only two, because the heart takes up some of the space in the left side of our chest. The lungs can also be divided up into even smaller portions, called 'bronchopulmonary segments'.

These are pyramidal-shaped areas which are also separated from each other by membranes. There are about 10 of them in each lung. Each segment receives its own blood supply and air supply.

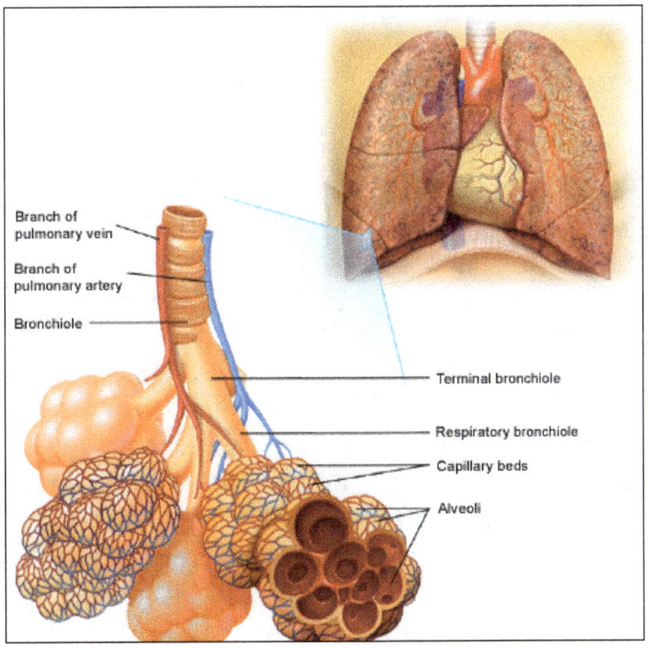

Blood Supply

The lungs are very vascular organs, meaning they receive a very large blood supply. This is because the pulmonary arteries, which supply the lungs, come directly from the right side of your heart. They carry blood which is low in oxygen and high in carbon dioxide into your lungs so that the carbon dioxide can be blown off, and more oxygen can be absorbed into the bloodstream. The newly oxygen-rich blood then travels back through the paired pulmonary veins into the left side of your heart. From there, it is pumped all around your body to supply oxygen to cells and organs.

The Pleurae

The lungs are covered by smooth membranes that we call pleurae. The pleurae have two layers, a 'visceral' layer which sticks closely to the outside surface of your lungs, and a 'parietal' layer which lines the inside of your chest wall (ribcage). The pleurae are important because they help you breathe in and out smoothly, without any friction. They also make sure that when your ribcage expands on breathing in, your lungs expand as well to fill the extra space.

The Diaphragm and Intercostal Muscles

When you breathe in (inspiration), your muscles need to work to fill your lungs with air. The diaphragm, a large, sheet-like muscle which stretches across your chest under the ribcage, does much of this work. At rest, it is shaped like a dome curving up into your chest. When you breathe in, the diaphragm contracts and flattens out, expanding the space in your chest and drawing air into your lungs. Other muscles, including the muscles between your ribs (the intercostal muscles) also help by moving your ribcage in and out. Breathing out (expiration) does not normally require your muscles to work. This is because your lungs are very elastic, and when your muscles relax at the end of inspiration your lungs simply recoil back into their resting position, pushing the air out as they go.

The Respiratory System Through the Ages

Breathing for the Premature Baby

When a baby is born, it must convert from getting all of its oxygen through the placenta to absorbing oxygen through its lungs. This is a complicated process, involving many changes in both air and blood pressures in the baby's lungs. For a baby born preterm (before 37 weeks gestation), the change is even harder. This is because the baby's lungs may not yet be mature enough to cope with the transition. The major problem with a preterm baby's lungs is a lack of something called 'surfactant'. This is a substance produced by cells in the lungs which helps keep the air sacs, or alveoli, open. Without surfactant, the pressures in the lungs change and the smaller alveoli collapse. This reduces the area across which oxygen and carbon dioxide can be exchanged, and not enough oxygen will be taken in. Normally, a foetus will begin producing surfactant from around 28-32 weeks gestation. When a baby is born before or around this age, it may not have enough surfactant to keep its lungs open. The baby may develop something called 'Neonatal Respiratory Distress Syndrome', or NRDS. Signs of NRDS include tachypnoea (very fast breathing), grunting, and cyanosis (blueness of the lips and tongue). Sometimes NRDS can be treated by giving the baby artificially made surfactant by a tube down into the baby's lungs.

The Respiratory System and Ageing

The normal process of ageing is associated with a number of changes in both the structure and function of the respiratory system. These include:

- Enlargement of the alveoli. The air spaces get bigger and lose their elasticity, meaning that there is less area for gases to be exchanged across. This change is sometimes referred to as 'senile emphysema'.

- The compliance (or springiness) of the chest wall decreases, so that it takes more effort to breathe in and out.
- The strength of the respiratory muscles (the diaphragm and intercostal muscles) decreases. This change is closely connected to the general health of the person.

Respiratory System of Marine Mammals

The respiratory tract of marine mammals begins with the external nares of pinnipeds and sirenians and blowhole(s) of cetaceans and ends with the lungs. Opening of the nares in all of these marine mammal groups is accomplished by contraction of skeletal muscle, whereas closure is a passive process. The active opening of the nostrils or blowholes is an energy-conserving adaptation to spending life in an aquatic medium, avoiding the need for continuous muscle contraction to keep water out of the respiratory tract.

The position of the nares in pinnipeds is not exceptional for mammals and no valve is present in the external nares of pinnipeds. Instead, the flat medial surfaces of two throat (arytenoid) cartilages lie in close proximity, which also abut the posterior surface of the epiglottis, providing a tight seal against the entry of water into the trachea. The powerful muscles of the larynx also help to keep this entrance closed.

Cetaceans breathe through the blowholes (external nares), which have migrated to a position high on the top of the head. The sperm whale, in which the blowhole is on the anterior top end of the head, slightly left of center, is the exception. Mysticetes possess two blowholes in contrast to the single blowhole of odontocetes.

Manatees breathe through two-valved nostrils situated at the tip of the rostrum. The nostrils of the dugong lie close together and are situated anterodorsally, thereby allowing the dugong to breathe with most of its body submerged. The nostrils are closed during diving by anteriorly hinged valves.

Deep-diving marine mammals have flexible chest walls and other structures that are capable of sufficient collapse to render the lungs virtually airless. The trachea in most pinnipeds is supported by cartilaginous rings that either form complete circles or are incomplete and overlap dorsally. This morphology permits the alveoli to collapse while the proximal airways are still open, ensuring that pulmonary gas exchange is eliminated and pulmonary entrapment of air is avoided when the animal is at depth.

The tracheal rings are reduced to ventral bars in Ross, Weddell, and leopard seals. In the ribbon seal, a longitudinal slit occurs between the ends of the incomplete cartilage rings on the dorsal surface of the trachea just before it bifurcates to form the bronchi. A membranous sac of unknown function extends both anteriorly and posteriorly from the trachea on the right side of the body. Male ribbon seals possess this sac, and it increases in size with age suggesting that it may have a role in sound production associated with mating behavior. The tracheal slit is present in females, but the sac is small or absent. In phocids and the walrus, the trachea divides into two primary bronchi immediately outside the substance of the lung. In otariids this bifurcation is located more anteriorly, at approximately the level of the first rib, and the two elongated bronchi run parallel until they diverge to enter the lungs dorsal to the heart.

The cetacean larynx is composed of a cartilaginous framework held together by a series of muscles. To keep inhaled air separate from food, the larynx of odontocetes (but not mysticetes) has two elongate cartilages. This provides a more direct connection between the trachea and blowhole than is found in mysticetes.

In cetaceans the trachea is short and broad and consists of several cartilaginous rings (varying from 5 to 7 rings in the beluga and sperm whale to 13 to 15 in the fin whale) that are interconnected with each other. Unlike baleen whales, the tracheal rings of toothed whales are closed and form a noncollapsing tube.

The dugong trachea is short (only four cartilage rings) and it is deeply divided by a medial septum. The manatee trachea is longer and is supported by 8–12 tracheal rings.

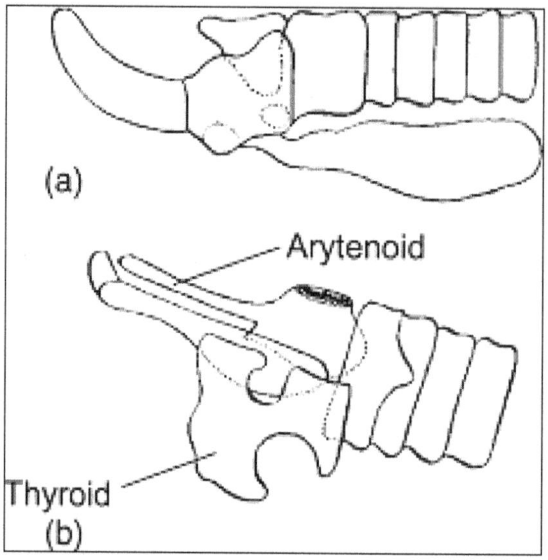

Lateral view of cetacean larynx in (a) mysticete (fin whale) and (b) odontocete (narwhal). Note the elongated arytenoid and thyroid cartilages in the odontocete.

Lungs

The lungs of marine mammals are not larger than those of terrestrial mammals, but some important differences exist between the lungs of marine mammals and those of terrestrial mammals. The left and right lungs of pinnipeds are approximately equal in size and are lobulated as in terrestrial carnivores; both lungs have three main lobes, but the right lung has an additional small intermediate lobe. There appears to be a tendency to reduce lobulation in the lungs of some pinnipeds, such as the walrus and ribbon, harp, and spotted seals. The latter species has been reported as having little or no lobulation. The bronchi subdivide within the lungs to form bronchioles and eventually end in alveoli, but the details of this transition vary among the three families.

The lungs of cetaceans are distinct from those of all other mammals in their overall sacculate shape and lack of lobes. Only occasionally is the apical part of the right lung somewhat prominent, resembling the apical lobe of the lungs of other mammals. The right lung is usually larger, longer, and heavier than the left. Such asymmetry of lung size, which is related in whales as in other mammals to the somewhat asymmetric position of the heart in the chest cavity, is seen in virtually all studied cetaceans: rorquals, various dolphins, and sperm and beluga whales.

The lungs of cetaceans, compared to those of terrestrial mammals, show greater rigidity and elasticity because of increased cartilaginous support. In baleen, sperm, and bottlenose whales,the septa projecting into the proximal portion of the air sacs contain heavy myoelastic bundles. In the smaller toothed whales these bundles are atrophied, but there are a series of myoelastic sphincters in the smallest bronchioles. In both the bundles and the sphincters, the muscular part may act to close the air sacs, whereas the elastic part may facilitate rapid expiration.Airway closure should delay alveolar collapse during dives and allow additional pulmonary gas exchange to occur for those odontocete families known to possess myoelastic terminal airway sphincters.

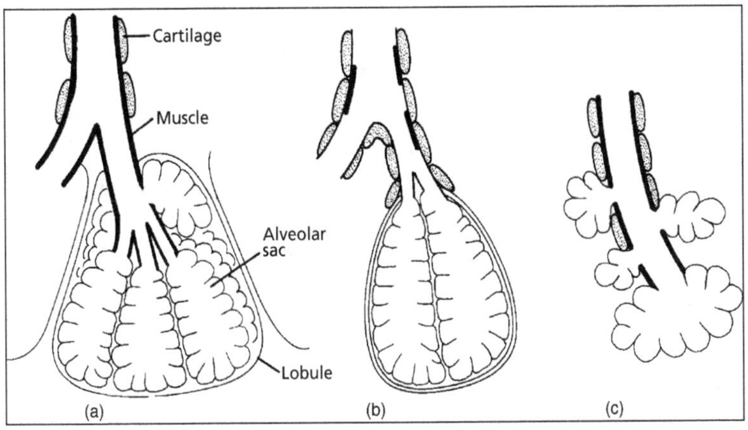

Diagram of the structure of alveoli and associated cartilage and muscle in pinnipeds. (a) Phocid. (b) Otariid. (c) Odobenid.

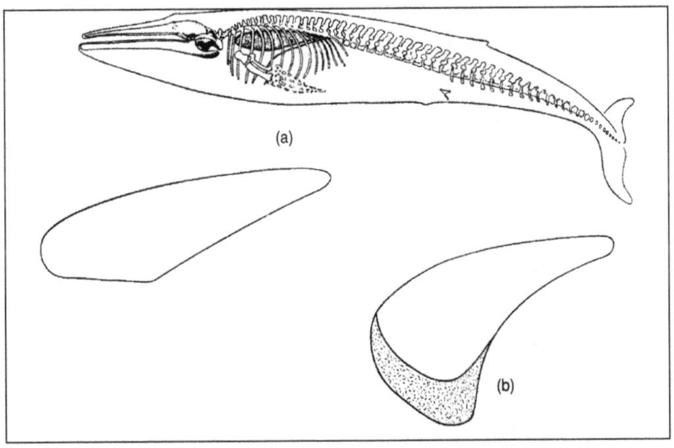

Diagram showing position of the cetacean lung and comparison of (a) mysticete (fin whale), (b) and odontocete (bottlenose dolphin) lung. Stippled area indicates particularly thin section of lung.

Relative lung volume is lower for cetaceans than terrestrial mammals. The relatively small lung volume of deep-diving species is a logical consequence of the inability of the respiratory tract to store gas because of the risks of an embolism and other difficulties that divers that breathe gas under pressure experience when surfacing rapidly. The unusual oblique position of the diaphragm permits the abdominal contents to occupy part of the thorax when the animal is under pressure. Deep divers also have a very small residual lung volumes, which means that the lung empties more completely and gas exchange can occur more fully in a respiratory cycle.

The lungs of sirenians are long and extend posteriorly almost as far back as the kidneys. They are separated from the abdominal viscera by a large obliquely sloped diaphragm, bronchial tree, and respiratory tissue. The primary bronchi run almost the length of the lungs with only a few smaller side branches or secondary bronchi. The secondary bronchi pass into smaller tubes, which in turn give rise to minute tubules supplying the respiratory vesicles. Another unique sirenian feature is that these vesicles arise laterally along the length of the bronchioles rather than from their ends, as is the typical mammalian condition. The bronchioles are very muscular and may function to close off respiratory vesicles when desired. For example, the dugong may use this technique to compress the volume and density of air in the lungs, thus enabling it to surface or sink without the use of flippers or tail and without expelling air. Cartilage occurs throughout the length of the air passages.

The thoracic cavity of the sea otter is large and the diaphragm is positioned obliquely. The right lung has four lobes and the left has two lobes. The lungs are large in relation to body size, nearly 2.5 times that found in other mammals of similar size. Large lungs serve more to regulate buoyancy than to store oxygen. The polar bear respiratory system is not unlike that of other bears. Although they are powerful swimmers they are not known to have any special physiological adaptations specific to diving.

Breathing

Breathing patterns of marine mammals vary. Pinnipeds breathe vigorously and frequently during the recovery phase after prolonged diving, but many species are typically periodic breathers under other circumstances. Particularly when resting or sleeping it is normal for seals to perform quite long apneas with short periods of rapid respiration between these nonbreathing periods. Although pinnipeds commonly exhale prior to diving, Hooker found that Antarctic fur seals consistently dive with full lungs and exhale during the latter stage of the ascent portion of a dive.

Towering blow of a blue whale.

Cetaceans exhale and inhale singly but very rapidly on surfacing. Whale blows represent the rapid emptying or expiration of whales'lungs through their blowholes in preparation for the next inspiration. A blow is one of the most visible behaviors of whales when they are observed at the sea surface. A particularly large amount of water may be spouted in baleen whales, whose blowholes are located in rather deep folds. The visibility of a blow is due to a mixture of vapor and seawater entrained into the exhaled column of air at the sea surface. When a blow occurs below the sea surface, as it sometimes does as bubble-blasts in gray whales or bubble trains in humpback whales, it may be intended as an audible or visual signal to other nearby whales. The size, shape, and orientation of a blow can help to identify some species of whales from a distance.

Vapor formed by contact between air warmed in the lungs and the cold external air sometimes enhances the visibility of the blow. The expired air of a blow also contains surfactantfrom the lungs. Surfactant is a complex a mixture of lipoproteins that reduce surface tension in pulmonary fluids and facilitate easy reinflation of collapsed lungs on surfacing. Pulmonary surfactant is secreted by alveolar type II cells and is necessary for normal mammalian lung function. Spragg et al. have found differences in surfactant of nondiving and diving mammals and suggest that these differences are associated with the repetitive collapse and reinflation of the lungs of divers.

A complete breathing cycle typically consists of a very rapid expiration (the blow) immediately followed by a slightly longer and much less obvious inspiration, then an extended yet variable period of breath-holding, or apnea. The rapid expiration of a typical whale blow provides more time to complete the next inspiration as the blowholes of a swimming animal breaks through the sea surface and results in little delay before submerging again.

The rapidity of the blow is accomplished by maintaining high flow rates throughout almost the entire expiration, which is in strong contrast to humans and other land mammals. The high expiratory flow rates of cetaceans are enhanced by very flexible chest walls and by cartilage reinforcement of the smallest terminal air passages of the lungs to prevent them from collapsing until the lungs are almost completely emptied. Small dolphins, for instance, expire and inspire in about 0.1 s, then hold their breath for 20 to 30 s before taking another breath. Even adult blue whales can empty their lungs of 1500 liters of air and refill them in as little as 2 s.

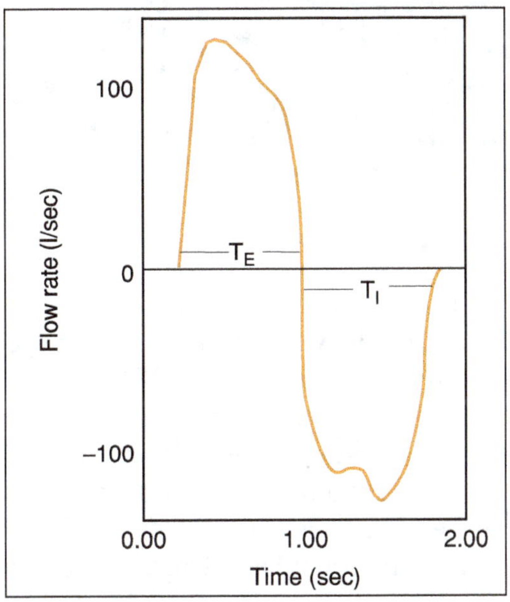

Typical ventilatory flow rates of a single expiratory (TE)/inspiratory (TI) event from a rehabilitating gray whale calf.

Blow patterns of whales vary, depending on their behaviors. In small cetaceans, swimming at low speeds, blowhole exposure during a blow is minimal and gradually changes to porpoising above the sea surface at higher swimming speeds. When migrating or feeding, larger baleen whales typically surface to blow several times in rapid succession, then make an extended dive of several minutes duration.

During inspiration, extensive elastic tissue in the lungs and diaphragm is stretched by diaphragm and intercostal musculature. These fibers recoil during expiration to rapidly and nearly completely empty the lungs. Oxygen uptake within the alveoli of the lungs may be enhanced as lung air is moved into contact with the walls of the alveoli by the action of small myoelastic bundles scattered throughout the lungs. In some species, the alveoli are highly vascularized to promote rapid uptake of oxygen. Bottlenose dolphins, for example, can remove nearly 90% of the oxygen available in each breath. In comparison, humans and most other terrestrial mammals use only about 20% of the inspired oxygen. It has been proposed that foreign particulate matter in inspired air or seawater may result in the formation of biomineral concretions, or calculi, discovered in the nasal sacs of some delphinids.

Manatees exhale after surfacing and, like cetaceans, can renew about 90% of the air in the lungs in a single breath. By comparison, humans at rest renew about 10% of the air in the lungs in a single breath.

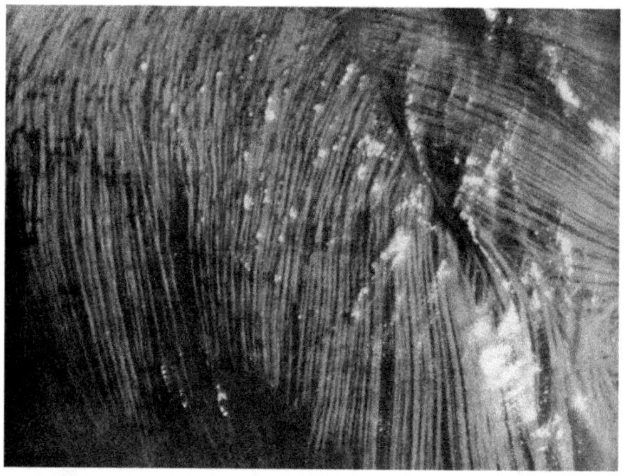

Extensive elastic fibers of diaphragm tissue of dolphin Stenella.

Respiratory Systems and Diving

Cetaceans typically dive with full lungs, whereas pinnipeds often exhale prior to diving. These differences support the contention that the volume of lung air at the beginning of a dive is of little importance in supplying oxygen during a dive, but it may be adjusted to achieve neutral buoyancy during some types of diving. Moreover, the lungs and their protective rib cage are modified to allow the lungs to collapse as the water pressure increases with depth. Complete lung collapse occurs at depths of 25–50 m for Weddell seals, 70 m for the bottlenose dolphin, and probably occurs in the first 50–100 m for most marine mammals. Any air remaining in the lungs below that depth is squeezed out of the alveoli and into the bronchi and trachea of the lungs.

By tolerating complete lung collapse, these animals avoid the need for respiratory structures capable of resisting the extreme water pressure experienced during deep dives. They also receive an additional bonus. As the air is forced out of the collapsing alveoli during a dive, the compressed air still within the larger air passages is blocked from contact with the thin, gas-exchanging walls of the alveoli. Consequently, little of this compressed gas is absorbed by the blood during dives and marine mammals avoid the potentially serious diving problems of decompression sickness (also called the bends) and nitrogen narcosis that sometimes plague human divers breathing compressed air.

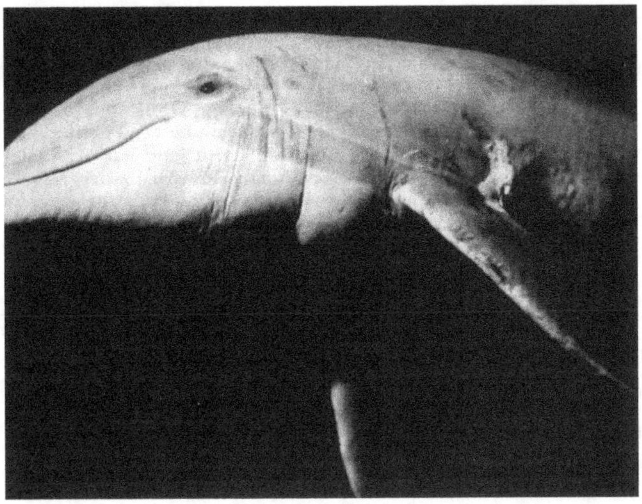

Bottlenose dolphin at a depth of 300 m, experiencing obvious thoracic collapse visible behind the left flipper.

Circulation in Monotremes

Circulatrory System of Echidna

Monotremes have two completely separate circulations and the systemic aorta is derived from the left fourth arch, as in the other mammals The left ventricular valve is tricuspid, though its movement is not controlled by chordae tendiniae as in other mammals, but by extensions of papillary muscles from the ventricular wall which are inserted into the membranous flaps of the valve. The right atrioventricular valve is also controlled by papillary muscles. The two coronary arteries 'have a typical mammalian course on the surface of the heart'. The venous drainage of the heart walls is achieved by the great cardiac vein which, as in mammals and birds, opens directly into the right atrium. In the Monotremata, however, venous blood is also carried to the right atrium by a coronary vein which also opens directly into the atrium. Coronary veins do not occur in other mammals, nor in birds.

The fibres of the sinus and atrioventricular nodes appear as discrete entities since they are paler in colour and have rounder nuclei than those of the contractile fibres. Both nodes are richly innervated with some of the axons spiralling around the nodal fibres; boutons terminaux are prominent features among the fibres.

The cardiac output is 59 ml/kg/min, pulse rate 115 beats per minute, arterial blood pressure systolic/diastolic 123/96 mm Hg and stroke volume 1.4 ml.

The erythrocytes in Tachyglossus are non-nucleated biconcave discs, as they are in all mammals except camels. The mean haemoglobin level is 17.4 g/100 ml, a value confirmed by Parer & Metcalfe.

A mysterious feature in red blood cells in Tachyglossus and Ornithorhynchus is their minuscule levels of adenosine triphosphate (ATP), the primary source of energy for utilisation of glucose in adult mammalian red cells. Monotreme erythrocytes can nevertheless metabolise glucose to lactate at a rate comparable to those of other mammals. It will be interesting to characterise the monotreme glycolytic enzymes which apparently can carry out hexose phosphorylation with a trace amount of ATP.

Bolliger & Backhouse found that the granular leucocytes are polymorphonuclear as in other mammals. Of the white blood cells, neutrophils account for 12%, monocytes 1% and lymphocytes 87% of the total; basophils and eosinophils are practically absent. These two kinds of white cells, however, are present in Platypus blood. The ultrastructural study carried out on Platypus blood cells has yet to be extended to Echidnas.

Tachyglossid serum exhibits seven clearly defined fractions as shown by cellulose acetate electrophoresis: α_1-, α_2-, α_3-, β_1-, β_2-, and γ-globulins and albumin. The mobilities of these fractions were found to be similar to those in human serum. The β_1 globulin (transferrin) binds iron and is essential for transport to a variety of cells, such as reticulocytes (developing red corpuscles) and bone marrow cells. Its aminoacid composition and sequence are the subjects of a study into the structural evolution of the serum and milk whey proteins in various mammals.

Atwell, Marchalonis & Ealey have further characterised the γ globulins or immunoglobulins observed by Morgan & Jordan from the serum of a Shortbeaked Echidna immunised with Salmonella adelaide flagellae. Atwell et al. isolated IgM and IgG, the two kinds of immunoglobulin found in the sera of mammals, including man. In terms of electrophoretic mobility, size and polypeptide structure, they resemble the immunoglobulins of man in particular. IgG predominates, and there is striking similarity between the γ heavy chain of human IgG and that of echidnas. The predominance of IgG immunoglobulin in the serum of T. aculeatus resembles more closely the mammalian pattern than that of reptiles, including the Tuatara, Sphenodon punctatus.

Circulatory System of Platypus

In its gross anatomy, the circulatory system of the Platypus exhibits the typically mammalian pattern of a closed double circulation, where the systemic arch is derived during development from the left side of the fourth gill arch. This anatomy was described in detail by and. described the vascularisation of the stomach and the alimentary system and Dowd presented a detailed study of the heart, its coronary vessels and its conducting system in both the Platypus and the Echidna. Apart from the presence of a coronary vein not found in other mammals and the indistinct sinus node, the heart of the Platypus appears to be typically mammalian in its anatomy.

Noted the presence of a complex of venous and arterial vessels in the pelvic area with small parallel veins and arteries supplying the muscles of the rear limbs and tail. No such complex was found in the axilla, but vessels in the manus were reasonably large with much smaller ones in the extensions of the web. In both areas of the forelimb, veins (venae comitantes) tend to parallel arteries.

Confirmed these findings and suggest that both the complex rete system arising from the short iliac arteries and the venae comitantes of the forelimb may serve to restrict heat loss from the unfurred extremities of the animal. The underside of the tail, the forefeet and the hind feet constitute

25.7% of the total surface area of the body. Measurements of low skin temperatures on these extremities at cold ambient temperatures substantiated this suggestion.

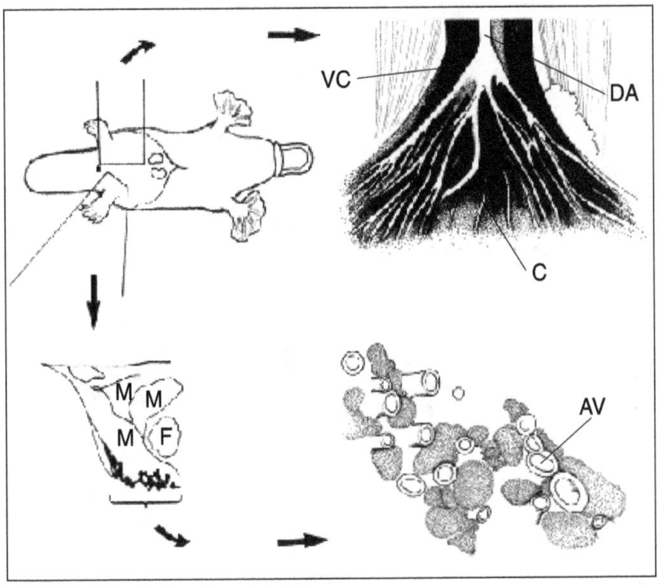

The pelvic rete system of the Platypus: (a) ramifications of the dorsal aorta and inferior vena cava in the pelvic region of the Platypus; (b) venous vessels paralleling arterial on the surface of muscles of the thigh region. Transverse section cut slightly distal to the branching of the iliac arteries. C = caudal artery; F = femur; DA = dorsal aorta; VC = inferior vena cava; M = muscle; AV = arterial vessels.

Vascularisation of the bill is poorly documented. However, large venous vessels are present along its margins and these have been used to obtain blood samples from the species.

Like those of other mammals, the red blood cells of the Platypus are nonnucleate and biconcave. The platelets are also typically mammalian. Unusual features of the leukocytes of the species, however, include the consistent presence of Döhle bodies in the segmented neutrophils and non-segmented nuclei in the eosinophils and basophils (Canfield & Whittington 1983). The species has a high haemoglobin level (175 ± 4.0 g/L) and haematocrit (52 ± 1.0%). These authors also found a low mean corpuscular volume (54 ± 1.5 μm3) and a high mean corpuscular haemoglobin concentration (353 ± 6.0 g/L), and reported a variety of white cell and blood biochemistry parameters. Similar haematological values were reported by Isaacks et al. in Tasmanian specimens. These authors also recorded extremely low levels of ATP in red blood cells of both the Platypus and the Echidna (0.06 and 0.03 μmol/ml, respectively). Such trace amounts of ATP appear still sufficient to maintain active cation transport across the red cell membranes in the two species.

Circulation in Placantal Mammals

Circulatory System of Humans

The circulatory system is an organ system that permits blood to circulate and transport nutrients (such as amino acids and electrolytes), oxygen, carbon dioxide, hormones, and blood cells to and from the cells in the body to provide nourishment and help in fighting diseases, stabilize temperature and pH, and maintain homeostasis.

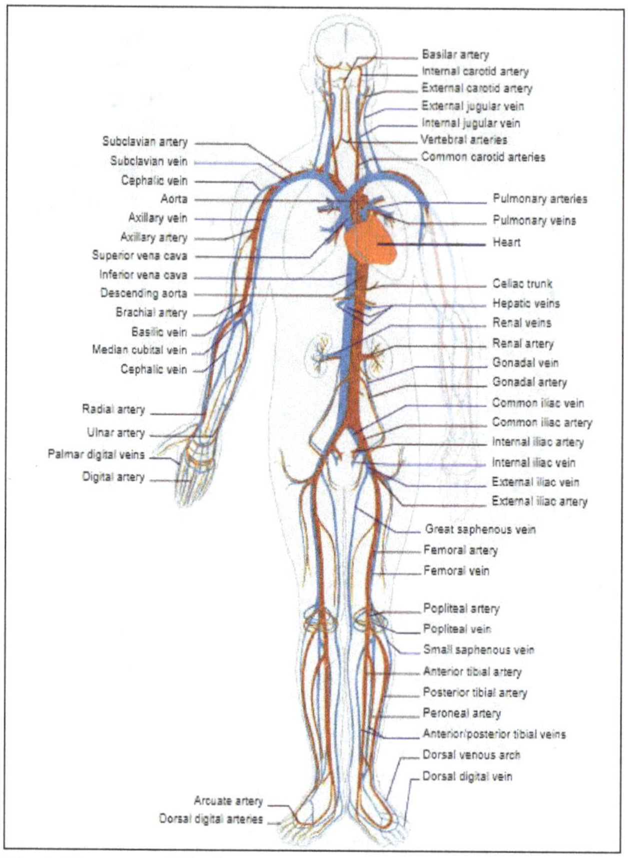

Simplified diagram of the human Circulatory system in anterior view.

Main Features of the Human Circulatory System

- A liquid, blood, to transport nutrients, wastes, oxygen and carbon dioxide, and hormones.
- Two pumps (in a single heart): one to pump deoxygenated blood to the lungs and the other to pump oxygenated blood to all the other organs and tissues of the body.
- A system of blood vessels to distribute blood throughout the body.
- Specialized organs for exchange of materials between the blood and the external environment; for example organs like the lungs and intestine that add materials to the blood and organs like the lungs and kidneys that remove materials from the blood and deposit them back in the external environment.

The Heart and Pulmonary System

The heart is located roughly in the center of the chest cavity. It is covered by a protective membrane, the pericardium.

- Deoxygenated blood from the body enters the right atrium.

- It flows through the tricuspid valve into the right ventricle. The term tricuspid refers to the three flaps of tissue that make up the valve.
- Contraction of the ventricle then closes the tricuspid valve and forces open the pulmonary valve.
- Blood flows into the pulmonary artery.
- This branches immediately, carrying blood to the right and left lungs.
- Here the blood gives up carbon dioxide and takes on a fresh supply of oxygen.
- The capillary beds of the lungs are drained by venules that are the tributaries of the pulmonary veins.
- Four pulmonary veins, two draining each lung, carry oxygenated blood to the left atrium of the heart.

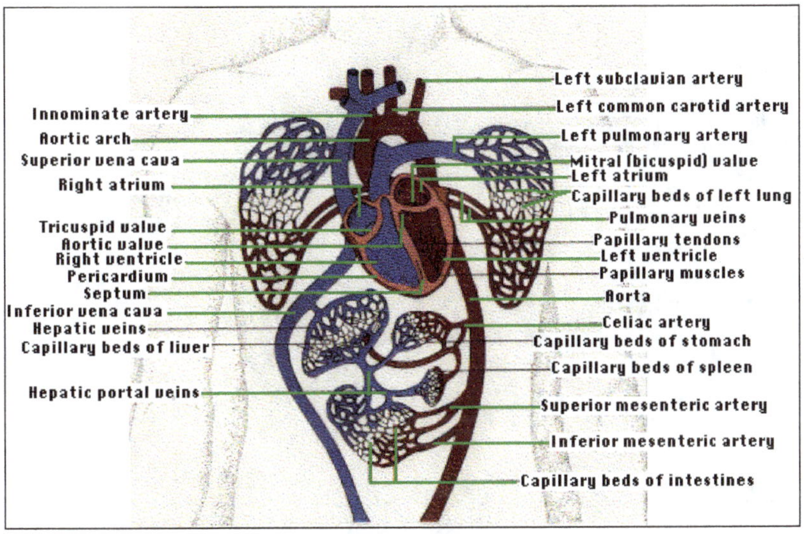

Human heart.

The above figure shows the human heart, with a schematic view of the pathway of blood through the lungs and internal organs. Oxygenated blood is shown in red; deoxygenated blood in blue. Note that the blood draining the stomach, spleen, and intestines passes through the liver before it is returned to the heart. Here surplus or harmful materials picked up from those organs can be removed before the blood returns to the general circulation.

The Coronary System

From the left atrium:

- Blood flows through the mitral valve (also known as the bicuspid valve) into the left ventricle.
- Contraction of the ventricle closes the mitral valve and opens the aortic valve at the entrance to the aorta.
- The first branches from the aorta occur just beyond the aortic valve still within the heart.

- Two openings lead to the right and left coronary arteries, which supply blood to the heart itself. Although the coronary arteries arise within the heart, they pass directly out to the surface of the heart and extend down across it. They supply blood to the network of capillaries that penetrate every portion of the heart.
- The capillaries drain into two coronary veins that empty into the right atrium.

The Systemic Circulation

The remainder of the system is known as the systemic circulation. The graphic shows the major arteries (in bright red) and veins (dark red) of the system. Blood from the aorta passes into a branching system of arteries that lead to all parts of the body. It then flows into a system of capillaries where its exchange functions take place.

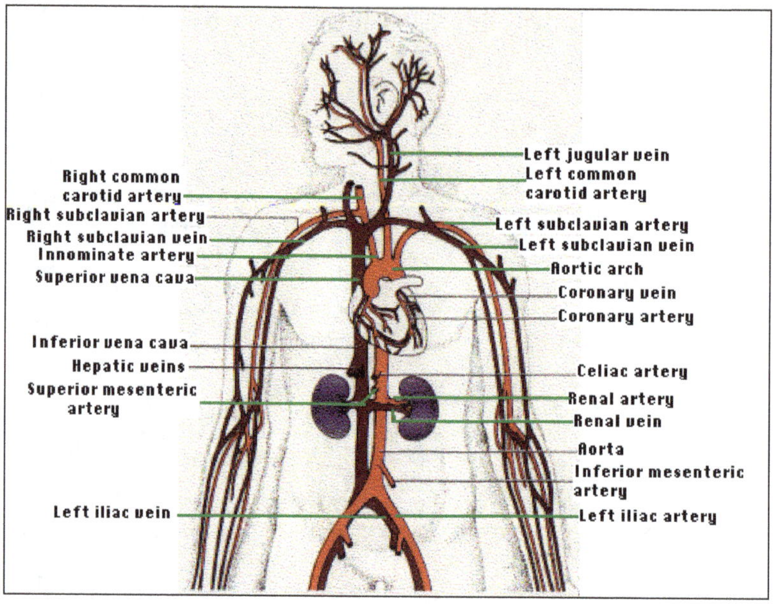

Human circulation system.

Blood from the capillaries flows into venules which are drained by veins.

- Veins draining the upper portion of the body lead to the superior vena cava.
- Veins draining the lower part of the body lead to the inferior vena cava.
- Both empty into the right atrium.

Circulatory System of Marine Mammals

I. General Anatomy

Heart

The basic structure and size of hearts in pinnipeds and cetaceans are typical of mammals. The four-chambered heart, with right ventricular outflow to the lungs and left ventricular output to the systemic circulation, weighs 0.5-1% of body mass in most pinnipeds and small cetaceans. In the great whales, relative heart mass is smaller, about 0.3-0.5% of body mass. Chamber size, stroke volume, and resting cardiac output and heart rate (where measured) are also in the general mammalian range and in agreement with mammalian allometric equations. Both the foramen ovale and the ductus arteriosus are closed in adult seals and cetaceans as in other mammals. Therefore, utilization of an intermittent fetal circulatory pathway does not appear to be a mechanism to bypass a potential increase in pulmonary vascular resistance during diving.

Arterial/Venous Systems

General aspects of the arterial and venous systems in marine mammals are remarkable for several features. First, dense sympathetic nerve innervation of proximal as well as distal arteries in seals may represent a mechanism by which the intense sympathetic vasoconstriction of the dive response can be maintained independent of local tissue metabolite-induced vasodilatation in the periphery. Angiography during forced submersions of seals supports this model. Second, venous capacitance is highly developed, especially in phocid seals and whales. This includes a large hepatic sinus and inferior vena cava, the latter of which in seals has been estimated to be capable of storing a fifth of the seals blood volume. Presumably, this large venous capacitance is related to the large blood volume of seals.

In some species, the spleen appears to be a significant storage organ for red blood cells. Increased splenic volumes in several pinniped species, and extensive sympathetic nerve innervation and smooth muscle development in the splenic capsule, are consistent with this storage role. Fluctuations in hematocrit between resting and diving states, or anesthetized and stressed states, also support such a role for the spleen in seals. It has been estimated that 30% of the blood volume can be stored in the spleen in Weddell seals (Leptomjchotes weddellii).

Another feature of the venous system, again well developed in both seals and whales, is the extradural venous system. This system, located within the vertebral canal and above the spinal cord, receives blood flow from the brain, back, and pelvic regions. It is linked with both the inferior and the superior vena cava via paravertebral communicating veins. The direction and magnitude of flow within the extradural vein is complex, varies with the respiratory cycle, and bears further investigation. In seals and cetaceans, the extradural vein is the primary venous drainage of the brain; the internal jugular vein is poorly developed or absent. The function of such a prominent vertebral venous system in these animals is unclear. In humans, it has been noted that extradural vein flow may participate in brain temperature regulation, and that, in the upright posture, the vertebral veins, kept open by attachment to the bony walls of the vertebral canal, are the primary cerebral venous drainage as blood flow decreases in the jugular veins due to venous collapse in the upright posture.

Blood Volume

Blood volumes are elevated and contribute to increased blood oxygen stores in marine mammals. On a mass-specific basis, most measurements indicate that blood volumes are two to three times the 70-ml kg-1 human value and that blood volume is greater in more active and in longer-diving species. The largest blood volumes (200-260 ml kg-1) have been found in some of the best divers, including elephant seals (Miraunga spp.), Weddell seals, and sperm whale (Phijseter macrocephalus).

II. Structural Adaptations

Vascular Thermoregulatory Adaptations

The parallel pattern of counterflowing arteries and veins, characteristic of counter current exchange units, is present in the flukes and flippers of cetaceans. Such arrangements, characteristic of blood vessel patterns in the limbs of many animals, are considered to conserve body heat by transferring heat from warm, outgoing arterial blood to cool venous blood returning from the limb. A superficial venous system, which does not return in conjunction with outgoing arteries, also occurs in the skin. These veins, which have well-developed muscular walls, have been considered to represent a route by which heat could be dissipated to the environment during periods of thermal stress.

Another structural adaptation observed in pinnipeds is the presence of numerous arteriovenous (a-v) anastomoses in the skin. These structures are connections between distal arteries and veins through which blood bypasses tissue capillary networks, and instead shunts directly from the arterial to venous system. The a-v anastomoses are distributed uniformly over the body surface of phocid seals, but, in otariids, are found in greater densities in the flippers. It is presumed that flow through these vessels allows heat exchange at the skin surface.

More recently, countercurrent anatomy has been observed in the reproductive organs of dolphins and pinnipeds. It has been proposed that the return of blood from the skin via vascular anastomoses allows relatively cool venous blood to prevent overheating of these organs. Temperature patterns along the length of the colon in the dolphin have been consistent with this hypothesis.

Aortic Bulbs/Windkessels

In pinnipeds, again particularly in phocid seals, the aortic root (ascending aorta) is dilated, forming the so-called aortic bulb. The bulb can accommodate the stroke volume ejected by the heart, and it is more distensible than the distal aorta. It has been proposed that the aortic bulb acts as a windkessel: the bulb expands and accepts blood ejected from the heart; then, gradual contraction of the bulb due to elastic fibers within its wall contributes to the maintenance of blood flow, especially to the brain and heart during diastole (relaxation phase of the cardiac cycle).

The ascending aorta, aortic arch, and proximal carotid arteries of whales are also very compliant and have also been hypothesized to act as a windkessel and preserve blood flow during diastole. This is especially important in whales because long diastoles accompany slow heart rates. Low heart rates can occur in whales due to the slow heart rates found in animals of large body mass and due to the bradycardias (decreased heart rates) that occur during diving.

The maintenance of blood flow and pressure during a long diastole is of course critical to the brain, but also to the heart. This is because coronaiy perfusion occurs during diastole when the heart is relaxed. Myocardial flow is dependent on the diastolic blood pressure as the driving pressure. Thus, species that are either large or have more profound diving responses are likely to have some form of an aortic windkessel.

A compliant ascending aorta may also contribute to a reduction in the impedance that the left ventricle must pump against during the peripheral vasoconstriction of the diving response. This reduction in afterload will decrease the work and oxygen consumption of the heart, which is of course beneficial to a diver with a limited oxygen supply.

Vascular Retia

The retia mirabilia (wonderful nets) of cetaceans have long been noted by anatomists. These plexuses of anastomosing arteries and veins occur along the vertebrae and base of the skull and are especially prominent in the thorax. The vascular retia are well developed in dolphins, in fact, more so than in large whales; they are also found in sirenians.

The thoracic rete is supplied by vessels from the aorta, which anastomose to form a complex, spongiform structure beneath the dorsal thoracic wall. This vascular tissue extends around the vertebrae into the vertebral canal and forms the primaiy arterial blood supply to the brain in cetaceans. The carotid arteries are vestigial or absent. The spinal meningeal artery in dolphins extends from the rete to the brain. Although mean blood pressure in the spinal meningeal artery of dolphins is equal to the aortic pressure, it is notable that the pressure is nonpulsatile; there is no systolic peak or diastolic trough. Thus, the cetacean brain appears to receive nonpulsatile blood flow. The significance of such a flow pattern, as well as the function of the retia, is unknown.

Large venous retia have also been reported in the abdomens of whales. Hypotheses about the role of the retia have included windkessel functions, intrathoracic vascular engorgement to prevent "lung squeeze " during diving, thermoregulation, and modification of composition of the blood.

Inferior Vena Caval Sphincter

In most pinnipeds, the inferior vena cava is associated with a striated muscle sphincter at the level of the diaphragm. Again, this is most well developed in phocid seals. The sphincter is innervated by the right phrenic nerve and is located cranial to the large hepatic sinus and inferior vena cava. Relaxation/contraction of the sphincter has been observed angiographically during forced submersions, and it is assumed that this is a mechanism to regulate venous return to the heart during diving bradycardias. Vena caval sphincters are also described in whales; they presumably regulate blood return from the large venous capacitance vessels in the abdomen (inferior vena cava and venous rete).

Another venous stnicture, especially developed in phocid seals, is the pericardial venous plexus. This extensive venous network drains the pericardium (membrane surrounding the heart) and empties into the inferior vena cava just cranial to the vena caval sphincter. It is especially developed in the better diving seals such as elephant seals. Its significance is unknown, although it has been reported to be associated with brown fat and has been hypothesized to function in thermoregulation.

Digestion in Monotremes

Digestive System of Echidna

The tongue is the sole means of catching and ingesting prey. If use of this organ is impaired in any way the animal starves. The tongue is long, vermiform and sticky and can be protruded up to 180 mm from the end of the snout. The stickiness is due to lubrication with a mucus composed largely of glycoprotein secreted by paired sub-lingual glands. The viscous mucus is passed via a duct from the anterior end of each gland to the floor of the buccal cavity. Each duct subdivides into a number of smaller ones and from these the mucus passes to the tongue. When the tongue is protruded, any ants or termites coming into contact with it are drawn back into the buccal cavity. When the tongue starts to protrude again, the insects are scraped off by rows of backwardly directed keratinous 'teeth' located along the roof of the cavity.

Retraction is brought about by contraction of the sternoglossi, two internal longitudinal muscles, that extend from near the tip through the fixed caudal end of the tongue to the sternum, where they take their origin at the xiphoid process. Protrusion results partly by contraction of circular muscles and partly by contraction of two genioglossal muscles attached to the caudal end of the tongue and to the mandibles. Since the base is fixed, contraction of the circular muscles deforms the tongue so that it is forced forwards. This is facilitated by the presence of a vascular stiffening mechanism discovered by Doran & Baggett. It lies anterior to the section shown in figure and consists of a centrally placed artery, a nerve ventral to the artery and vascular spaces located dorsally and ventrally. These spaces are supplied with blood from the central artery via a series of bulbous extensions, and small channels. The tongue is stiffened by engorgement of the vascular spaces with blood from the central artery. In this condition it is strong enough to penetrate termite-ridden wood. This stiffening of the tongue has no apparent effect on protrusion and retraction, which occur at a rate of about 100 actions per minute. has shown that the sinusoids of the erectile tissue unite in the mid-line anterior to a transverse sinus, forming a large space lined with endothelium and containing arterial blood. Striated fibres from the circular muscles are inserted into the wall of the sinus. Contraction of the circular muscles compresses and distorts the large sinus raising the pressure of the blood contained therein, and displacing blood into the dorsal and ventral sinusoids, leading to rapid protrusion of the tongue.

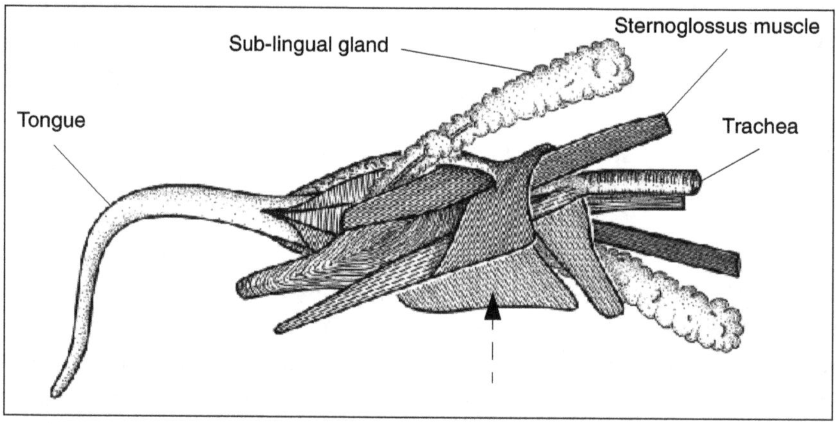

Dissection of the musculature of the tongue of the Short-beaked Echidna.

In addition to rapid protrusion and retraction, the end of the tongue is capable of movement left, right, up, down and of recurving on itself. This ability helps to apprehend ants and termites scurrying in many directions at once. It is achieved by the contraction of radially arranged musculature and three pairs of longitudinal muscles, all of which take their origin in the dermis of the anterior end of the tongue.

As well as apprehending the prey, the tongue masticates it. The insects are passed to the posterior end of the buccal cavity where they are comminuted by the grinding action of a set of spines, located on the dorsal surface of the base of the tongue. The cuticular teeth, as Home termed the latter, are directed backwards and project about 2.5 mm above the surface of the pad; they are covered with keratin. Located above the pad, along the palate, are seven transverse rows of keratinous teeth also directed backwards.

When prey arrive at the posterior end of the buccal cavity, the pad is pressed up against the palate and is worked backwards and forwards by the action of musculature described by Fewkes and Doran & Baggett. Since both sets of teeth are directed backwards they can slide past one another and their oscillations grind the insect prey into fragments. An illustration of a sagittal section of a head in Doran & Baggett showing the lingual teeth forwardly directed is incorrect. The trituration process is quite audible and the sound, a scratching noise, has been recorded along with slow-motion photography of the tongue in action.

The epithelium bearing the palatal spines is applied to a concave surface of bone, made up of the posterior ends of the palatines and of the robust ectopterygoids. This offers the necessary stout support for the grinding action of lingual pad against the palatal spines.

The triturated insects are passed through the oesophagus to a distensible stomach lined by stratified squamous epithelium.

Apart from a set of Brunner's glands at the posterior end, the stomach contains no other glands, and hence no peptic digestion occurs. The pH of the stomach 2–4 hours after ingestion of a meal of 200 g of termites (Nasutitermes exitiosus) does not fall below 6.2. Except for the possibility of self-digestion of the triturated insects in the stomach, all digestion of proteins, lipids and carbohydrates takes place in the small and large intestines. These exhibit the usual villi and crypts of Lieberkühn found in the intestines of other vertebrates. The villi are clothed distally with a simple columnar epithelium, centrally with goblet cells secreting mucigens. The deeper parts or crypts exhibit Paneth cells loaded with secretion granules. At the anterior end of the small intestine a combined pancreatic and bile duct passes digestive secretions of the pancreas and liver into the lumen of the intestine. The succus entericus exhibits lipase, amylase and trypsin-like activities. Sodium taurocholate is the emulsifying salt in the bile of Tachyglossus.

The amylase activity of the succus entericus would presumably contribute to digestion of glycogen, but as far as digestion of carbohydrates of shorter chain length is concerned,) has shown that the intestinal mucosa of echidnas has maltase, isomaltase and trehalase activity but no sucrase and practically no lactase activity, a finding confirmed by Stewart et al. The presence of trehalase is interesting since trehalose (a disaccharide of two molecules of glucose) is the main sugar found in insects. The latter authors also found quantities of the enzymes fucosidase and N-acetyl-D-glucosaminidase. This is in contrast with the enzymes present in the intestinal mucosae of suckling echidnas, which have appreciable lactase, maltase, isomaltase, neuraminidase and practically no sucrase activities, but relatively enormous amounts of trehalase, fucosidase and N-acetyl-D-glucosaminidase. This doubtless reflects the need to digest the two main oligosaccharides found in the milk: sialyllactose (4-o-acetyl-N-acetylneuraminyllactose) and fucosyllactose. Stewart et al. adduced evidence that digestion of these oligosaccharides does not take place at the border of the villi (as happens in the Eutheria), but intracellularly by the release of the digestive enzymes from lysosomes. The concept of lysosomal digestion entails a mechanism for entry of the milk oligosaccharides into the mucosal cells. Such a route has been described by Krause as an extensive network of tubules in the apical cytoplasm of the cells of the intestinal mucosa in suckling echidnas. These tubules are in communication with invaginations of the apical cell membrane and they lie close to small spherical vacuoles which are probably lysosomes.

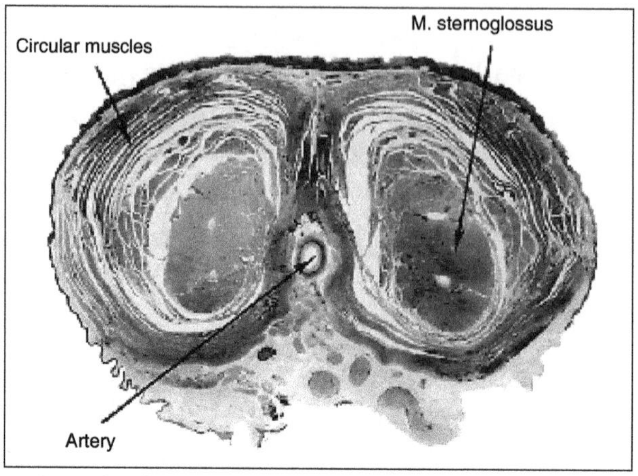

Cross section of proximal region of the tongue of the Shortbeaked Echidna x 5.6.

As an apparent compensation for the lack of peptide digestion, the adult has a very long intestine (some 3.5 m), which imposes a slow rate of passage. Clearance of a meal of termites takes upwards of 2 days providing the pancreatic and intestinal digestive enzymes with ample time to act effectively on their substrates.

Digestive System of Platypus

The teeth are replaced by keratinous grinding pads in the adult. Food material, which consists mainly of benthic invertebrate species, is stored in the cheek pouches which lie adjacent to these pads. Serrations in the skin of the lower jaw are thought to be involved in sorting the masticated food particles.

The digestive tract is very simple and relatively short, 1400 mm long in a female of 440 mm. The stomach is small and thin-walled, has no gastric glands and, therefore, no peptic digestion, although Brunner's glands are present. There are no villi in the short small intestine, but there are numerous surface folds. Groups of intestinal glands drain into common ducts which empty into the lumen between these folds. The large intestine is also short. The rectum is of greater diameter than the rest of the tract. There is also a small caecum.

Nothing is known of the digestive processes in the species, although its diet would suggest high proteolytic activity in the secretions of both the pancreas and the intestinal wall. The food material is so finely ground once it enters the digestive tract that analysis of fragments in the cheek pouches is the only method of food recognition which has been used successfully to date.

Digestion in Placantal Mammals

Digestive System of Humans

The human digestive system consists of the gastrointestinal tract plus the accessory organs of digestion (the tongue, salivary glands, pancreas, liver, and gallbladder). Digestion involves the breakdown of food into smaller and smaller components, until they can be absorbed and assimilated into the body. The process of digestion has three stages. The first stage is the cephalic phase of digestion which begins with gastric secretions in response to the sight and smell of food. This stage includes the mechanical breakdown of food by chewing, and the chemical breakdown by digestive enzymes, that takes place in the mouth.

Saliva contains digestive enzymes called amylase, and lingual lipase, secreted by the salivary glands and serous glands on the tongue. The enzymes start to break down the food in the mouth. Chewing, in which the food is mixed with saliva, begins the mechanical process of digestion. This produces a bolus which can be swallowed down the esophagus to enter the stomach. In the stomach the gastric phase of digestion takes place. The food is further broken down by mixing with gastric acid until it passes into the duodenum, in the third intestinal phase of digestion, where it is mixed with a number of enzymes produced by the pancreas. Digestion is helped by the chewing of food carried out by the muscles of mastication, the tongue, and the teeth, and also by the contractions of peristalsis, and segmentation. Gastric acid, and the production of mucus in the stomach, are essential for the continuation of digestion.

Peristalsis is the rhythmic contraction of muscles that begins in the esophagus and continues along the wall of the stomach and the rest of the gastrointestinal tract. This initially results in the production of chyme which when fully broken down in the small intestine is absorbed as chyle into the lymphatic system. Most of the digestion of food takes place in the small intestine. Water and some minerals are reabsorbed back into the blood in the colon of the large intestine. The waste products of digestion (feces) are defecated from the anus via the rectum.

Components

There are several organs and other components involved in the digestion of food. The organs known as the accessory digestive organs are the liver, gall bladder and pancreas. Other components include the mouth, salivary glands, tongue, teeth and epiglottis.

The largest structure of the digestive system is the gastrointestinal tract (GI tract). This starts at the mouth and ends at the anus, covering a distance of about nine (9) metres.

The largest part of the GI tract is the colon or large intestine. Water is absorbed here and the remaining waste matter is stored prior to defecation.

Most of the digestion of food takes place in the small intestine which is the longest part of the GI tract.

A major digestive organ is the stomach. Within its mucosa are millions of embedded gastric glands. Their secretions are vital to the functioning of the organ.

There are many specialised cells of the GI tract. These include the various cells of the gastric glands, taste cells, pancreatic duct cells, enterocytes and microfold cells.

Some parts of the digestive system are also part of the excretory system, including the large intestine.

Mouth

The mouth is the first part of the upper gastrointestinal tract and is equipped with several structures that begin the first processes of digestion. These include salivary glands, teeth and the tongue. The mouth consists of two regions; the vestibule and the oral cavity proper. The vestibule is the area between the teeth, lips and cheeks, and the rest is the oral cavity proper. Most of the oral cavity is lined with oral mucosa, a mucous membrane that produces a lubricating mucus, of which only a small amount is needed. Mucous membranes vary in structure in the different regions of the body but they all produce a lubricating mucus, which is either secreted by surface cells or more usually by underlying glands. The mucous membrane in the mouth continues as the thin mucosa which lines the bases of the teeth. The main component of mucus is a glycoprotein called mucin and the type secreted varies according to the region involved. Mucin is viscous, clear, and clinging. Underlying the mucous membrane in the mouth is a thin layer of smooth muscle tissue and the loose connection to the membrane gives it its great elasticity.[] It covers the cheeks, inner surfaces of the lips, and floor of the mouth, and the mucin produced is highly protective against tooth decay.

The roof of the mouth is termed the palate and it separates the oral cavity from the nasal cavity. The palate is hard at the front of the mouth since the overlying mucosa is covering a plate of bone; it is softer and more pliable at the back being made of muscle and connective tissue, and it can move to swallow food and liquids. The soft palate ends at the uvula. The surface of the hard palate allows for the pressure needed in eating food, to leave the nasal passage clear. The opening between the lips is termed the oral fissure, and the opening into the throat is called the fauces.

At either side of the soft palate are the palatoglossus muscles which also reach into regions of the tongue. These muscles raise the back of the tongue and also close both sides of the fauces to enable food to be swallowed. Mucus helps in the mastication of food in its ability to soften and collect the food in the formation of the bolus.

Salivary Glands

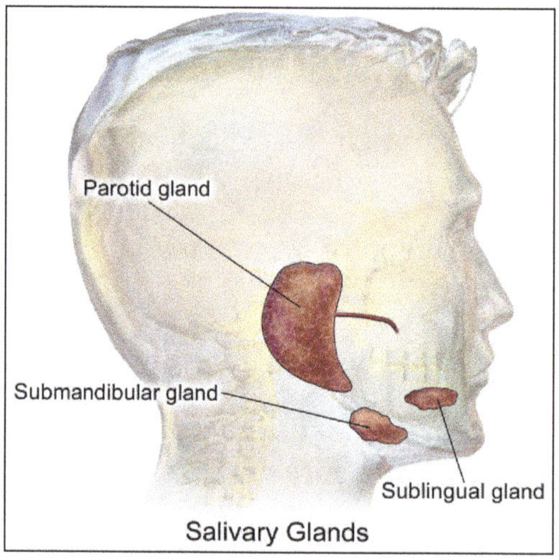

The main salivary glands.

There are three pairs of main salivary glands and between 800 and 1,000 minor salivary glands, all of which mainly serve the digestive process, and also play an important role in the maintenance of dental health and general mouth lubrication, without which speech would be impossible. The main glands are all exocrine glands, secreting via ducts. All of these glands terminate in the mouth. The largest of these are the parotid glands—their secretion is mainly serous. The next pair are underneath the jaw, the submandibular glands, these produce both serous fluid and mucus. The serous fluid is produced by serous glands in these salivary glands which also produce lingual lipase. They produce about 70% of the oral cavity saliva. The third pair are the sublingual glands located underneath the tongue and their secretion is mainly mucous with a small percentage of saliva.

Within the oral mucosa, and also on the tongue, palates, and floor of the mouth, are the minor salivary glands; their secretions are mainly mucous and they are innervated by the facial nerve (CN7). The glands also secrete amylase a first stage in the breakdown of food acting on the carbohydrate in the food to transform the starch content into maltose. There are other serous glands on the surface of the tongue that encircle taste buds on the back part of the tongue and these also produce lingual lipase. Lipase is a digestive enzyme that catalyses the hydrolysis of lipids (fats). These glands are termed Von Ebner's glands which have also been shown to have another function in the secretion of histatins which offer an early defense (outside of the immune system) against microbes in food, when it makes contact with these glands on the tongue tissue. Sensory information can stimulate the secretion of saliva providing the necessary fluid for the tongue to work with and also to ease swallowing of the food.

Saliva

Saliva moistens and softens food, and along with the chewing action of the teeth, transforms the food into a smooth bolus. The bolus is further helped by the lubrication provided by the saliva in its passage from the mouth into the esophagus. Also of importance is the presence in saliva of the digestive enzymes amylase and lipase. Amylase starts to work on the starch in carbohydrates, breaking it down into the simple sugars of maltose and dextrose that can be further broken down in the small intestine. Saliva in the mouth can account for 30% of this initial starch digestion. Lipase starts to work on breaking down fats. Lipase is further produced in the pancreas where it is released to continue this digestion of fats. The presence of salivary lipase is of prime importance in young babies whose pancreatic lipase has yet to be developed.

As well as its role in supplying digestive enzymes, saliva has a cleansing action for the teeth and mouth. It also has an immunological role in supplying antibodies to the system, such as immunoglobulin A. This is seen to be key in preventing infections of the salivary glands, importantly that of parotitis.

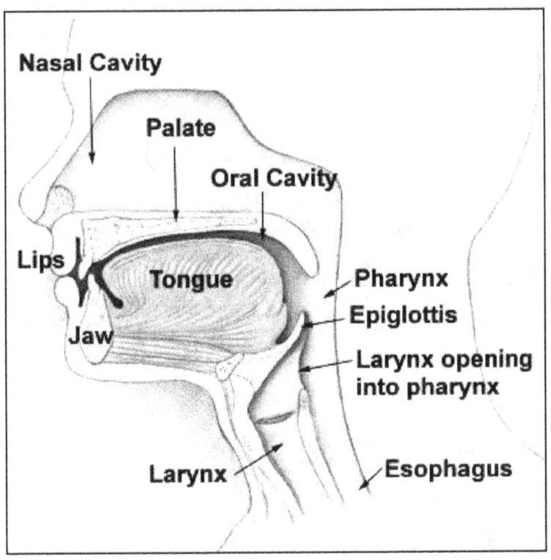

Saliva also contains a glycoprotein called haptocorrin which is a binding protein to vitamin B.. It binds with the vitamin in order to carry it safely through the acidic content of the stomach. When it reaches the duodenum, pancreatic enzymes break down the glycoprotein and free the vitamin which then binds with intrinsic factor.

Tongue

Food enters the mouth where the first stage in the digestive process takes place, with the action of the tongue and the secretion of saliva. The tongue is a fleshy and muscular sensory organ, and the very first sensory information is received via the taste buds in the papillae on its surface. If the taste is agreeable, the tongue will go into action, manipulating the food in the mouth which stimulates the secretion of saliva from the salivary glands. The liquid quality of the saliva will help in the softening of the food and its enzyme content will start to break down the food whilst it is still in the mouth. The first part of the food to be broken down is the starch of carbohydrates (by the enzyme amylase in the saliva).

The tongue is attached to the floor of the mouth by a ligamentous band called the frenum[1] and this gives it great mobility for the manipulation of food (and speech); the range of manipulation is optimally controlled by the action of several muscles and limited in its external range by the stretch of the frenum. The tongue's two sets of muscles, are four intrinsic muscles that originate in the tongue and are involved with its shaping, and four extrinsic muscles originating in bone that are involved with its movement.

Taste

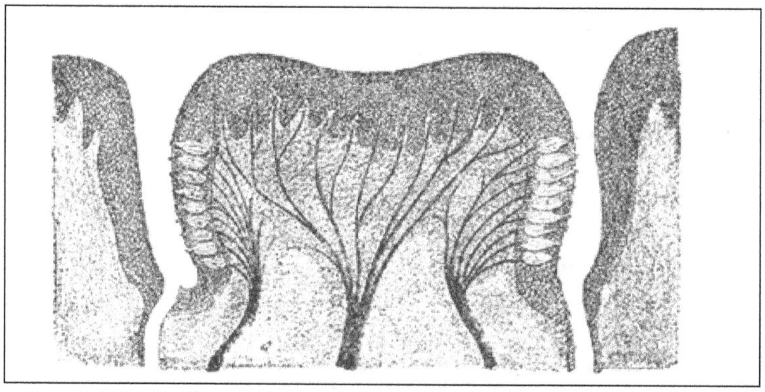

Cross section of circumvallate papilla showing arrangement of nerves and taste buds.

Taste is a form of chemoreception that takes place in the specialised taste receptors, contained in structures called taste buds in the mouth. Taste buds are mainly on the upper surface (dorsum) of the tongue. The function of taste perception is vital to help prevent harmful or rotten foods from being consumed. There are also taste buds on the epiglottis and upper part of the esophagus. The taste buds are innervated by a branch of the facial nerve the chorda tympani, and the glossopharyngeal nerve. Taste messages are sent via these cranial nerves to the brain. The brain can distinguish between the chemical qualities of the food. The five basic tastes are referred to as those of saltiness, sourness, bitterness, sweetness, and umami. The detection of saltiness and sourness enables the control of salt and acid balance. The detection of bitterness warns of poisons—many of a plant's defences are of poisonous compounds that are bitter. Sweetness guides to those foods that will supply energy; the initial breakdown of the energy-giving carbohydrates by salivary amylase creates the taste of sweetness since simple sugars are the first result. The taste of umami is thought to signal protein-rich food. Sour tastes are acidic which is often found in bad food. The brain has to decide very quickly whether the food should be eaten or not. It was the findings in 1991, describing the first olfactory receptors that helped to prompt the research into taste. The olfactory receptors are located on cell surfaces in the nose which bind to chemicals enabling the detection of smells. It is assumed that signals from taste receptors work together with those from the nose, to form an idea of complex food flavours.

Teeth

Teeth are complex structures made of materials specific to them. They are made of a bone-like material called dentin, which is covered by the hardest tissue in the body—enamel. Teeth have different shapes to deal with different aspects of mastication employed in tearing and chewing pieces of food into smaller and smaller pieces. This results in a much larger surface area for the action of digestive enzymes. The teeth are named after their particular roles in the process of mastication—incisors are used for cutting or biting off pieces of food; canines, are used for tearing, premolars and molars are used for chewing and grinding. Mastication of the food with the help of saliva and mucus results in the formation of a soft bolus which can then be swallowed to make its way down the upper gastrointestinal tract to the stomach. The digestive enzymes in saliva also help in keeping the teeth clean by breaking down any lodged food particles.

Epiglottis

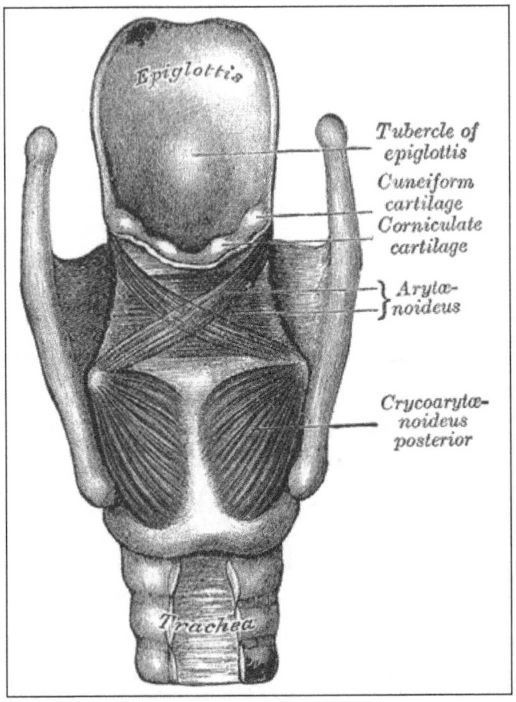

The epiglottis is a flap of elastic cartilage attached to the entrance of the larynx. It is covered with a mucous membrane and there are taste buds on its lingual surface which faces into the mouth. Its laryngeal surface faces into the larynx. The epiglottis functions to guard the entrance of the glottis, the opening between the vocal folds. It is normally pointed upward during breathing with its underside functioning as part of the pharynx, but during swallowing, the epiglottis folds down to a more horizontal position, with its upper side functioning as part of the pharynx. In this manner it prevents food from going into the trachea and instead directs it to the esophagus, which is behind. During swallowing, the backward motion of the tongue forces the epiglottis over the glottis' opening to prevent any food that is being swallowed from entering the larynx which leads to the lungs; the larynx is also pulled upwards to assist this process. Stimulation of the larynx by ingested matter produces a strong cough reflex in order to protect the lungs.

Pharynx

The pharynx is a part of the conducting zone of the respiratory system and also a part of the digestive system. It is the part of the throat immediately behind the nasal cavity at the back of the mouth and above the esophagus and larynx. The pharynx is made up of three parts. The lower two parts—the oropharynx and the laryngopharynx are involved in the digestive system. The laryngopharynx connects to the esophagus and it serves as a passageway for both air and food. Air enters the larynx anteriorly but anything swallowed has priority and the passage of air is temporarily blocked. The pharynx is innervated by the pharyngeal plexus of the vagus nerve. Muscles in the pharynx push the food into the esophagus. The pharynx joins the esophagus at the oesophageal inlet which is located behind the cricoid cartilage.

Esophagus

The esophagus, commonly known as the foodpipe or gullet, consists of a muscular tube through which food passes from the pharynx to the stomach. The esophagus is continuous with the laryngopharynx. It passes through the posterior mediastinum in the thorax and enters the stomach through a hole in the thoracic diaphragm—the esophageal hiatus, at the level of the tenth thoracic vertebra (T10). Its length averages 25 cm, varying with an individual's height. It is divided into cervical, thoracic and abdominal parts. The pharynx joins the esophagus at the esophageal inlet which is behind the cricoid cartilage.

At rest the esophagus is closed at both ends, by the upper and lower esophageal sphincters. The opening of the upper sphincter is triggered by the swallowing reflex so that food is allowed through. The sphincter also serves to prevent back flow from the esophagus into the pharynx. The esophagus has a mucous membrane and the epithelium which has a protective function is continuously replaced due to the volume of food that passes inside the esophagus. During swallowing, food passes from the mouth through the pharynx into the esophagus. The epiglottis folds down to a more horizontal position to direct the food into the esophagus, and away from the trachea.

Once in the esophagus, the bolus travels down to the stomach via rhythmic contraction and relaxation of muscles known as peristalsis. The lower esophageal sphincter is a muscular sphincter surrounding the lower part of the esophagus. The gastroesophageal junction between the esophagus and the stomach is controlled by the lower esophageal sphincter, which remains constricted at all times other than during swallowing and vomiting to prevent the contents of the stomach from entering the esophagus. As the esophagus does not have the same protection from acid as the stomach, any failure of this sphincter can lead to heartburn.

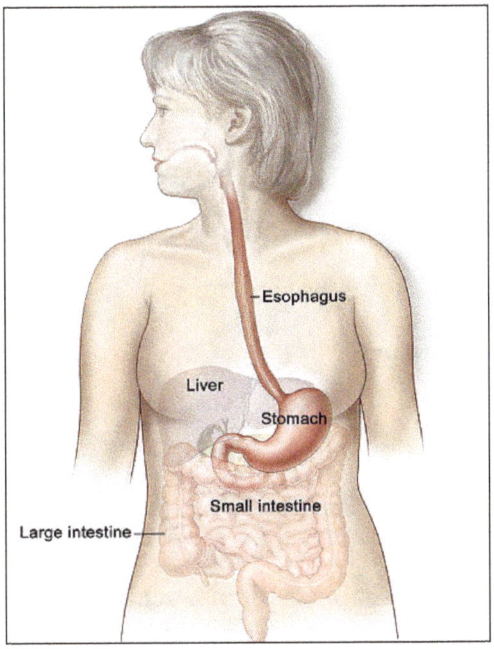

Adult digestive system.

Diaphragm

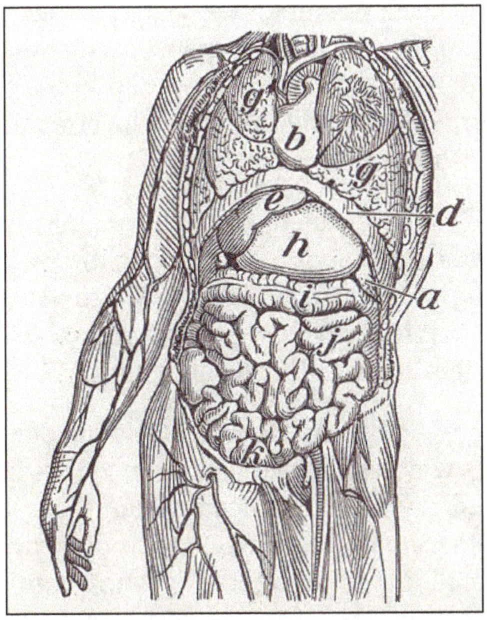

The diaphragm is an important part of the body's digestive system. The muscular diaphragm separates the thoracic cavity from the abdominal cavity where most of the digestive organs are located. The suspensory muscle attaches the ascending duodenum to the diaphragm. This muscle is thought to be of help in the digestive system in that its attachment offers a wider angle to the duodenojejunal flexure for the easier passage of digesting material. The diaphragm also attaches to, and anchors the liver at its bare area. The esophagus enters the abdomen through a hole in the diaphragm at the level of T10.

Stomach

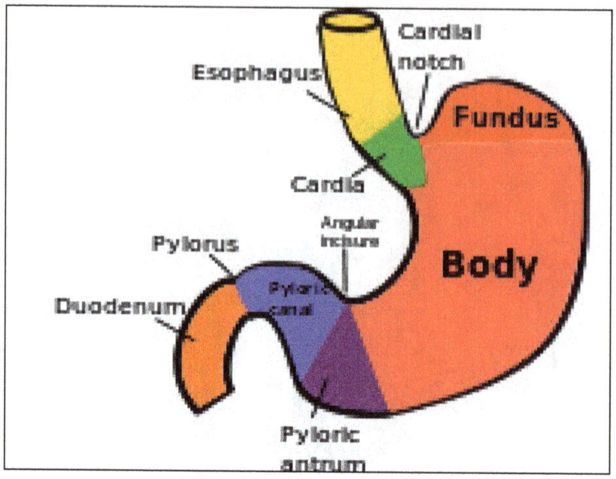

Areas of the stomach.

The stomach is a major organ of the gastrointestinal tract and digestive system. It is a consistently J-shaped organ joined to the esophagus at its upper end and to the duodenum at its lower end. Gastric acid (informally *gastric juice*), produced in the stomach plays a vital role in the digestive process, and mainly contains hydrochloric acid and sodium chloride. A peptide hormone, gastrin, produced by G cells in the gastric glands, stimulates the production of gastric juice which activates the digestive enzymes. Pepsinogen is a precursor enzyme (zymogen) produced by the gastric chief cells, and gastric acid activates this to the enzyme pepsin which begins the digestion of proteins. As these two chemicals would damage the stomach wall, mucus is secreted by innumerable gastric glands in the stomach, to provide a slimy protective layer against the damaging effects of the chemicals on the inner layers of the stomach.

At the same time that protein is being digested, mechanical churning occurs through the action of peristalsis, waves of muscular contractions that move along the stomach wall. This allows the mass of food to further mix with the digestive enzymes. Gastric lipase secreted by the chief cells in the fundic glands in the gastric mucosa of the stomach, is an acidic lipase, in contrast with the alkaline pancreatic lipase. This breaks down fats to some degree though is not as efficient as the pancreatic lipase.

The pylorus, the lowest section of the stomach which attaches to the duodenum via the pyloric canal, contains countless glands which secrete digestive enzymes including gastrin. After an hour or two, a thick semi-liquid called chyme is produced. When the pyloric sphincter, or valve opens, chyme enters the duodenum where it mixes further with digestive enzymes from the pancreas, and then passes through the small intestine, where digestion continues. When the chyme is fully digested, it is absorbed into the blood. 95% of absorption of nutrients occurs in the small intestine. Water and minerals are reabsorbed back into the blood in the colon of the large intestine, where the environment is slightly acidic. Some vitamins, such as biotin and vitamin K produced by bacteria in the gut flora of the colon are also absorbed.

The parietal cells in the fundus of the stomach, produce a glycoprotein called intrinsic factor which is essential for the absorption of vitamin B12. Vitamin B12 (cobalamin), is carried to, and through the stomach, bound to a glycoprotein secreted by the salivary glands - transcobalamin I also called haptocorrin, which protects the acid-sensitive vitamin from the acidic stomach contents. Once in the more neutral duodenum, pancreatic enzymes break down the protective glycoprotein. The freed vitamin B12 then binds to intrinsic factor which is then absorbed by the enterocytes in the ileum.

The stomach is a distensible organ and can normally expand to hold about one litre of food. This expansion is enabled by a series of gastric folds in the inner walls of the stomach. The stomach of a newborn baby will only be able to expand to retain about 30 ml.

Spleen

The spleen is the largest lymphoid organ in the body but has other functions. It breaks down both red and white blood cells that are *spent*. This is why it is sometimes known as the 'graveyard of red blood cells'. A product of this *digestion* is the pigment bilirubin, which is sent to the liver and secreted in the bile. Another product is iron, which is used in the formation of new blood cells in the bone marrow. Medicine treats the spleen solely as belonging to the lymphatic system, though it is acknowledged that the full range of its important functions is not yet understood.

Liver

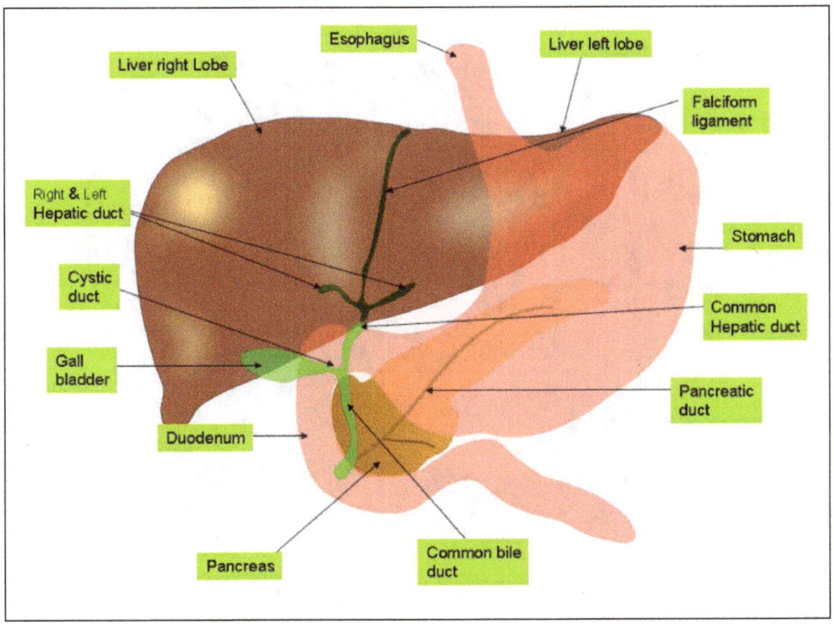

The liver is the second largest organ (after the skin) and is an accessory digestive gland which plays a role in the body's metabolism. The liver has many functions some of which are important to digestion. The liver can detoxify various metabolites; synthesise proteins and produce biochemicals needed for digestion. It regulates the storage of glycogen which it can form from glucose (glycogenesis). The liver can also synthesise glucose from certain amino acids. Its digestive functions are largely involved with the breaking down of carbohydrates. It also maintains protein metabolism in its synthesis and degradation. In lipid metabolism it synthesises cholesterol. Fats are also produced in the process of lipogenesis. The liver synthesises the bulk of lipoproteins. The liver is located in the upper right quadrant of the abdomen and below the diaphragm to which it is attached at one part, the bare area of the liver. This is to the right of the stomach and it overlies the gall bladder. The liver synthesises bile acids and lecithin to promote the digestion of fat.

Bile

Bile produced by the liver is made up of water (97%), bile salts, mucus and pigments, 1% fats and inorganic salts. Bilirubin is its major pigment. Bile acts partly as a surfactant which lowers the surface tension between either two liquids or a solid and a liquid and helps to emulsify the fats in the chyme. Food fat is dispersed by the action of bile into smaller units called micelles. The breaking down into micelles creates a much larger surface area for the pancreatic enzyme, lipase to work on. Lipase digests the triglycerides which are broken down into two fatty acids and a monoglyceride. These are then absorbed by villi on the intestinal wall. If fats are not absorbed in this way in the small intestine problems can arise later in the large intestine which is not equipped to absorb fats. Bile also helps in the absorption of vitamin K from the diet. Bile is collected and delivered through the common hepatic duct. This duct joins with the cystic duct to connect in a common bile duct with the gallbladder. Bile is stored in the gallbladder for release when food is discharged into the duodenum and also after a few hours.

Gallbladder

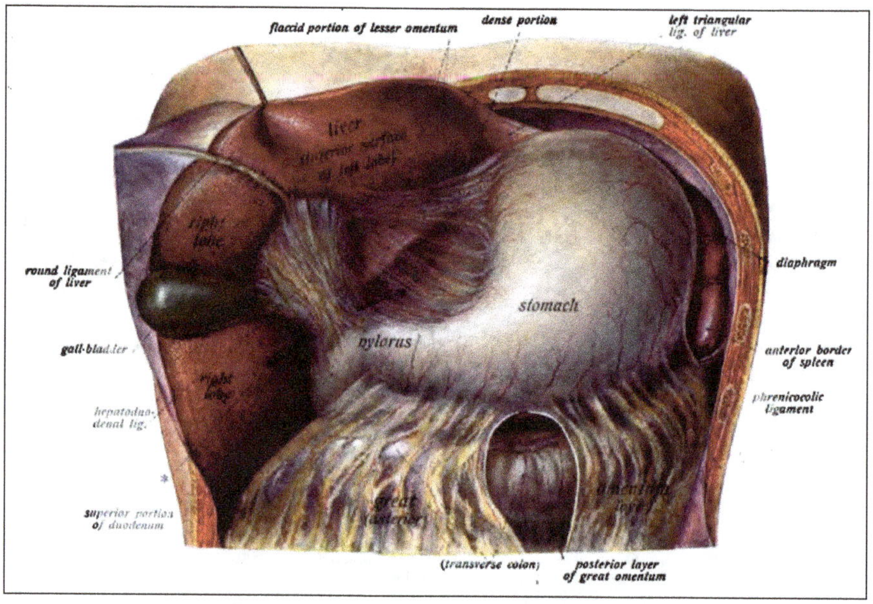

Gallbladder shown in green below the liver.

The gallbladder is a hollow part of the biliary tract that sits just beneath the liver, with the gallbladder body resting in a small depression. It is a small organ where the bile produced by the liver is stored, before being released into the small intestine. Bile flows from the liver through the bile ducts and into the gall bladder for storage. The bile is released in response to cholecystokinin (CCK) a peptide hormone released from the duodenum. The production of CCK (by endocrine cells of the duodenum) is stimulated by the presence of fat in the duodenum.

It is divided into three sections, a fundus, body and neck. The neck tapers and connects to the biliary tract via the cystic duct, which then joins the common hepatic duct to form the common bile duct. At this junction is a mucosal fold called *Hartmann's pouch*, where gallstones commonly get stuck. The muscular layer of the body is of smooth muscle tissue that helps the gallbladder contract, so that it can discharge its bile into the bile duct. The gallbladder needs to store bile in a natural, semi-liquid form at all times. Hydrogen ions secreted from the inner lining of the gallbladder keep the bile acidic enough to prevent hardening. To dilute the bile, water and electrolytes from the digestion system are added. Also, salts attach themselves to cholesterol molecules in the bile to keep them from crystallising. If there is too much cholesterol or bilirubin in the bile, or if the gallbladder doesn't empty properly the systems can fail. This is how gallstones form when a small piece of calcium gets coated with either cholesterol or bilirubin and the bile crystallises and forms a gallstone. The main purpose of the gallbladder is to store and release bile, or *gall*. Bile is released into the small intestine in order to help in the digestion of fats by breaking down larger molecules into smaller ones. After the fat is absorbed, the bile is also absorbed and transported back to the liver for reuse.

Pancreas

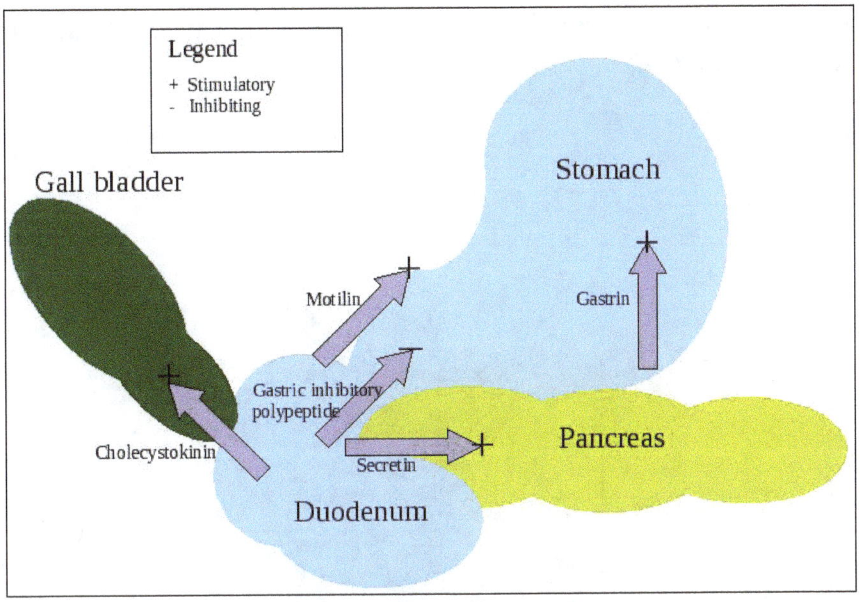

Action of digestive hormones.

The pancreas is a major organ functioning as an accessory digestive gland in the digestive system. It is both an endocrine gland and an exocrine gland. The endocrine part secretes insulin when the blood sugar becomes high; insulin moves glucose from the blood into the muscles and other tissues for use as energy. The endocrine part releases glucagon when the blood sugar is low; glucagon allows stored sugar to be broken down into glucose by the liver in order to re-balance the sugar levels. The pancreas produces and releases important digestive enzymes in the pancreatic juice that it delivers to the duodenum. The pancreas lies below and at the back of the stomach. It connects to the duodenum via the pancreatic duct which it joins near to the bile duct's connection where both the bile and pancreatic juice can act on the chyme that is released from the stomach into the duodenum. Aqueous pancreatic secretions from pancreatic duct cells contain bicarbonate ions which are alkaline and help with the bile to neutralise the acidic chyme that is churned out by the stomach.

The pancreas is also the main source of enzymes for the digestion of fats and proteins. Some of these are released in response to the production of CKK in the duodenum. (The enzymes that digest polysaccharides, by contrast, are primarily produced by the walls of the intestines.) The cells are filled with secretory granules containing the precursor digestive enzymes. The major proteases, the pancreatic enzymes which work on proteins, are trypsinogen and chymotrypsinogen. Elastase is also produced. Smaller amounts of lipase and amylase are secreted. The pancreas also secretes phospholipase A2, lysophospholipase, and cholesterol esterase. The precursor zymogens, are inactive variants of the enzymes; which avoids the onset of pancreatitis caused by autodegradation. Once released in the intestine, the enzyme enteropeptidase present in the intestinal mucosa activates trypsinogen by cleaving it to form trypsin; further cleavage results in chymotripsin.

Lower Gastrointestinal Tract

The lower gastrointestinal tract (GI), includes the small intestine and all of the large intestine. The intestine is also called the bowel or the gut. The lower GI starts at the pyloric sphincter of the stomach and finishes at the anus. The small intestine is subdivided into the duodenum, the jejunum and the ileum. The cecum marks the division between the small and large intestine. The large intestine includes the rectum and anal canal.

Small Intestine

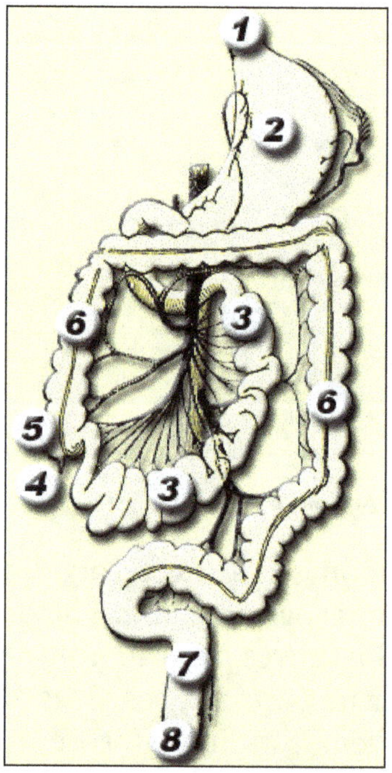

Lower GI tract - 3) Small intestine; 5) Cecum; 6) Large intestine.

Partially digested food starts to arrive in the small intestine as semi-liquid chyme, one hour after it is eaten. After two hours the stomach has emptied.

In the small intestine, the pH becomes crucial; it needs to be finely balanced in order to activate digestive enzymes. The chyme is very acidic, with a low pH, having been released from the stomach and needs to be made much more alkaline. This is achieved in the duodenum by the addition of bile from the gall bladder combined with the bicarbonate secretions from the pancreatic duct and also from secretions of bicarbonate-rich mucus from duodenal glands known as Brunner's glands. The chyme arrives in the intestines having been released from the stomach through the opening of the pyloric sphincter. The resulting alkaline fluid mix neutralises the gastric acid which would damage the lining of the intestine. The mucus component lubricates the walls of the intestine.

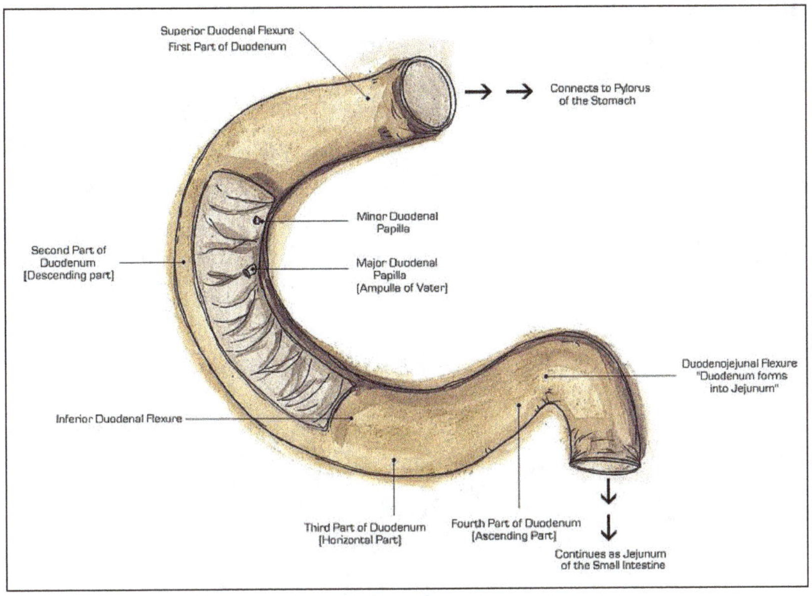

Duodenum.

When the digested food particles are reduced enough in size and composition, they can be absorbed by the intestinal wall and carried to the bloodstream. The first receptacle for this chyme is the duodenal bulb. From here it passes into the first of the three sections of the small intestine, the duodenum. The duodenum is the first and shortest section of the small intestine. It is a hollow, jointed C-shaped tube connecting the stomach to the jejunum. It starts at the duodenal bulb and ends at the suspensory muscle of duodenum. The attachment of the suspensory muscle to the diaphragm is thought to help the passage of food by making a wider angle at its attachment.

Most food digestion takes place in the small intestine. Segmentation contractions act to mix and move the chyme more slowly in the small intestine allowing more time for absorption (and these continue in the large intestine). In the duodenum, pancreatic lipase is secreted together with a co-enzyme, colipase to further digest the fat content of the chyme. From this breakdown, smaller particles of emulsified fats called chylomicrons are produced. There are also digestive cells called enterocytes lining the intestines (the majority being in the small intestine). They are unusual cells in that they have villi on their surface which in turn have innumerable microvilli on their surface. All these villi make for a greater surface area, not only for the absorption of chyme but also for its further digestion by large numbers of digestive enzymes present on the microvilli.

The chylomicrons are small enough to pass through the enterocyte villi and into their lymph capillaries called lacteals. A milky fluid called chyle, consisting mainly of the emulsified fats of the chylomicrons, results from the absorbed mix with the lymph in the lacteals. Chyle is then transported through the lymphatic system to the rest of the body.

The suspensory muscle marks the end of the duodenum and the division between the upper gastrointestinal tract and the lower GI tract. The digestive tract continues as the jejunum which continues as the ileum. The jejunum, the midsection of the small intestine contains circular folds, flaps of doubled mucosal membrane which partially encircle and sometimes completely encircle the lumen of the intestine. These folds together with villi serve to increase the surface area of the jejunum enabling an increased absorption of digested sugars, amino acids and fatty acids into the bloodstream. The circular folds also slow the passage of food giving more time for nutrients to be absorbed.

The last part of the small intestine is the ileum. This also contains villi and vitamin B12; bile acids and any residue nutrients are absorbed here. When the chyme is exhausted of its nutrients the remaining waste material changes into the semi-solids called feces, which pass to the large intestine, where bacteria in the gut flora further break down residual proteins and starches.

Cecum

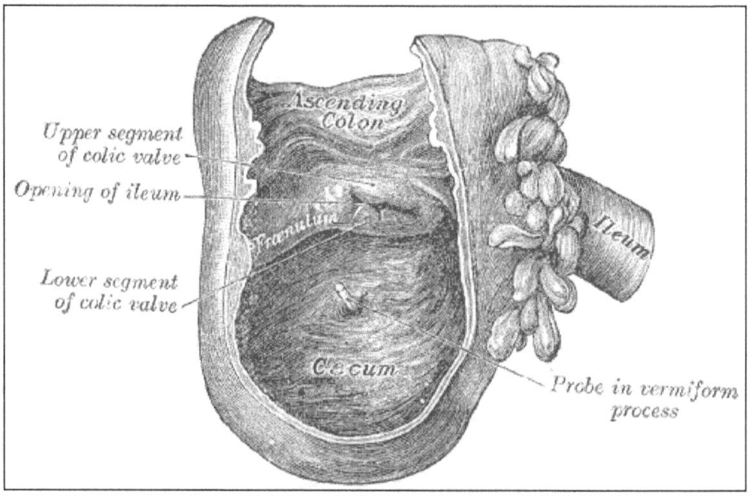

Cecum and beginning of ascending colon.

The cecum is a pouch marking the division between the small intestine and the large intestine. It lies below the ileocecal valve in the lower right quadrant of the abdomen. The cecum receives chyme from the last part of the small intestine, the ileum, and connects to the ascending colon of the large intestine. At this junction there is a sphincter or valve, the ileocecal valve which slows the passage of chyme from the ileum, allowing further digestion. It is also the site of the appendix attachment.

Large Intestine

In the large intestine, the passage of the digesting food in the colon is a lot slower, taking from 12 to 50 hours until it is removed by defecation. The colon mainly serves as a site for the fermentation of digestible matter by the gut flora. The time taken varies considerably between individuals. The remaining semi-solid waste is termed feces and is removed by the coordinated contractions of the intestinal walls, termed peristalsis, which propels the excreta forward to reach the rectum and exit via defecation from the anus. The wall has an outer layer of longitudinal muscles, the taeniae coli, and an inner layer of circular muscles. The circular muscle keeps the material moving forward and also prevents any back flow of waste. Also of help in the action of peristalsis is the basal electrical rhythm that determines the frequency of contractions. The taeniae coli can be seen and are responsible for the bulges (haustra) present in the colon. Most parts of the GI tract are covered with serous membranes and have a mesentery. Other more muscular parts are lined with adventitia.

Blood Supply

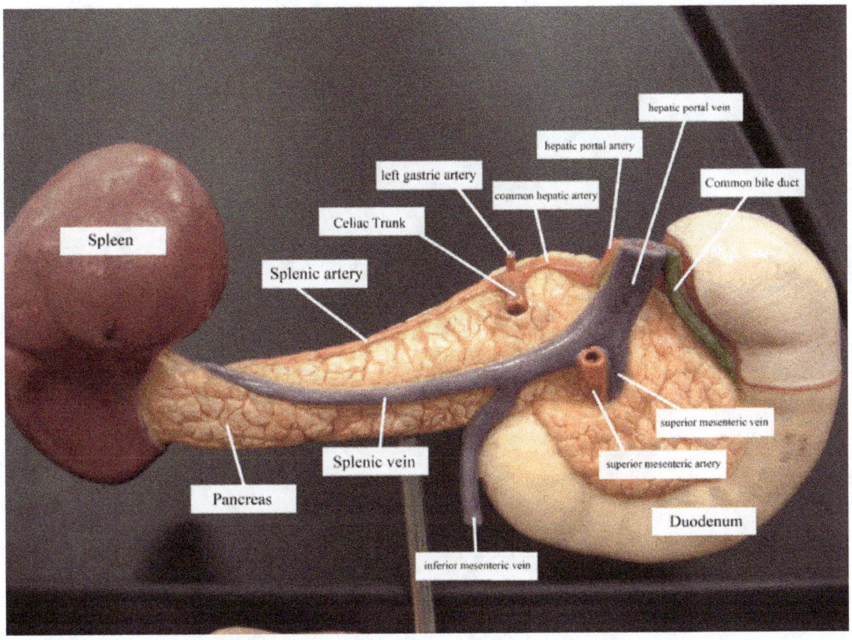

Arteries and veins around the pancreas and spleen.

The digestive system is supplied by the celiac artery. The celiac artery is the first major branch from the abdominal aorta, and is the only major artery that nourishes the digestive organs.

There are three main divisions – the left gastric artery, the common hepatic artery and the splenic artery.

The celiac artery supplies the liver, stomach, spleen and the upper 1/3 of the duodenum (to the sphincter of Oddi) and the pancreas with oxygenated blood. Most of the blood is returned to the liver via the portal venous system for further processing and detoxification before returning to the systemic circulation via the hepatic veins.

The next branch from the abdominal aorta is the superior mesenteric artery, which supplies the regions of the digestive tract derived from the midgut, which includes the distal 2/3 of the duodenum, jejunum, ileum, cecum, appendix, ascending colon, and the proximal 2/3 of the transverse colon.

The final branch which is important for the digestive system is the inferior mesenteric artery, which supplies the regions of the digestive tract derived from the hindgut, which includes the distal 1/3 of the transverse colon, descending colon, sigmoid colon, rectum, and the anus above the pectinate line.

Nerve Supply

The enteric nervous system consists of some one hundred million neurons that are embedded in the peritoneum, the lining of the gastrointestinal tract extending from the esophagus to the anus. These neurons are collected into two plexuses - the myenteric (or Auerbach's) plexus that lies between the longitudinal and the smooth muscle layers, and the submucosal (or Meissner's) plexus that lies between the circular smooth muscle layer and the mucosa.

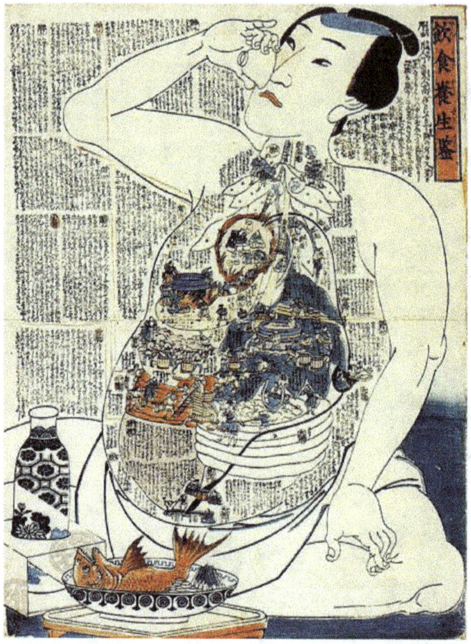

Dietary life rules, Japan, Edo period.

Parasympathetic innervation to the ascending colon is supplied by the vagus nerve. Sympathetic innervation is supplied by the splanchnic nerves that join the celiac ganglia. Most of the digestive tract is innervated by the two large celiac ganglia, with the upper part of each ganglion joined by the greater splanchnic nerve and the lower parts joined by the lesser splanchnic nerve. It is from these ganglia that many of the gastric plexuses arise.

Development

Early in embryonic development, the embryo has three germ layers and abuts a yolk sac. During the second week of development, the embryo grows and begins to surround and envelop portions of this sac. The enveloped portions form the basis for the adult gastrointestinal tract. Sections of this foregut begin to differentiate into the organs of the gastrointestinal tract, such as the esophagus, stomach, and intestines.

During the fourth week of development, the stomach rotates. The stomach, originally lying in the midline of the embryo, rotates so that its body is on the left. This rotation also affects the part of the gastrointestinal tube immediately below the stomach, which will go on to become the duodenum. By the end of the fourth week, the developing duodenum begins to spout a small outpouching on its right side, the hepatic diverticulum, which will go on to become the biliary tree. Just below this is a second outpouching, known as the *cystic diverticulum*, that will eventually develop into the gallbladder.

Clinical Significance

Each part of the digestive system is subject to a wide range of disorders many of which can be congenital. Mouth diseases can also be caused by pathogenic bacteria, viruses, fungi and as a side effect of some medications. Mouth diseases include tongue diseases and salivary gland diseases. A common gum disease in the mouth is gingivitis which is caused by bacteria in plaque. The most common viral infection of the mouth is gingivostomatitis caused by herpes simplex. A common fungal infection is candidiasis commonly known as *thrush* which affects the mucous membranes of the mouth.

There are a number of esophageal diseases such as the development of Schatzki rings that can restrict the passageway, causing difficulties in swallowing. They can also completely block the esophagus.

Stomach diseases are often chronic conditions and include gastroparesis, gastritis, and peptic ulcers.

A number of problems including malnutrition and anemia can arise from malabsorption, the abnormal absorption of nutrients in the GI tract. Malabsorption can have many causes ranging from infection, to enzyme deficiencies such as exocrine pancreatic insufficiency. It can also arise as a result of other gastrointestinal diseases such as coeliac disease. Coeliac disease is an autoimmune disorder of the small intestine. This can cause vitamin deficiencies due to the improper absorption of nutrients in the small intestine. The small intestine can also be obstructed by a volvulus, a loop of intestine that becomes twisted enclosing its attached mesentery. This can cause mesenteric ischemia if severe enough.

A common disorder of the bowel is diverticulitis. Diverticula are small pouches that can form inside the bowel wall, which can become inflamed to give diverticulitis. This disease can have complications if an inflamed diverticulum bursts and infection sets in. Any infection can spread further to the lining of the abdomen (peritoneum) and cause potentially fatal peritonitis.

Crohn's disease is a common chronic inflammatory bowel disease (IBD), which can affect any part of the GI tract, but it mostly starts in the terminal ileum.

Ulcerative colitis an ulcerative form of colitis, is the other major inflammatory bowel disease which is restricted to the colon and rectum. Both of these IBDs can give an increased risk of the development of colorectal cancer. Ulcerative coliltis is the most common of the IBDs·

Irritable bowel syndrome (IBS) is the most common of the functional gastrointestinal disorders. These are idiopathic disorders that the Rome process has helped to define.

Giardiasis is a disease of the small intestine caused by a protist parasite *Giardia lamblia*. This does not spread but remains confined to the lumen of the small intestine. It can often be asymptomatic, but as often can be indicated by a variety of symptoms. Giardiasis is the most common pathogenic parasitic infection in humans.

There are diagnostic tools mostly involving the ingestion of barium sulphate to investigate disorders of the GI tract. These are known as upper gastrointestinal series that enable imaging of the pharynx, larynx, oesophagous, stomach and small intestine and lower gastrointestinal series for imaging of the colon.

In Pregnancy

Gestation can predispose for certain digestive disorders. Gestational diabetes can develop in the mother as a result of pregnancy and while this often presents with few symptoms it can lead to pre-eclampsia.

Digestive System of Marine Mammals

The digestive system of marine mammals consists of the mouth, esophagus, stomach, small and large intestine, colon, and rectum. Accessory organs are also critical for digestive function including the liver and pancreas.

Most marine mammals (except sea otters) don't chew their food. Rather they pluck it out of the water and swallow it whole. The fish travel down the esophagus to the stomach. In cetaceans, the stomach has multiple compartments. Stomach anatomy in cetaceans is diverse with the number of chambers varying across species. There is also significant diversity in the anatomical names that have been given to these various compartments. In harbor porpoises, the compartment into which the esophagus terminates is called the forestomach or esophageal stomach. It is lined with thick ridges (rugae) and is mainly a storage compartment for food though some protein digestion does occur here. The next chamber is called the fundic or mainstomach and is a dark red in color. This is the chamber where HCl is secreted along with digestive enzymes. The next chamber is the pyloric stomach which is lined with mucus cells and is usually yellow to brown in color. After moving through the pyloric sphincter, the food, now called chyme, moves into the duodenal ampulla which is usually smooth in texture and is often full of bile (brownish fluid) which is secreted into this region from the liver via the hepatopancreatic duct. The chyme next travels to the duodenum which is the most anterior part of the small intestine where it gets mixed with the products of the pancreas including bicarbonate to increase the pH. The chyme then makes its way to the rest of the small intestine which consists of the jejunum and ileum. Most of the absorption of nutrients occurs in the first quarter of the small intestine. The large intestine follows the small intestine and then finally reaches the colon and rectum and waste is excreted from the anus.

Nervous System of Monotremes

Nervous System of Echidna

The brain has been described in great detail by. These works have been summarised by Griffiths. Only some matters of phylogenetic and physiological interest are discussed here.

Monotreme brains exhibit many typically mammalian structures. The cerebellum is large and divided into a number of lobes whose surfaces are thrown into folds. The lateral aspects of the cerebellum are joined by the pons Varolii, a massive band of ventrally directed nerve fibres found only in brains of mammals. The roof of the midbrain is also distinctively mammalian in that there are four prominences, the corpora quadrigemina or superior and inferior colliculi, instead of the two optic lobes of the sauropsidan brain. The presence of huge trigeminal nerves is a feature of prototherian brains. The trigeminal is a mixed sensory and motor nerve, but the sensory component is by far the greater, conveying impulses from the plethora of tactile organs in the snout to the cortex via a 'way station' in the thalamus.

The forebrain bears large olfactory bulbs which, enigmatically, have no mitral cell monolayer. The forebrain is divisible into a pyriform cortex, hippocampal cortex and neocortex. The latter, as in the marsupial brain, lacks the corpus callosum which interconnects the two halves of the eutherian neocortex.

The neocortex or neopallium, the hallmark of a mammalian brain, consists of a hypertrophy of the lateral and dorsal aspects of the cerebral hemispheres. This brings about a ventral displacement of the pyriform cortex and where it and the neocortex meet, a longitudinal fissure, the rhinal sulcus, occurs. The latter is a landmark for palaeontologists studying brain endocasts of fossil mammals. Such an endocast of the brain of the mammal-like reptile, Probainognathus, indicated that a neocortex had evolved at least once in the Middle Triassic. A neocortex does not occur in the brains of the Sauropsida.

The neocortex of Echidnas is markedly gyrencephalic, whereas the brains of Platypus and many marsupials are smooth or lissencephalic. The huge gyrencephalic neocortex of Echidnas has been a source of wonderment to neuroanatomists. Elliott Smith wrote:

> "The most obtrusive feature of this brain is the relatively enormous development of the cerebral hemispheres which are much larger, both actually and relatively, than those of the platypus. In addition the extent of the cortex is very considerably increased by numerous deep sulci. The meaning of this large neopallium is quite incomprehensible. The factors which the study of other mammalian brains has shown to be the determinants of the extent of the cortex fail completely to explain how it is that a small animal of the lowliest status in the mammalian series comes to possess this large cortical apparatus".

This mammal 'of the lowliest status' is, however, the equal of the Wistar rat in position-habit and habit-reversal learning and in its ability to learn to operate a simple piece of machinery. The latter authors state:

> "Further studies of learning will undoubtedly disclose important facts about the intelligence of these remarkable animals and modify the quaint, explicitly or tacitly-held views that echidnas are little more than animated pin cushions or, at the best glorified reptiles".

In all mammals the neocortex is organised for reception and analysis of sensory impulses and for dispatch of motor impulses. Lende maps auditory, visual, somatosensory and motor areas in the cortex of Echidnas. The somatosensory, visual and auditory areas are clustered at the caudal pole of the cortex. Coincident and exactly coextensive with the somatic sensory area is the motor cortex, but rostral to this sensorimotor area is an area with motor function only. There is considerable representation of the snout and tongue in the sensorimotor area, but nowhere near as great as that of the muzzle in the Playtpus cortex.

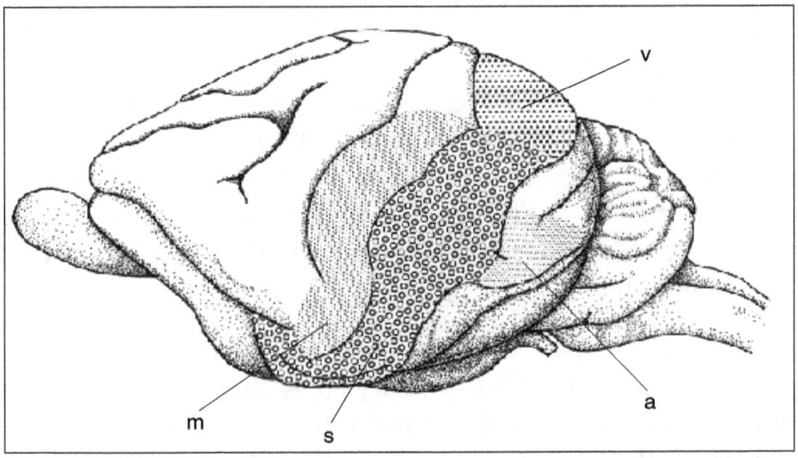

Somatosensory and motor (s); motor (m); visual (v) and auditory (a) areas of the cortex of the Short-beaked Echidna.

Pathways of projection of sensory impulses to the cortex from various parts of the body and sense organs are reviewed by Griffiths. The final way stations for relaying impulses to the cortex are in the thalamus. From here visual, auditory, tactile, temperature and proprioceptive impulses are projected onto the cortex. Two architectonic fields in the sensorimotor amalgam and the thalamic afferents to them have been identified. These fields are possibly homologous areas in eutherian cortices known as 3a and 3b.

Only mammals have direct connection between the motor cortex and the lower reaches of the spinal cord. As might be expected, these corticospinal or pyramidal tracts occur in the brains of monotremes. They have been studied experimentally in Tachyglossus by.

Special Senses

The eye of Tachyglossus has been described by As Gates pointed out, there are inconsistencies in these descriptions.

The eyeball is roughly spherical, about 9.5 mm in diameter. The fibrous sclera exhibits a cartilaginous cup which is characteristic of the eyes of reptiles but not of other mammals, except the Platypus. The retina shows the layers typical of the retinas of all vertebrates: a thin layer of nerve fibres, a ganglion cell layer and a receptor layer. Stone has shown unequivocally that the receptors exhibit oil droplets and Locket examined well-fixed retinas by light and electron microscopy and found both rods and cones in a ratio of about 5:1.

The nerve fibres from the retina form the optic nerve and the number of fibres therein has been estimated at about 15 000. This is small compared with that in monkeys, for example, but it is double that estimated for the bat, Eptesicus fuscus, a mammal capable of making a single visual discrimination. On this basis, the Echidna has a reasonable degree of visual function. has shown that the Short-beaked Echidna is indeed very good at learning by visual discrimination, to a level at least comparable to that of the rat.

The primary optic pathways from eye to the thalamus have been elucidated by Campbell & Hayhow. There is almost complete decussation of the fibres via the optic chiasma, only about 1% of fibres going to the ipselateral side. This may confer a modicum of binocular vision, since the eyes are directed forward to a small degree. The path of visual afferents from thalamus to the cortex has not yet been determined.

Cartilaginous ear trumpets carry sound waves from the pinnae, downward, medially and dorsally to the tympanic membrane. The stapes, standing on a membrane in the fenestra ovalis, conveys the sound impulses to the cochlea where the mechanical energy is transduced to electrical energy in the organ of Corti. The cochlea is not coiled as in the metatherian and eutherian middle ear but the organ of Corti is similar, but not identical, to that in other mammals. Uncoiled cochleas are also found in Mesozoic mammals.

The electrical potentials (cochlear microphonics) produced by the organ of Corti can be measured and are an indication of the sensitivity of the auditory apparatus to different sound frequencies. The CM audiogram for the Echidna's cochlea shows maximum response close to sound at 5 kHz. This frequency is low compared to the range of frequencies found to stimulate maximally the cochleae of other mammals, but is probably ideal for detecting sounds emitted by termites and ants underground.

The ear ossicles of Tachyglossus are peculiar in that the incus is firmly ankylosed to the malleus, a process which in turn is tightly attached to the petrosal bone. This implies that the ear would be sensitive to sounds conducted by the body. Johnstone found that tapping the body at various points elicited very small CM potentials, but when the snout was tapped a large potential was observed. Echidnas probably get auditory information via mechanoreceptors in the snout when this organ is resting on or buried in the ground.

The snout has several tactile functions. Iggo, have shown by electrophysiological techniques that the leathery skin of the dorsal surface of the snout is densely packed with specialised sensory end organs innervated by trigeminal nerve fibres. Four kinds of mechanoreceptors and two kinds of thermoreceptors are present. The latter are either cold sensitive, excited by a fall in skin temperature, or warmth sensitive, fired persistently at temperatures above the normal TB of 32°C. Iggo et al. conclude that the snout provides detailed information about the immediate physical environment.

The olfactory organs, unlike those in the Platypus, are very well developed. The area covered by the olfactory epithelium is enormous; it consists of seven vertical endoturbinals dependent from the cribriform plate; a large number of ectoturbinals (ethmoturbinals); and sets of nasoturbinals and maxilloturbinals. The cribriform plate is pierced with a large number of pores through which the branches of the olfactory nerves pass to the olfactory bulbs.

There is no published information on physiological or behavioural studies of olfaction, however, it is likely that sense of smell is of importance to echidnas in at least three ways. Tachyglossids are solitary animals for most of the year but may form groups in the breeding season; olfaction probably plays a part in bringing individuals together. Olfaction probably assists in detection of ants and termites, since to humans, at least, both kinds of organisms possess distinctive odours. The sense of smell is probably the one special sense that an Echidna has at the time of hatching, since fully developed olfactory sense receptors are present in the cavum nasum. This development presumably enables the hatchling to find the milk areola since other senses are lacking (the malleus is still part of the jaw and the eyes are covered with a thin layer of skinlike epithelium).

Two slits forming a V are located at the dorsal surface of the hind end of the dental pad of the tongue. These lead down into deeply seated circumvallate papillae in the trenches of which taste buds occur. In addition, sets of foliate papillae furnished with taste buds are located laterally and posterior to the dental pad. These arrangements would appear to be well suited for taste since the receptors occur where the prey is being crushed.

Nervous System of Platypus

The brain is relatively large and is essentially mammalian in its organisation. The neocortex is very large in relation to the pyriform and hippocampal cortices. Unlike that of the Echidna, the brain of the Platypus is not highly fissured, but is lissencephalic. There is no corpus callosum joining the two hemispheres, though there is communication between the two halves of the telencephalon through the anterior and hippocampal commissures.

The olfactory bulb is well developed, but, as in the Echidna, it lacks the well-defined layer of mitral cells found in all other mammals. Nothing is known of olfaction in the animal and the physiological importance of the absence of these cells is open to question. Hines found the olfactory pathways in the brain to be similar to those in other mammals. As the nostrils are closed by a valve in the bill during diving, olfaction probably is not important in foraging behaviour. Griffiths suggested, however, that olfaction may be important in courtship and in the reflex induction of milk ejection in lactating females. It may be important in other social behaviour patterns. Both sexes have well-developed sternal scent glands. Burrell noted marking of burrows by males with a 'foxy odour'. Strahan & Thomas reported that a captive male deposited mucous-like secretions over underwater objects, but noted that the female did not appear to react to these secretions. This latter behaviour is, therefore, difficult to interpret.

Jacobson's Organs are present, opening into the buccal cavity, and are probably associated with smelling food in the mouth.

The vallate foliate papilla complex, associated with taste in mammals, is found in both the Echidna and the Platypus in grooves on either side of the dorsal surface of the tongue.

The eye of the Platypus is supported by a cartilaginous cup, a structure not found in any metatherian or eutherian species. There is a nictitating membrane, lachrymal and Harderian glands are present and both lids have tarsal plates. The pupil is round, with a sphincter of smooth muscle, the choroid is unusually thick and deeply pigmented and the retina has rods and double cones with colourless oil droplets. Although the internal structure of the eye is mammalian, the scleral cartilage, the double cones and the oil droplets are sauropsidan characters. The eye is small (6.0 mm diameter), spherical, and has a flat lens which would appear to give it little accommodation ability, but to adapt it more to vision in water than in air. indicated that this is of little importance, as the Platypus closes its eyes when underwater. photographed two captive specimens momentarily opening their eyes while diving. For a species which forages for its food mainly in turbid waters and mainly at night, a reliance on visual sensory input would seem a paradox.

Moreover, Pridmore (personal communication) found that captive animals could navigate and feed with their eyes covered. Although it has poor capacities for accommodation, the eye is extremely sensitive to movements while the animal is on land or on the surface of the water. Patently, more investigation on the nature of vision is required.

Primary optic pathways have been investigated in the central nervous system. There are few fibres in the optic chiasma which pass to ipsilateral sides of the brain. Positioning of the eyes in the grooves (which also house the ear openings) on each side of the head and posterior to the extensions of the frontal shield suggests that the species is unlikely to possess stereoscopic vision.

Studied cochlear microphonic potentials of the ear of the Platypus in response to sounds of differing frequencies. The middle ear bones consist of the malleus, incus and stapes, as in other mammals, but the articulation between the malleus and incus is unusual. The stapes is columelliform as it is in the Echidna, marsupials and pangolins. Although the cochlea is not coiled and resembles that of Mesozoic mammals, its internal structure is mammalian, as is the organisation of its organ of Corti.

Although the cochlea was found to respond to a reasonably wide range of frequencies of sound, it was less sensitive than the cochlea of most other mammals, including the Echidna. This lack of sensitivity compared with other mammals was attributed to differences in structure. The slight reduction in sensitivity below that of the Echidna is thought to be due to dissimilarities in the middle ears of the two monotremes. The conclusion, however, is that audition in both of these animals is more similar to that found in marsupials and eutherians than it is to reptiles.

For an animal which closes its ears, eyes and nostrils when diving, underwater navigation and location of food would seem to present a problem. Home noted the massive innervation of the bill, while Poulton recorded the occurrence of numerous peripheral nerve terminals within the skin of the muzzle. Until neurophysiological measurements were made by Bohringer & Rowe, little of the nature of the tactile sense of this organ was established.

These workers found a large area of representation of nerves from the numerous nerve terminals, or rod organs, in the sensory cortex. This area was much larger than those which represented either the visual or auditory inputs. Within the large area of bill representation, individual neurones were found to have very restricted receptive fields. Bohringer further described the rod organs associated with the many small pits in the bill (distinct from the larger ones associated with sweat glands) and suggests they function to permit tactile reception of a very sensitive nature, but still allow physical protection of the muzzle by the keratinised epidermis.

The sequential stimulation of rod organs in a specific order may permit the Platypus to localise its position or the position of its food by the evaluation of both static and moving components of tactile stimuli. Andres & von During reconstructed the fine detail of the rod organs and suggest that their innervation makes them a possible three-dimensional detection system. These workers also indicate the possibility of the transmission of water movement direction and pressure through the secretion films and tonofibrils to the nerves associated with the gland ducts in the bill. How the Platypus manages to move about underwater, avoiding objects and finding its prey without actually having to touch them with its bill, is not completely clear.

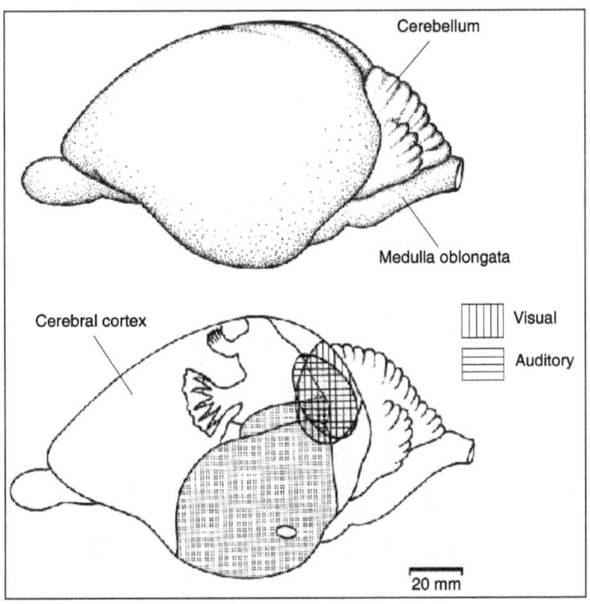

Organisation of the sensory cortex: (a) brain of the Platypus showing the lissencephalic nature of the cortex; (b) the brain has a Platypus figurine (stippled) drawn in proportion to the amount of representation of each area in the cortex. Areas receiving sensory input from the eyes and ears are also shown.

Captives respond to small electrical fields in water. Individuals are attracted more frequently to submerged miniature 1.5 V batteries and to fields generated between two carbon electrodes than to spent batteries or when fields were turned off. Animals also respond with reflex head and tail movements to fields in water between aluminium plates placed 3 m apart, at threshold values from between 50 μV/cm and 500 μV/cm. These authors also record cortical evoked potentials from electrical stimulus to wet cotton on the bill and suggested electroreception mediated via the trigeminal system, with representation in the cortex next to the trigeminal map of bill mechanoreceptors mapped by Bohringer & Rowe. These latter authors found that motor areas of the cortex overlapped sensory areas.

Nervous System of Placantal Mammals

Nervous System of Humans

The nervous system can be divided into two major regions: the central and peripheral nervous systems. The central nervous system (CNS) is the brain and spinal cord, and the peripheral nervous system (PNS) is everything else. The brain is contained within the cranial cavity of the skull, and the spinal cord is contained within the vertebral cavity of the vertebral column. It is a bit of an oversimplification to say that the CNS is what is inside these two cavities and the peripheral nervous system is outside of them, but that is one way to start to think about it. In actuality, there are some elements of the peripheral nervous system that are within the cranial or vertebral cavities. The peripheral nervous system is so named because it is on the periphery—meaning beyond the brain and spinal cord. Depending on different aspects of the nervous system, the dividing line between central and peripheral is not necessarily universal.

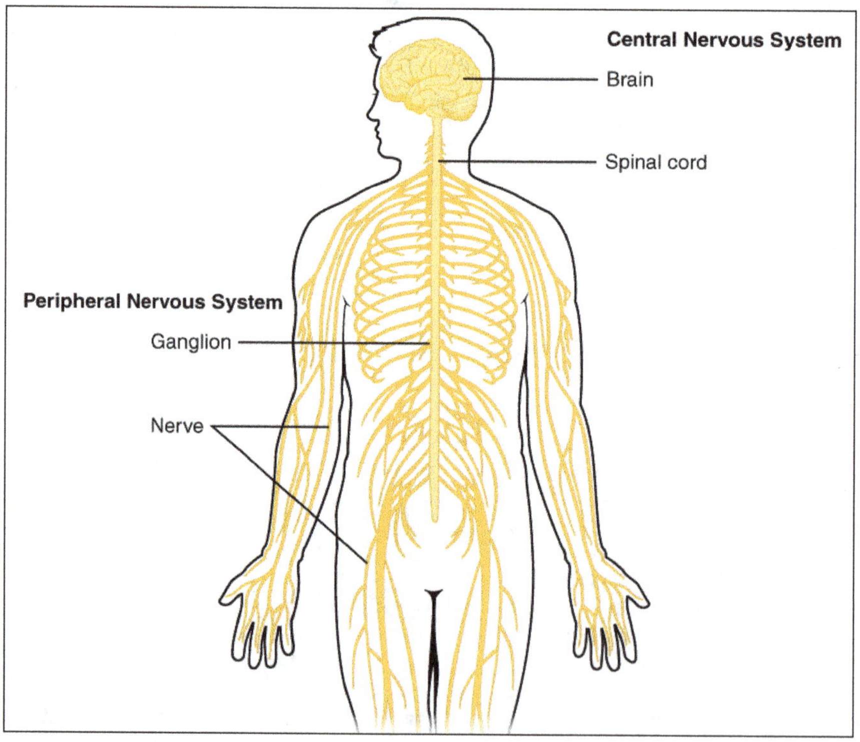

Central and Peripheral Nervous System. The structures of the PNS are referred to as ganglia and nerves, which can be seen as distinct structures. The equivalent structures in the CNS are not obvious from this overall perspective and are best examined in prepared tissue under the microscope.

Nervous tissue, present in both the CNS and PNS, contains two basic types of cells: neurons and glial cells. A glial cell is one of a variety of cells that provide a framework of tissue that supports the neurons and their activities. The neuron is the more functionally important of the two, in terms of the communicative function of the nervous system. To describe the functional divisions of the nervous system, it is important to understand the structure of a neuron. Neurons are cells and therefore have a soma, or cell body, but they also have extensions of the cell; each extension is generally referred to as a process. There is one important process that every neuron has called an axon, which is the fiber that connects a neuron with its target. Another type of process that branches off from the soma is the dendrite. Dendrites are responsible for receiving most of the input from other neurons. Looking at nervous tissue, there are regions that predominantly contain cell bodies and regions that are largely composed of just axons. These two regions within nervous system structures are often referred to as gray matter (the regions with many cell bodies and dendrites) or white matter (the regions with many axons). Figure demonstrates the appearance of these regions in the brain and spinal cord. The colors ascribed to these regions are what would be seen in "fresh," or unstained, nervous tissue. Gray matter is not necessarily gray. It can be pinkish because of blood content, or even slightly tan, depending on how long the tissue has been preserved. But white matter is white because axons are insulated by a lipid-rich substance called myelin. Lipids can appear as white ("fatty") material, much like the fat on a raw piece of chicken or beef. Actually, gray matter may have that color ascribed to it because next to the white matter, it is just darker—hence, gray.

The distinction between gray matter and white matter is most often applied to central nervous tissue, which has large regions that can be seen with the unaided eye. When looking at peripheral structures, often a microscope is used and the tissue is stained with artificial colors. That is not to say that central nervous tissue cannot be stained and viewed under a microscope, but unstained tissue is most likely from the CNS—for example, a frontal section of the brain or cross section of the spinal cord.

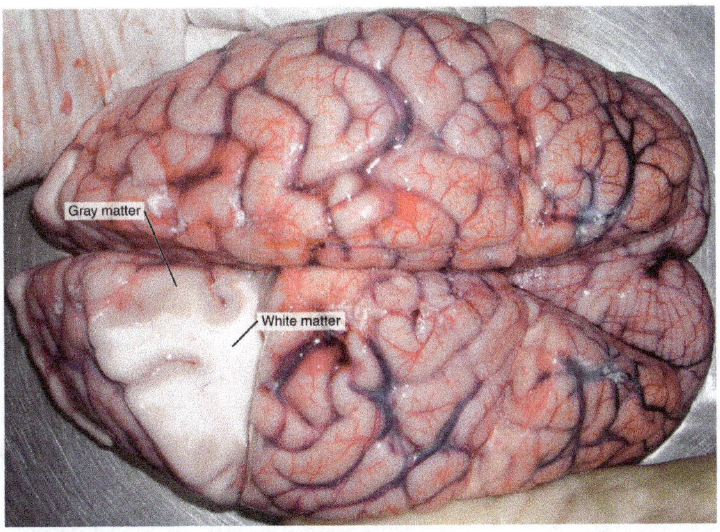

Gray Matter and White Matter. A brain removed during an autopsy, with a partial section removed, shows white matter surrounded by gray matter. Gray matter makes up the outer cortex of the brain.

Regardless of the appearance of stained or unstained tissue, the cell bodies of neurons or axons can be located in discrete anatomical structures that need to be named. Those names are specific to whether the structure is central or peripheral. A localized collection of neuron cell bodies in the CNS is referred to as a nucleus. In the PNS, a cluster of neuron cell bodies is referred to as a ganglion. Figure indicates how the term nucleus has a few different meanings within anatomy and physiology. It is the center of an atom, where protons and neutrons are found; it is the center of a cell, where the DNA is found; and it is a center of some function in the CNS. There is also a potentially confusing use of the word ganglion (plural = ganglia) that has a historical explanation. In the central nervous system, there is a group of nuclei that are connected together and were once called the basal ganglia before "ganglion" became accepted as a description for a peripheral structure. Some sources refer to this group of nuclei as the "basal nuclei" to avoid confusion.

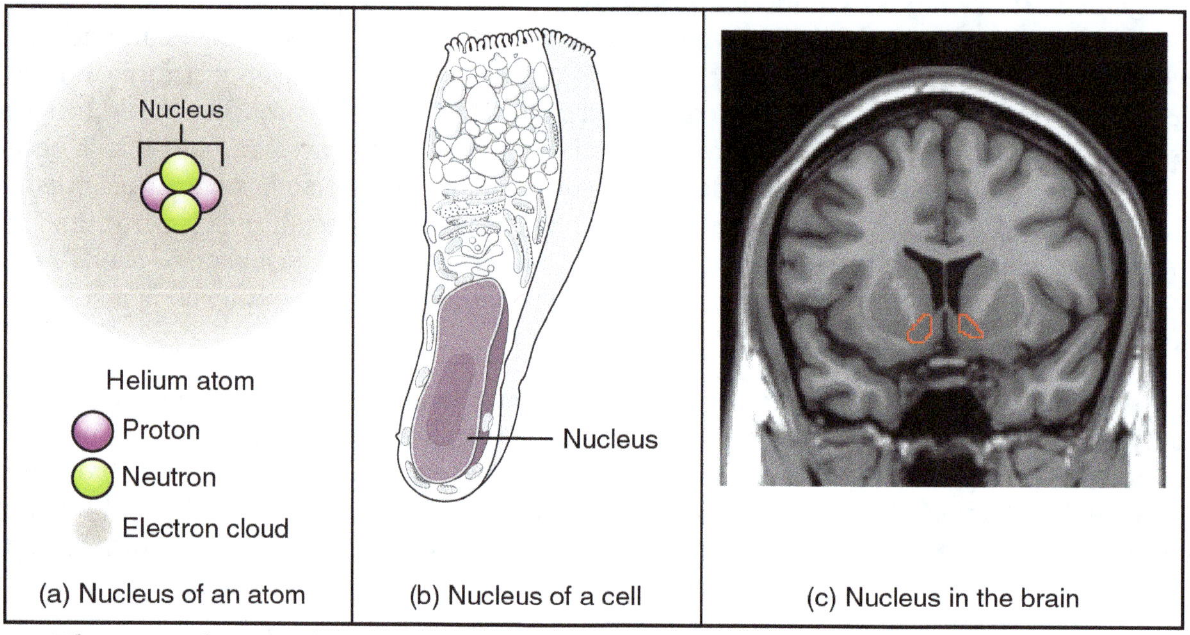

What Is a Nucleus? (a) The nucleus of an atom contains its protons and neutrons. (b) The nucleus of a cell is the organelle that contains DNA. (c) A nucleus in the CNS is a localized center of function with the cell bodies of several neurons, shown here circled in red.

Terminology applied to bundles of axons also differs depending on location. A bundle of axons, or fibers, found in the CNS is called a tract whereas the same thing in the PNS would be called a nerve. There is an important point to make about these terms, which is that they can both be used to refer to the same bundle of axons. When those axons are in the PNS, the term is nerve, but if they are CNS, the term is tract. The most obvious example of this is the axons that project from the retina into the brain. Those axons are called the optic nerve as they leave the eye, but when they are inside the cranium, they are referred to as the optic tract. There is a specific place where the name changes, which is the optic chiasm, but they are still the same axons. A similar situation outside of science can be described for some roads. Imagine a road called "Broad Street" in a town called "Anyville." The road leaves Anyville and goes to the next town over, called "Hometown." When the road crosses the line between the two towns and is in Hometown, its name changes to "Main Street." That is the idea behind the naming of the retinal axons. In the PNS, they are called the optic nerve, and in the CNS, they are the optic tract. Table helps to clarify which of these terms apply to the central or peripheral nervous systems.

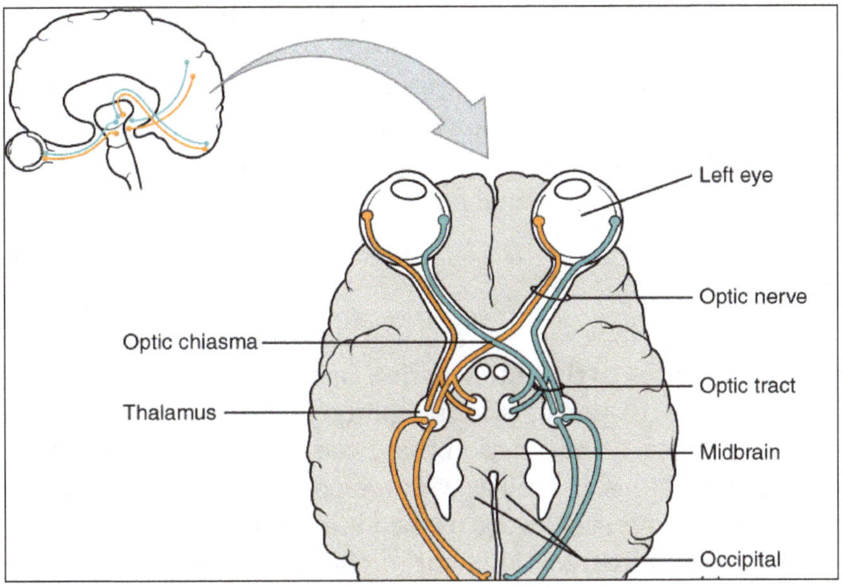

Optic Nerve Versus Optic Tract. This drawing of the connections of the eye to the brain shows the optic nerve extending from the eye to the chiasm, where the structure continues as the optic tract. The same axons extend from the eye to the brain through these two bundles of fibers, but the chiasm represents the border between peripheral and central.

In 2003, the Nobel Prize in Physiology or Medicine was awarded to Paul C. Lauterbur and Sir Peter Mansfield for discoveries related to magnetic resonance imaging (MRI). This is a tool to see the structures of the body (not just the nervous system) that depends on magnetic fields associated with certain atomic nuclei. The utility of this technique in the nervous system is that fat tissue and water appear as different shades between black and white. Because white matter is fatty (from myelin) and gray matter is not, they can be easily distinguished in MRI images. Also, the results from an MRI session are compared with images obtained from X-ray or computed tomography. How do the imaging techniques shown in this game indicate the separation of white and gray matter compared with the freshly dissected tissue shown earlier?

Structures of the CNS and PNS		
	CNS	PNS
Group of Neuron Cell Bodies (i.e., gray matter)	Nucleus	Ganglion
Bundle of Axons (i.e., white matter)	Tract	Nerve

Functional Divisions of the Nervous System

The nervous system can also be divided on the basis of its functions, but anatomical divisions and functional divisions are different. The CNS and the PNS both contribute to the same functions, but those functions can be attributed to different regions of the brain (such as the cerebral cortex or the hypothalamus) or to different ganglia in the periphery. The problem with trying to fit functional differences into anatomical divisions is that sometimes the same structure can be part of several functions. For example, the optic nerve carries signals from the retina that are either used for the conscious perception of visual stimuli, which takes place in the cerebral cortex, or for the reflexive responses of smooth muscle tissue that are processed through the hypothalamus.

There are two ways to consider how the nervous system is divided functionally. First, the basic functions of the nervous system are sensation, integration, and response. Secondly, control of the body can be somatic or autonomic—divisions that are largely defined by the structures that are involved in the response. There is also a region of the peripheral nervous system that is called the enteric nervous system that is responsible for a specific set of the functions within the realm of autonomic control related to gastrointestinal functions.

Basic Functions

The nervous system is involved in receiving information about the environment around us (sensation) and generating responses to that information (motor responses). The nervous system can be divided into regions that are responsible for sensation (sensory functions) and for the response (motor functions). But there is a third function that needs to be included. Sensory input needs to be integrated with other sensations, as well as with memories, emotional state, or learning (cognition). Some regions of the nervous system are termed integration or association areas. The process of integration combines sensory perceptions and higher cognitive functions such as memories, learning, and emotion to produce a response.

Sensation. The first major function of the nervous system is sensation—receiving information about the environment to gain input about what is happening outside the body (or, sometimes, within the body). The sensory functions of the nervous system register the presence of a change from homeostasis or a particular event in the environment, known as a stimulus. The senses we think of most are the "big five": taste, smell, touch, sight, and hearing. The stimuli for taste and smell are both chemical substances (molecules, compounds, ions, etc.), touch is physical or mechanical stimuli that interact with the skin, sight is light stimuli, and hearing is the perception of sound, which is a physical stimulus similar to some aspects of touch. There are actually more senses than just those, but that list represents the major senses. Those five are all senses that receive stimuli from the outside world, and of which there is conscious perception. Additional sensory stimuli might be from the internal environment (inside the body), such as the stretch of an organ wall or the concentration of certain ions in the blood.

Response. The nervous system produces a response on the basis of the stimuli perceived by sensory structures. An obvious response would be the movement of muscles, such as withdrawing a hand from a hot stove, but there are broader uses of the term. The nervous system can cause the contraction of all three types of muscle tissue. For example, skeletal muscle contracts to move the skeleton, cardiac muscle is influenced as heart rate increases during exercise, and smooth muscle contracts as the digestive system moves food along the digestive tract. Responses also include the neural control of glands in the body as well, such as the production and secretion of sweat by the eccrine and merocrine sweat glands found in the skin to lower body temperature.

Responses can be divided into those that are voluntary or conscious (contraction of skeletal muscle) and those that are involuntary (contraction of smooth muscles, regulation of cardiac muscle, activation of glands). Voluntary responses are governed by the somatic nervous system and involuntary responses are governed by the autonomic nervous system.

Integration. Stimuli that are received by sensory structures are communicated to the nervous system where that information is processed. This is called integration. Stimuli are compared with, or integrated with, other stimuli, memories of previous stimuli, or the state of a person at a particular time. This leads to the specific response that will be generated. Seeing a baseball pitched to a batter will not automatically cause the batter to swing. The trajectory of the ball and its speed will need to be considered. Maybe the count is three balls and one strike, and the batter wants to let this pitch go by in the hope of getting a walk to first base. Or maybe the batter's team is so far ahead, it would be fun to just swing away.

Controlling the Body

The nervous system can be divided into two parts mostly on the basis of a functional difference in responses. The somatic nervous system (SNS) is responsible for conscious perception and voluntary motor responses. Voluntary motor response means the contraction of skeletal muscle, but those contractions are not always voluntary in the sense that you have to want to perform them. Some somatic motor responses are reflexes, and often happen without a conscious decision to perform them. If your friend jumps out from behind a corner and yells "Boo!" you will be startled and you might scream or leap back. You didn't decide to do that, and you may not have wanted to give your friend a reason to laugh at your expense, but it is a reflex involving skeletal muscle contractions. Other motor responses become automatic (in other words, unconscious) as a person learns motor skills (referred to as "habit learning" or "procedural memory").

The autonomic nervous system (ANS) is responsible for involuntary control of the body, usually for the sake of homeostasis (regulation of the internal environment). Sensory input for autonomic functions can be from sensory structures tuned to external or internal environmental stimuli. The motor output extends to smooth and cardiac muscle as well as glandular tissue. The role of the autonomic system is to regulate the organ systems of the body, which usually means to control homeostasis. Sweat glands, for example, are controlled by the autonomic system. When you are hot, sweating helps cool your body down. That is a homeostatic mechanism. But when you are nervous, you might start sweating also. That is not homeostatic, it is the physiological response to an emotional state.

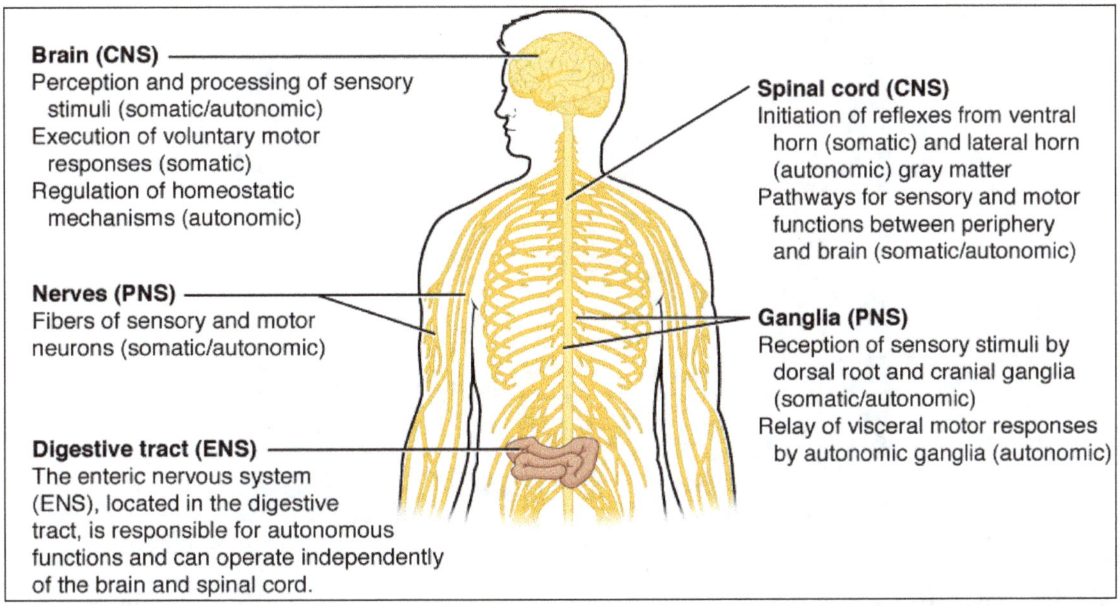

Somatic, Autonomic, and Enteric Structures of the Nervous System. Somatic structures include the spinal nerves, both motor and sensory fibers, as well as the sensory ganglia (posterior root ganglia and cranial nerve ganglia).

There is another division of the nervous system that describes functional responses. The enteric nervous system (ENS) is responsible for controlling the smooth muscle and glandular tissue in your digestive system. It is a large part of the PNS, and is not dependent on the CNS. It is sometimes valid, however, to consider the enteric system to be a part of the autonomic system because the neural structures that make up the enteric system are a component of the autonomic output that regulates digestion. There are some differences between the two, but for our purposes here there will be a good bit of overlap. Figure for examples of where these divisions of the nervous system can be found.

Autonomic structures are found in the nerves also, but include the sympathetic and parasympathetic ganglia. The enteric nervous system includes the nervous tissue within the organs of the digestive tract.

This leads to the discovery of a hereditary condition that affects the brain and spinal cord. The electromyography and MRI tests indicated deficiencies in the spinal cord and cerebellum, both of which are responsible for controlling coordinated movements. To what functional division of the nervous system would these structures belong?

How Much of Your Brain Do You Use? Have you ever heard the claim that humans only use 10 percent of their brains? Maybe you have seen an advertisement on a website saying that there is a secret to unlocking the full potential of your mind—as if there were 90 percent of your brain sitting idle, just waiting for you to use it.

An easy way to see how much of the brain a person uses is to take measurements of brain activity while performing a task. An example of this kind of measurement is functional magnetic resonance imaging (fMRI), which generates a map of the most active areas and can be generated and presented in three dimensions. This procedure is different from the standard MRI technique because it is measuring changes in the tissue in time with an experimental condition or event.

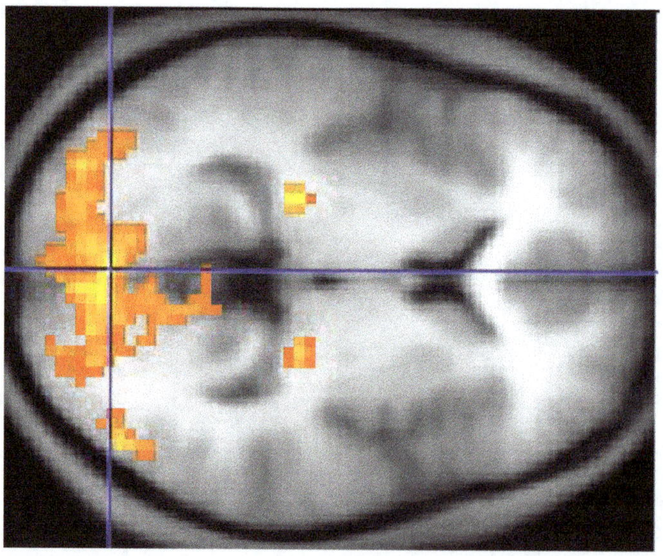

fMRI. This fMRI shows activation of the visual cortex in response to visual stimuli.

The underlying assumption is that active nervous tissue will have greater blood flow. By having the subject perform a visual task, activity all over the brain can be measured. Consider this possible experiment: the subject is told to look at a screen with a black dot in the middle (a fixation point). A photograph of a face is projected on the screen away from the center. The subject has to look at the photograph and decipher what it is. The subject has been instructed to push a button if the photograph is of someone they recognize. The photograph might be of a celebrity, so the subject would press the button, or it might be of a random person unknown to the subject, so the subject would not press the button.

In this task, visual sensory areas would be active, integrating areas would be active, motor areas responsible for moving the eyes would be active, and motor areas for pressing the button with a finger would be active. Those areas are distributed all around the brain and the fMRI images would show activity in more than just 10 percent of the brain (some evidence suggests that about 80 percent of the brain is using energy—based on blood flow to the tissue—during well-defined tasks similar to the one suggested above). This task does not even include all of the functions the brain performs. There is no language response, the body is mostly lying still in the MRI machine, and it does not consider the autonomic functions that would be ongoing in the background.

The nervous system can be separated into divisions on the basis of anatomy and physiology. The anatomical divisions are the central and peripheral nervous systems. The CNS is the brain and spinal cord. The PNS is everything else. Functionally, the nervous system can be divided into those regions that are responsible for sensation, those that are responsible for integration, and those that are responsible for generating responses. All of these functional areas are found in both the central and peripheral anatomy.

Considering the anatomical regions of the nervous system, there are specific names for the structures within each division. A localized collection of neuron cell bodies is referred to as a nucleus in the CNS and as a ganglion in the PNS. A bundle of axons is referred to as a tract in the CNS and as a nerve in the PNS. Whereas nuclei and ganglia are specifically in the central or peripheral divisions, axons can cross the boundary between the two. A single axon can be part of a nerve and a tract. The name for that specific structure depends on its location.

Nervous tissue can also be described as gray matter and white matter on the basis of its appearance in unstained tissue. These descriptions are more often used in the CNS. Gray matter is where nuclei are found and white matter is where tracts are found. In the PNS, ganglia are basically gray matter and nerves are white matter.

The nervous system can also be divided on the basis of how it controls the body. The somatic nervous system (SNS) is responsible for functions that result in moving skeletal muscles. Any sensory or integrative functions that result in the movement of skeletal muscle would be considered somatic. The autonomic nervous system (ANS) is responsible for functions that affect cardiac or smooth muscle tissue, or that cause glands to produce their secretions. Autonomic functions are distributed between central and peripheral regions of the nervous system. The sensations that lead to autonomic functions can be the same sensations that are part of initiating somatic responses. Somatic and autonomic integrative functions may overlap as well.

A special division of the nervous system is the enteric nervous system, which is responsible for controlling the digestive organs. Parts of the autonomic nervous system overlap with the enteric nervous system. The enteric nervous system is exclusively found in the periphery because it is the nervous tissue in the organs of the digestive system.

Interactive Link Questions

MRI uses the relative amount of water in tissue to distinguish different areas, so gray and white matter in the nervous system can be seen clearly in these images.

Nervous Tissue

Nervous tissue is composed of two types of cells, neurons and glial cells. Neurons are the primary type of cell that most anyone associates with the nervous system. They are responsible for the computation and communication that the nervous system provides. They are electrically active and release chemical signals to target cells. Glial cells, or glia, are known to play a supporting role for nervous tissue. Ongoing research pursues an expanded role that glial cells might play in signaling, but neurons are still considered the basis of this function. Neurons are important, but without glial support they would not be able to perform their function.

Neurons

Neurons are the cells considered to be the basis of nervous tissue. They are responsible for the electrical signals that communicate information about sensations, and that produce movements in response to those stimuli, along with inducing thought processes within the brain. An important part of the function of neurons is in their structure, or shape. The three-dimensional shape of these cells makes the immense numbers of connections within the nervous system possible.

Parts of a Neuron

the main part of a neuron is the cell body, which is also known as the soma (soma = "body"). The cell body contains the nucleus and most of the major organelles. But what makes neurons special is that they have many extensions of their cell membranes, which are generally referred to as processes. Neurons are usually described as having one, and only one, axon—a fiber that emerges from the cell body and projects to target cells. That single axon can branch repeatedly to communicate with many target cells. It is the axon that propagates the nerve impulse, which is communicated to one or more cells. The other processes of the neuron are dendrites, which receive information from other neurons at specialized areas of contact called synapses. The dendrites are usually highly branched processes, providing locations for other neurons to communicate with the cell body. Information flows through a neuron from the dendrites, across the cell body, and down the axon. This gives the neuron a polarity—meaning that information flows in this one direction. Figure shows the relationship of these parts to one another.

Where the axon emerges from the cell body, there is a special region referred to as the axon hillock. This is a tapering of the cell body toward the axon fiber. Within the axon hillock, the cytoplasm changes to a solution of limited components called axoplasm. Because the axon hillock represents the beginning of the axon, it is also referred to as the initial segment.

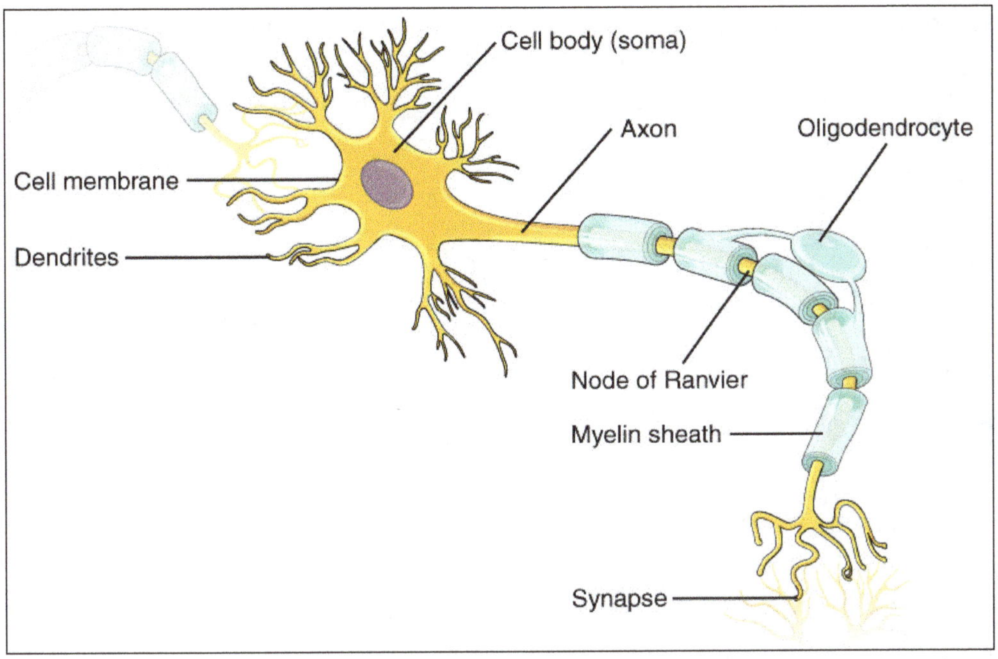

Parts of a Neuron. The major parts of the neuron are labeled on a multipolar neuron from the CNS.

Many axons are wrapped by an insulating substance called myelin, which is actually made from glial cells. Myelin acts as insulation much like the plastic or rubber that is used to insulate electrical wires. A key difference between myelin and the insulation on a wire is that there are gaps in the myelin covering of an axon. Each gap is called a node of Ranvier and is important to the way that electrical signals travel down the axon. The length of the axon between each gap, which is wrapped in myelin, is referred to as an axon segment. At the end of the axon is the axon terminal, where there are usually several branches extending toward the target cell, each of which ends in an enlargement called a synaptic end bulb. These bulbs are what make the connection with the target cell at the synapse.

Types of Neurons

There are many neurons in the nervous system—a number in the trillions. And there are many different types of neurons. They can be classified by many different criteria. The first way to classify them is by the number of processes attached to the cell body. Using the standard model of neurons, one of these processes is the axon, and the rest are dendrites. Because information flows through the neuron from dendrites or cell bodies toward the axon, these names are based on the neuron's polarity.

Unipolar cells have only one process emerging from the cell. True unipolar cells are only found in invertebrate animals, so the unipolar cells in humans are more appropriately called "pseudo-unipolar" cells. Invertebrate unipolar cells do not have dendrites. Human unipolar cells have an axon that emerges from the cell body, but it splits so that the axon can extend along a very long distance. At one end of the axon are dendrites, and at the other end, the axon forms synaptic connections with a target. Unipolar cells are exclusively sensory neurons and have two unique characteristics. First, their dendrites are receiving sensory information, sometimes directly from the stimulus itself. Secondly, the cell bodies of unipolar neurons are always found in ganglia. Sensory reception is a peripheral function (those dendrites are in the periphery, perhaps in the skin) so the cell body is in the periphery, though closer to the CNS in a ganglion. The axon projects from the dendrite endings, past the cell body in a ganglion, and into the central nervous system.

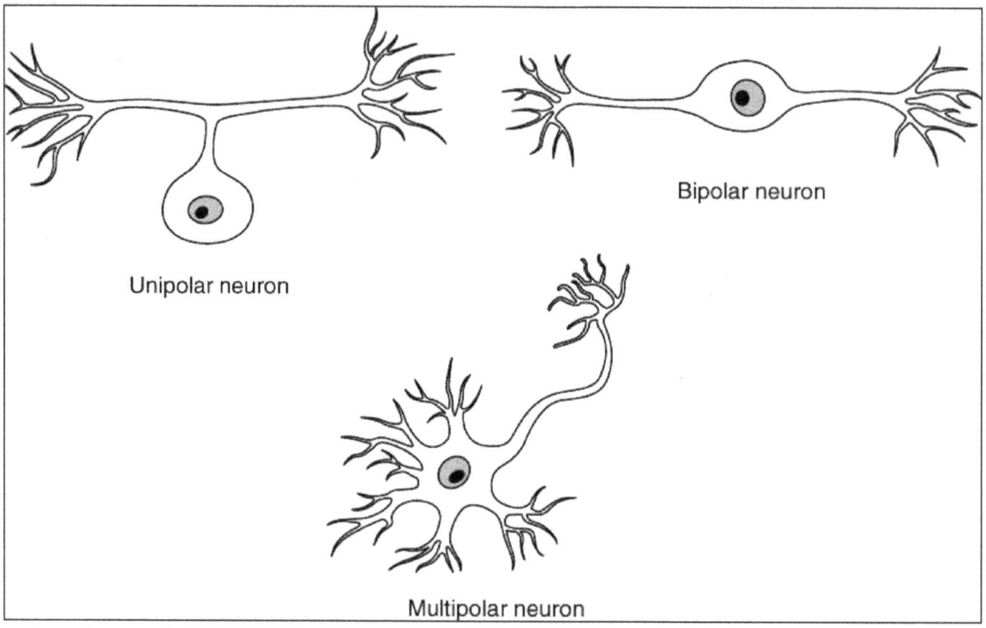

Neuron Classification by Shape. Unipolar cells have one process that includes both the axon and dendrite. Bipolar cells have two processes, the axon and a dendrite. Multipolar cells have more than two processes, the axon and two or more dendrites.

Bipolar cells have two processes, which extend from each end of the cell body, opposite to each other. One is the axon and one the dendrite. Bipolar cells are not very common. They are found mainly in the olfactory epithelium (where smell stimuli are sensed), and as part of the retina.

Multipolar neurons are all of the neurons that are not unipolar or bipolar. They have one axon and two or more dendrites (usually many more). With the exception of the unipolar sensory ganglion cells, and the two specific bipolar cells, all other neurons are multipolar. Some cutting edge research suggests that certain neurons in the CNS do not conform to the standard model of "one, and only one" axon. Some sources describe a fourth type of neuron, called an anaxonic neuron. The name suggests that it has no axon (an- = "without"), but this is not accurate. Anaxonic neurons are very small, and if you look through a microscope at the standard resolution used in histology (approximately 400X to 1000X total magnification), you will not be able to distinguish any process specifically as an axon or a dendrite. Any of those processes can function as an axon depending on the conditions at any given time. Nevertheless, even if they cannot be easily seen, and one specific process is definitively the axon, these neurons have multiple processes and are therefore multipolar.

Neurons can also be classified on the basis of where they are found, who found them, what they do, or even what chemicals they use to communicate with each other. Some neurons referred to in this topic on the nervous system are named on the basis of those sorts of classifications. For example, a multipolar neuron that has a very important role to play in a part of the brain called the cerebellum is known as a Purkinje (commonly pronounced per-KIN-gee) cell. It is named after the anatomist who discovered it.

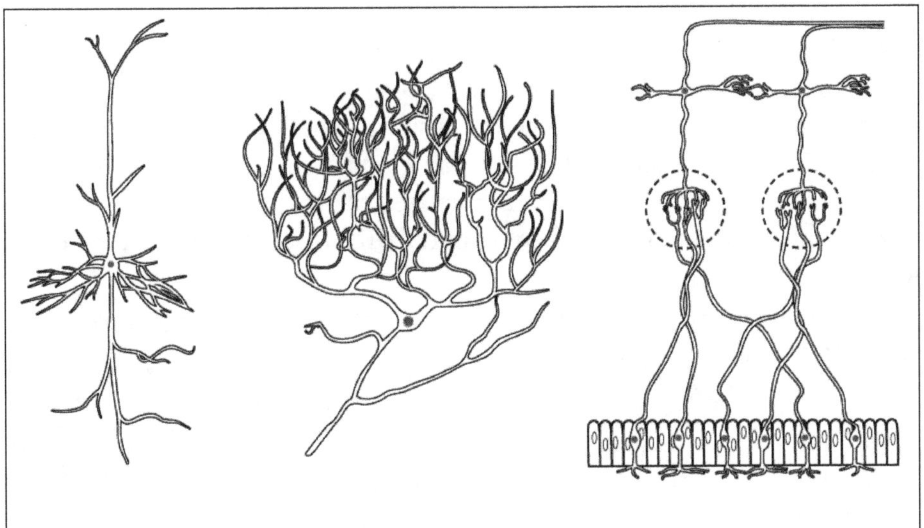

(a) Pyramidal cell of the cerebral cortex (b) Purkinje cell of the cerebellar cortex (c) Olfactory cells in the olfactory epithelium and olfactory bulbs

Other Neuron Classifications.

Three examples of neurons that are classified on the basis of other criteria. (a) The pyramidal cell is a multipolar cell with a cell body that is shaped something like a pyramid. (b) The Purkinje cell in the cerebellum was named after the scientist who originally described it. (c) Olfactory neurons are named for the functional group with which they belong.

Glial Cells

Glial cells, or neuroglia or simply glia, are the other type of cell found in nervous tissue. They are considered to be supporting cells, and many functions are directed at helping neurons complete their function for communication.

There are six types of glial cells. Four of them are found in the CNS and two are found in the PNS. Table outlines some common characteristics and functions.

Glial Cell Types by Location and Basic Function (Table)		
CNS glia	PNS glia	Basic function
Astrocyte	Satellite cell	Support
Oligodendrocyte	Schwann cell	Insulation, myelination
Microglia	–	Immune surveillance and phagocytosis
Ependymal cell	–	Creating CSF

Glial Cells of the CNS

One cell providing support to neurons of the CNS is the astrocyte, so named because it appears to be star-shaped under the microscope (astro- = "star"). Astrocytes have many processes extending from their main cell body (not axons or dendrites like neurons, just cell extensions). Those processes extend to interact with neurons, blood vessels, or the connective tissue covering the CNS that is called the pia mater. Generally, they are supporting cells for the neurons in the central nervous system. Some ways in which they support neurons in the central nervous system are by maintaining the concentration of chemicals in the extracellular space, removing excess signaling molecules, reacting to tissue damage, and contributing to the blood-brain barrier (BBB). The blood-brain barrier is a physiological barrier that keeps many substances that circulate in the rest of the body from getting into the central nervous system, restricting what can cross from circulating blood into the CNS. Nutrient molecules, such as glucose or amino acids, can pass through the BBB, but other molecules cannot. This actually causes problems with drug delivery to the CNS. Pharmaceutical companies are challenged to design drugs that can cross the BBB as well as have an effect on the nervous system.

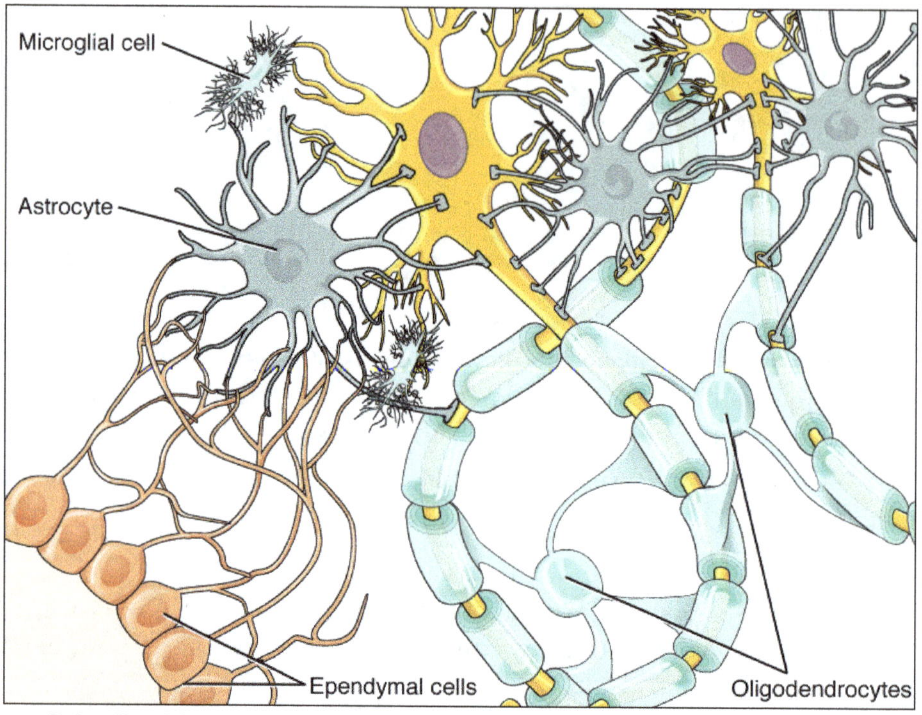

Glial Cells of the CNS. The CNS has astrocytes, oligodendrocytes, microglia, and ependymal cells that support the neurons of the CNS in several ways.

Like a few other parts of the body, the brain has a privileged blood supply. Very little can pass through by diffusion. Most substances that cross the wall of a blood vessel into the CNS must do so through an active transport process. Because of this, only specific types of molecules can enter the CNS. Glucose—the primary energy source—is allowed, as are amino acids. Water and some other small particles, like gases and ions, can enter. But most everything else cannot, including white blood cells, which are one of the body's main lines of defense. While this barrier protects the CNS from exposure to toxic or pathogenic substances, it also keeps out the cells that could protect the brain and spinal cord from disease and damage. The BBB also makes it harder for pharmaceuticals to be developed that can affect the nervous system. Aside from finding efficacious substances, the means of delivery is also crucial.

Also found in CNS tissue is the oligodendrocyte, sometimes called just "oligo," which is the glial cell type that insulates axons in the CNS. The name means "cell of a few branches" (oligo- = "few"; dendro- = "branches"; -cyte = "cell"). There are a few processes that extend from the cell body. Each one reaches out and surrounds an axon to insulate it in myelin. One oligodendrocyte will provide the myelin for multiple axon segments, either for the same axon or for separate axons.

Microglia are, as the name implies, smaller than most of the other glial cells. Ongoing research into these cells, although not entirely conclusive, suggests that they may originate as white blood cells, called macrophages, that become part of the CNS during early development. While their origin is not conclusively determined, their function is related to what macrophages do in the rest of the body. When macrophages encounter diseased or damaged cells in the rest of the body, they ingest and digest those cells or the pathogens that cause disease. Microglia are the cells in the CNS that can do this in normal, healthy tissue, and they are therefore also referred to as CNS-resident macrophages.

The ependymal cell is a glial cell that filters blood to make cerebrospinal fluid (CSF), the fluid that circulates through the CNS. Because of the privileged blood supply inherent in the BBB, the extracellular space in nervous tissue does not easily exchange components with the blood. Ependymal cells line each ventricle, one of four central cavities that are remnants of the hollow center of the neural tube formed during the embryonic development of the brain. The choroid plexus is a specialized structure in the ventricles where ependymal cells come in contact with blood vessels and filter and absorb components of the blood to produce cerebrospinal fluid. Because of this, ependymal cells can be considered a component of the BBB, or a place where the BBB breaks down. These glial cells appear similar to epithelial cells, making a single layer of cells with little intracellular space and tight connections between adjacent cells. They also have cilia on their apical surface to help move the CSF through the ventricular space. The relationship of these glial cells to the structure of the CNS is seen in figure.

Glial Cells of the PNS

One of the two types of glial cells found in the PNS is the satellite cell. Satellite cells are found in sensory and autonomic ganglia, where they surround the cell bodies of neurons. This accounts for the name, based on their appearance under the microscope. They provide support, performing similar functions in the periphery as astrocytes do in the CNS—except, of course, for establishing the BBB.

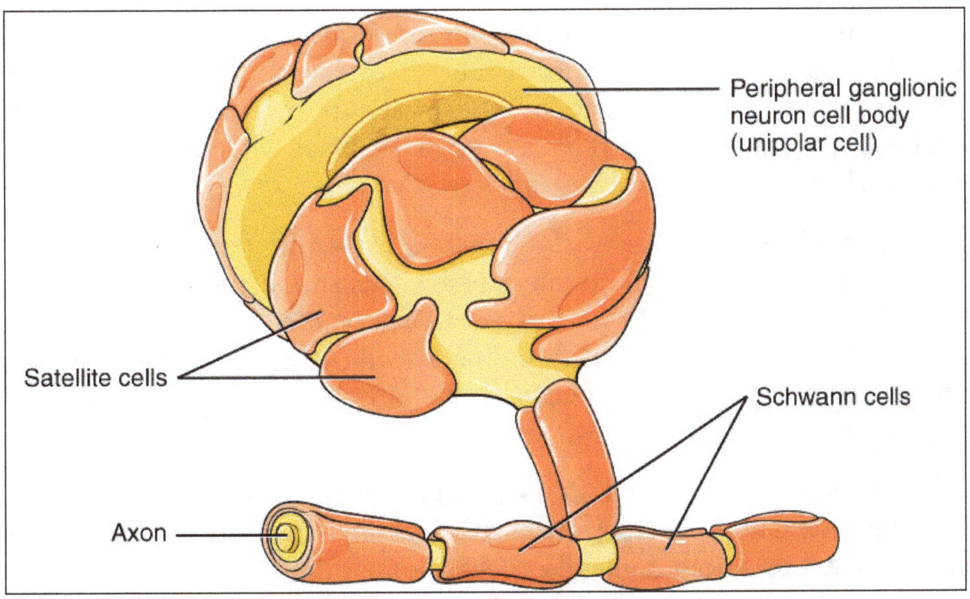

Glial Cells of the PNS. The PNS has satellite cells and Schwann cells.

The second type of glial cell is the Schwann cell, which insulate axons with myelin in the periphery. Schwann cells are different than oligodendrocytes, in that a Schwann cell wraps around a portion of only one axon segment and no others. Oligodendrocytes have processes that reach out to multiple axon segments, whereas the entire Schwann cell surrounds just one axon segment. The nucleus and cytoplasm of the Schwann cell are on the edge of the myelin sheath. The relationship of these two types of glial cells to ganglia and nerves in the PNS is seen in figure.

Myelin

The insulation for axons in the nervous system is provided by glial cells, oligodendrocytes in the CNS, and Schwann cells in the PNS. Whereas the manner in which either cell is associated with the axon segment, or segments, that it insulates is different, the means of myelinating an axon segment is mostly the same in the two situations. Myelin is a lipid-rich sheath that surrounds the axon and by doing so creates a myelin sheath that facilitates the transmission of electrical signals along the axon. The lipids are essentially the phospholipids of the glial cell membrane. Myelin, however, is more than just the membrane of the glial cell. It also includes important proteins that are integral to that membrane. Some of the proteins help to hold the layers of the glial cell membrane closely together.

The appearance of the myelin sheath can be thought of as similar to the pastry wrapped around a hot dog for "pigs in a blanket" or a similar food. The glial cell is wrapped around the axon several times with little to no cytoplasm between the glial cell layers. For oligodendrocytes, the rest of the cell is separate from the myelin sheath as a cell process extends back toward the cell body. A few other processes provide the same insulation for other axon segments in the area. For Schwann cells, the outermost layer of the cell membrane contains cytoplasm and the nucleus of the cell as a bulge on one side of the myelin sheath. During development, the glial cell is loosely or incompletely wrapped around the axon. The edges of this loose enclosure extend toward each other, and one end tucks under the other. The inner edge wraps around the axon, creating several layers, and the other edge closes around the outside so that the axon is completely enclosed.

The axon contains microtubules and neurofilaments that are bounded by a plasma membrane known as the axolemma. Outside the plasma membrane of the axon is the myelin sheath, which is composed of the tightly wrapped plasma membrane of a Schwann cell. What aspects of the cells in this image react with the stain to make them a deep, dark, black color, such as the multiple layers that are the myelin sheath?

Myelin sheaths can extend for one or two millimeters, depending on the diameter of the axon. Axon diameters can be as small as 1 to 20 micrometers. Because a micrometer is 1/1000 of a millimeter, this means that the length of a myelin sheath can be 100–1000 times the diameter of the axon. Figures show the myelin sheath surrounding an axon segment, but are not to scale. If the myelin sheath were drawn to scale, the neuron would have to be immense—possibly covering an entire wall of the room in which you are sitting.

Disorders of the Nervous Tissue Several diseases can result from the demyelination of axons. The causes of these diseases are not the same; some have genetic causes, some are caused by pathogens, and others are the result of autoimmune disorders. Though the causes are varied, the results are largely similar. The myelin insulation of axons is compromised, making electrical signaling slower.

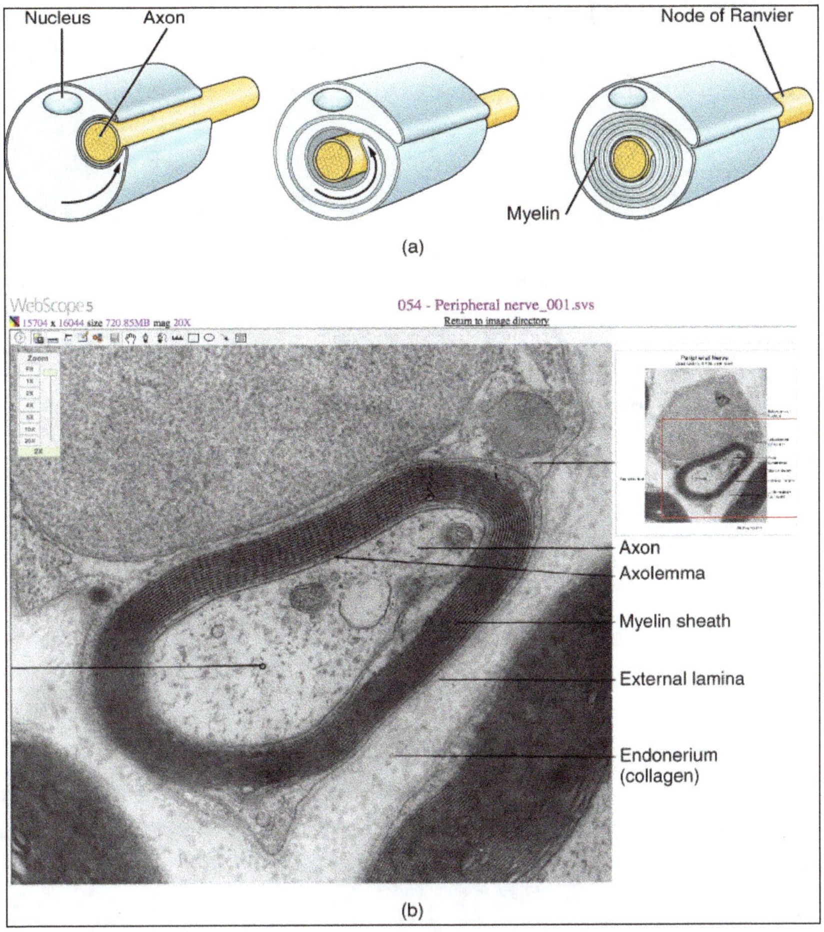

The Process of Myelination. Myelinating glia wrap several layers of cell membrane around the cell membrane of an axon segment. A single Schwann cell insulates a segment of a peripheral nerve, whereas in the CNS, an oligodendrocyte may provide insulation for a few separate axon segments. EM × 1,460,000.

Multiple sclerosis (MS) is one such disease. It is an example of an autoimmune disease. The antibodies produced by lymphocytes (a type of white blood cell) mark myelin as something that should not be in the body. This causes inflammation and the destruction of the myelin in the central nervous system. As the insulation around the axons is destroyed by the disease, scarring becomes obvious. This is where the name of the disease comes from; sclerosis means hardening of tissue, which is what a scar is. Multiple scars are found in the white matter of the brain and spinal cord. The symptoms of MS include both somatic and autonomic deficits. Control of the musculature is compromised, as is control of organs such as the bladder.

Guillain-Barré (pronounced gee-YAN bah-RAY) syndrome is an example of a demyelinating disease of the peripheral nervous system. It is also the result of an autoimmune reaction, but the inflammation is in peripheral nerves. Sensory symptoms or motor deficits are common, and autonomic failures can lead to changes in the heart rhythm or a drop in blood pressure, especially when standing, which causes dizziness.

The Function of Nervous Tissue

Having looked at the components of nervous tissue, and the basic anatomy of the nervous system, next comes an understanding of how nervous tissue is capable of communicating within the nervous system. Before getting to the nuts and bolts of how this works, an illustration of how the components come together will be helpful. An example is summarized in figure.

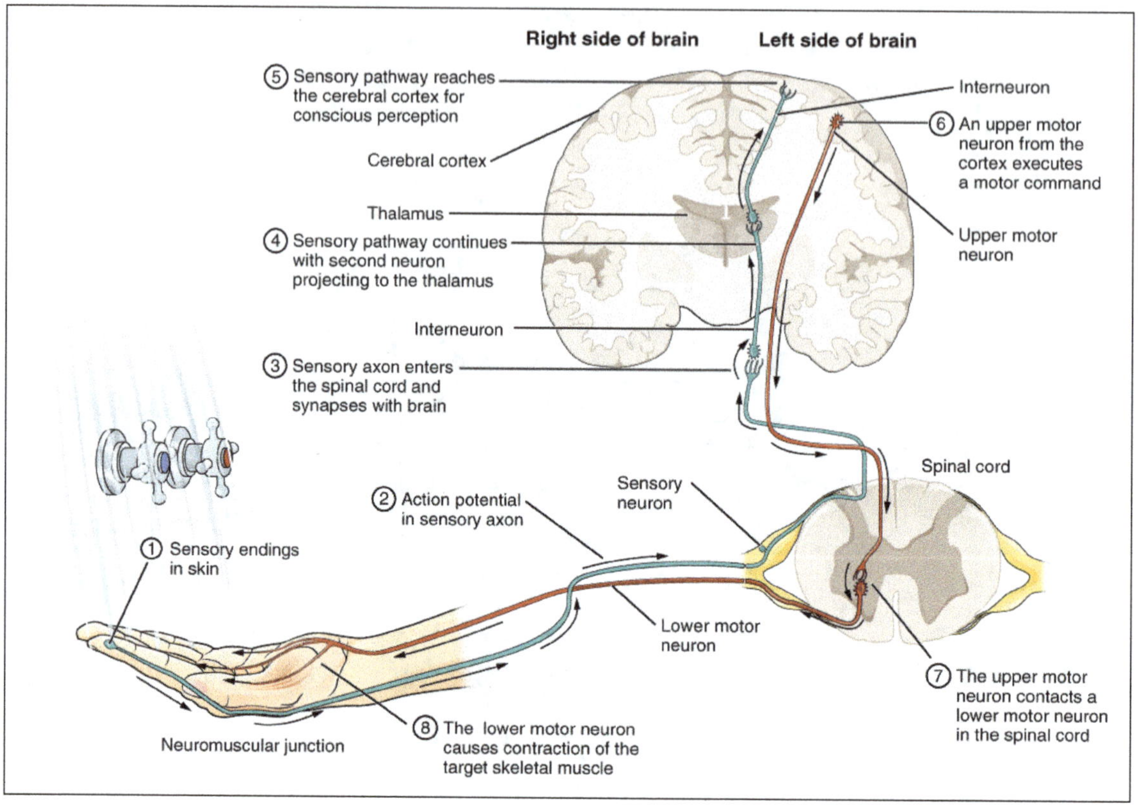

Testing the Water. (1) The sensory neuron has endings in the skin that sense a stimulus such as water temperature. The strength of the signal that starts here is dependent on the strength of the stimulus. (2) The graded potential from the sensory endings, if strong enough, will initiate an action potential at the initial segment of the axon (which is immediately adjacent to the sensory endings in the skin). (3) The axon of the peripheral sensory neuron enters the spinal cord and contacts another neuron in the gray matter. The contact is a synapse where another graded potential is caused by the release of a chemical signal from the axon terminals. (4) An action potential is initiated at the initial segment of this neuron and travels up the sensory pathway to a region of the brain called the thalamus. Another synapse passes the information along to the next neuron. (5) The sensory pathway ends when the signal reaches the cerebral cortex. (6) After integration with neurons in other parts of the cerebral cortex, a motor command is sent from the precentral gyrus of the frontal cortex. (7) The upper motor neuron sends an action potential down to the spinal cord. The target of the upper motor neuron is the dendrites of the lower motor neuron in the gray matter of the spinal cord. (8) The axon of the lower motor neuron emerges from the spinal cord in a nerve and connects to a muscle through a neuromuscular junction to cause contraction of the target muscle.

Imagine you are about to take a shower in the morning before going to school. You have turned on the faucet to start the water as you prepare to get in the shower. After a few minutes, you expect the water to be a temperature that will be comfortable to enter. So you put your hand out into the spray of water. What happens next depends on how your nervous system interacts with the stimulus of the water temperature and what you do in response to that stimulus.

Found in the skin of your fingers or toes is a type of sensory receptor that is sensitive to temperature, called a thermoreceptor. When you place your hand under the shower, the cell membrane of the thermoreceptors changes its electrical state (voltage). The amount of change is dependent on the strength of the stimulus (how hot the water is). This is called a graded potential. If the stimulus is strong, the voltage of the cell membrane will change enough to generate an electrical signal that will travel down the axon. You have learned about this type of signaling before, with respect to the interaction of nerves and muscles at the neuromuscular junction. The voltage at which such a signal is generated is called the threshold, and the resulting electrical signal is called an action potential. In this example, the action potential travels—a process known as propagation—along the axon from the axon hillock to the axon terminals and into the synaptic end bulbs. When this signal reaches the end bulbs, it causes the release of a signaling molecule called a neurotransmitter.

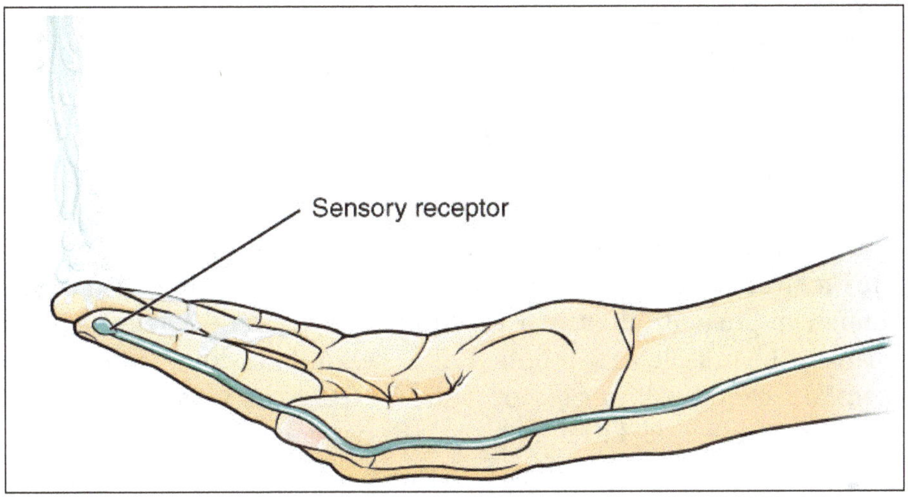

The Sensory Input. Receptors in the skin sense the temperature of the water.

The neurotransmitter diffuses across the short distance of the synapse and binds to a receptor protein of the target neuron. When the molecular signal binds to the receptor, the cell membrane of the target neuron changes its electrical state and a new graded potential begins. If that graded potential is strong enough to reach threshold, the second neuron generates an action potential at its axon hillock. The target of this neuron is another neuron in the thalamus of the brain, the part of the CNS that acts as a relay for sensory information. At another synapse, neurotransmitter is released and binds to its receptor. The thalamus then sends the sensory information to the cerebral cortex, the outermost layer of gray matter in the brain, where conscious perception of that water temperature begins.

Within the cerebral cortex, information is processed among many neurons, integrating the stimulus of the water temperature with other sensory stimuli, with your emotional state (you just aren't ready to wake up; the bed is calling to you), memories. Finally, a plan is developed about what to do, whether that is to turn the temperature up, turn the whole shower off and go back to bed, or step into the shower. To do any of these things, the cerebral cortex has to send a command out to your body to move muscles.

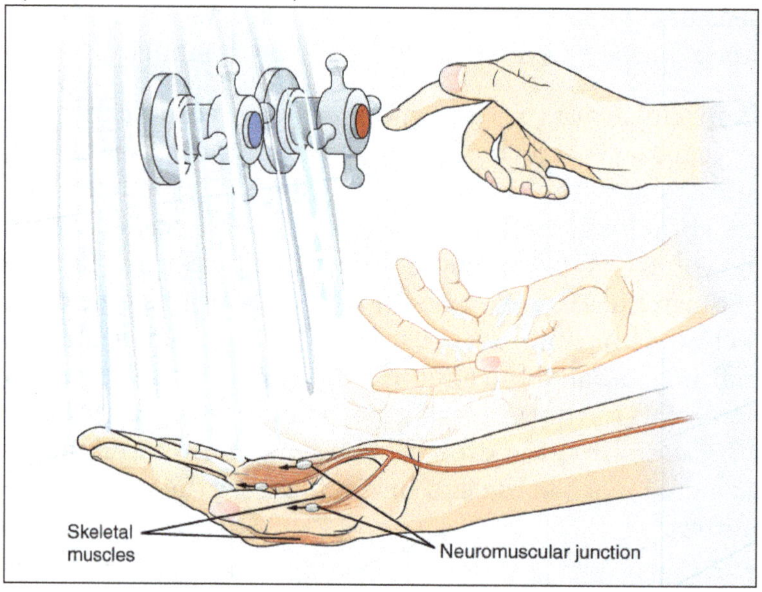

The Motor Response. On the basis of the sensory input and the integration in the CNS, a motor response is formulated and executed.

A region of the cortex is specialized for sending signals down to the spinal cord for movement. The upper motor neuron is in this region, called the precentral gyrus of the frontal cortex, which has an axon that extends all the way down the spinal cord. At the level of the spinal cord at which this axon makes a synapse, a graded potential occurs in the cell membrane of a lower motor neuron. This second motor neuron is responsible for causing muscle fibers to contract. The axon terminates on muscle fibers at the neuromuscular junction. Acetylcholine is released at this specialized synapse, which causes the muscle action potential to begin, following a large potential known as an end plate potential. When the lower motor neuron excites the muscle fiber, it contracts. All of this occurs in a fraction of a second, but this story is the basis of how the nervous system functions.

Sensory Nervous System of Marine Mammals

Sensory systems evolved to allow animals to receive and process information from their surroundings. To understand how sensory systems operate in any given environment, we must understand how the physical characteristics ofthat environment affect the available information and its propagation and reception. In a very real sense we need to look at both the medium and the message. Signals in the marine environment can be substantially differentfrom those in air, and the oceanicmedium itselfchangesthe message in a number of ways.

When their evolutionary paths took them into the oceans, marine mammals had to adaptsensory systems that had evolved in air into ones that were able to detect and process Signals in water. The sensory systems of marine .Mammals are functionally similar to those of terrestrial mammals in that they act as highly selective filters. If every environmental cue available received equal attention, the brain would be barraged by sensory inputs. Instead, sensory organs are filters, selecting and attending to signalsthat, evolutionarily, proved to be important. Consider how predator and prey are driven to be both similar and different sensorially. Because their activities intersect in place and time, they need to have similar visual sensitivities, but different fields of view. The predator usually has binocular overlap that provides a precise judgment of distance to the prey. The prey may forego binocular vision and accurate visual depth judgments in favor ofgreater lateral visual fields to detect a predator. Thus, two species may have overlapping sensory ranges, but no two have identicalsensory capacities. Consequently, each animal's perceived world is only a subset ofthe real phySical world, that is, it is a species-specific model, constructed from the blocks ofdata its senses can capture.

In animal behavior, this conceptis called the Umwelt. As a technical term, Umwelt means an animal's perceptually limited construct ofthe world. In common usage, it simply means the environment. This dual meaning reflects the complex interaction ofsensory adaptations and habitat. Senses are tuned to relevantstimuliby evolution but are limited by the physical parameters ofthe habitat. For example, human sensory systems are geared to diurnal, airborne cues. Humans are highly developed visually, with 38 times more optic nerve fibers than auditory nerve fibers, and a hearingrange (20 to 20,000 Hz) that is narrower than that ofmany other animals. By observingspecies adapted to different habitats and analyzing their sensory biology, we can learn how they detect and use phYSical cues that are normally imperceptible to us. Ifwe develop technology that translates those cues into our sensory ranges, we can glimpse at the world as otherspecies perceive it. Marine mammals offer us a very special glimpse. In aquatic environments, our air-adapted senses are out oftheir element and are effectively detuned. By studying marine mammal sensory systems and abilities, we can understand how land mammal senses were evolutionarily retuned to operate in water. From that knowledge, we gain a valuable window into the oceans, the most extensive and unexplored environment on earth.

Generally, the term sensory system refers to the peripheral, as opposed to the brain, or central, components an animal uses to detect and analyze a signal. There are four essential functions for any sensory system: (1) capture an environmental signal, (2) filter it, (3) transduce it to a neural impulse, and (4) send processed information to the central nervous system. Each function may involve more than one form ofreceptor orperipheral processor. The blockdiagram in Figure compares a generic sensory receptor system with equivalent stages for mammalian eyes and ears. In vision, the first step, signal capture, is accomplished by the refraction oflight at the cornea andthe pupil's ability to control the light intensity entering the eye. Second, the lens focuses light on the retina while also acting as a first-stage filter, passing only some portions of the full spectrum of light. The tapetum, a reflective layerbehind the retina, reflectsthe photons not captured on the first passage through the retina back through for a second chance at absorption. Third, pigments within the rod and cone receptor cells absorb each wavelength with a different efficiency. Fourth, the rods or cones pass a chemicalsignal to horizontal orbipolar cellsthat modify and transmit the Signal to amacrine or ganglion cells. Axons ofganglion cells make up the optic nerve, which passes signals to midbrain structures and ultimately to the cortex. In hearing, the first step is the capture ofsound by the external ear. The external ear and the ear canal act as first-stage filters, attenuating some sounds based on their direction (pinnal shadowing) and amplifying others according to the resonance characteristics ofthe outer ear and canal. Second, the middle ear components act as second stage filters. The middle ear bones mechanically transmit vibrations of the eardrum, or tympanic membrane, to the oval window, which is the acoustic entrance to the fluid-filled inner ear. Thisbony chain acts as a series ofleversthat provides a nearly 40-dB boost to the incoming signal, which compensates for the loss ofacoustic powerthat would normally occurfrom a simple transmission of sound in air into flUid. The mass, stiffuess, and shape ofthe middle ear cavity and ofthe middle ear ossicular chain also influence the efficiency with which different frequencies are transmitted to the inner ear. Third, at the level ofthe inner ear, the basilar membrane acts as a bank offilters that determine the range offrequencies the brainwill ultimately process. The detectable sound or "hearing" range is dictated by the stiffness andmass characteristics ofthis membrane. Fourth, when sensory cells with flexible cilia, the hair cells, are bentthrough the motion ofthe basilar membrane, a chemicalsignal is transmitted via the auditory afferent (inward) fibers to the brainstem. Thus, in both the eye and ear, the'signal goes through a minimum ofthree and as many as five layers ofsignal processing before it is transformed into a neural impulse.

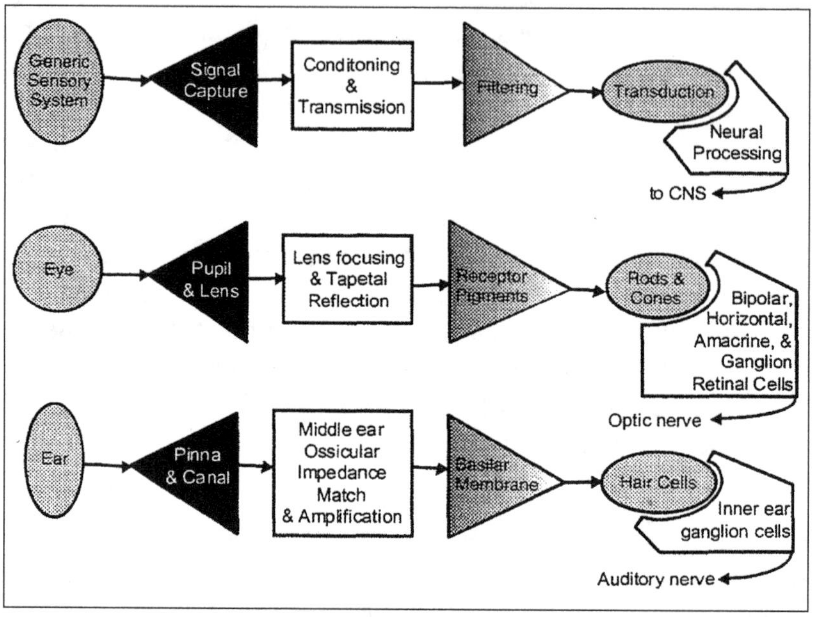

A generic sensory system is shown with parallelstages for mammalian visual and auditory systems.

For both sensory systems, there is extensive central processing as well as efferent (outward) feedback signals from the central nervous system that affect the responses at the receptor. Depending on the stimulus, the behavioral and physiologic state ofthe animal, and the type ofreceptor, a stimulus can be perceived but elicit no action, or it can prompt a set ofsignals to be sent to an effector that modulates the stimulus intensity, as in withdrawal from pain, pupillary contraction in bright light, or rotation ofthe head or pinna to enhance detection ofa particularsound. Nowwe turn to a more detailed look at individual sensory systems.

References

- Comparative-anatomy, science: britannica.com, Retrieved 15 April, 2019
- Respiratory-system-reptiles-anatomy-13100: ourreptileforum.com, Retrieved 19 March, 2019
- Reptilian-cardiovascular-anatomy-and-physiology-evaluation-and-monitoring-proceedings: veterinarycalendar.dvm360.com, Retrieved 10 January, 2019
- Circulatory-system, physiology-body-systems: poultryhub.org, Retrieved 31 August, 2019
- Circulatory-system-marine-mammals, marine-mammals: what-when-how.com, Retrieved 21 July, 2019
- Basic-structure-and-function-of-the-nervous-system, anatomyandphysiology-12-1: opentextbc.ca, Retrieved 08 June, 2019
- The-function-of-nervous-tissue anatomyandphysiology-12-3: opentextbc.ca, Retrieved 28 April, 2019

Reproductive System of Vertebrates

4 CHAPTER

A system consisting of reproductive organs, necessary to carry-on the generation of any living species is referred to as reproductive system. There are two modes of reproduction, sexual and asexual. This chapter has been carefully written to provide an easy understanding of the varied facts of reproduction in pisces, reptiles, amphibians, aves and mammals.

REPRODUCTION IN PISCES

The methods of reproduction in fishes are varied, but most fishes lay a large number of small eggs, fertilized and scattered outside of the body. The eggs of pelagic fishes usually remain suspended in the open water. Many shore and freshwater fishes lay eggs on the bottom or among plants. Some have adhesive eggs. The mortality of the young and especially of the eggs is very high, and often only a few individuals grow to maturity out of hundreds, thousands, and in some cases millions of eggs laid.

Males produce sperm, usually as a milky white substance called milt, in two (sometimes one) testes within the body cavity. In bony fishes a sperm duct leads from each testis to a urogenital opening behind the vent or anus. In sharks and rays and in cyclostomes the duct leads to a cloaca. Sometimes the pelvic fins are modified to help transmit the milt to the eggs at the female's vent or on the substrate where the female has placed them. Sometimes accessory organs are used to fertilize females internally-for example, the claspers of many sharks and rays.

In the females the eggs are formed in two ovaries (sometimes only one) and pass from the ovaries to the urogenital opening and to the outside. In some fishes the eggs are fertilized internally but shed before development takes place. Members of about a dozen families each of bony fishes (teleosts) and sharks bear live young. Many skates and rays also bear live young. In some bony fishes the eggs simply develop within the female, the young emerging when the eggs hatch (ovoviviparous).

Others develop within the ovary and are nourished by ovarian tissues after hatching (viviparous). There are also other methods utilized by fishes to nourish young within the female. In all live-bearers the young are born at a relatively large size and are few in number. In one family of primarily marine fishes, the surfperches from the Pacific coast of North America, Japan, and Korea, the males of at least one species appear to be born sexually mature, although they are not fully grown.

Some fishes are hermaphroditic, an individual producing both sperm and eggs, usually at different stages of its life. Self-fertilization, however, is probably rare. Successful reproduction and in many cases defense of the eggs and young is assured by rather stereotyped but often elaborate courtship and parental behaviour, either by the male or the female or both. Some fishes prepare nests by hollowing out depressions in the sand bottom (cichlids, for example), build nests with plant materials and sticky threads excreted by the kidneys (sticklebacks), or blow a cluster of mucus-covered bubbles at the water surface (gouramis). The eggs are laid in these structures. Some varieties of cichlids and catfishes incubate eggs in their mouths.

Some fishes, such as salmon, undergo long migrations from the ocean and up large rivers to spawn in gravel beds where they themselves hatched (anadromous fishes). Others undertake shorter migrations from lakes into streams or in other ways enter for spawning habitats that they do not ordinarily occupy.

A pair of bettas spawning under a bubble nest.

Fish reproductive organs include testes and ovaries. In most species, gonads are paired organs of similar size, which can be partially or totally fused. There may also be a range of secondary organs that increase reproductive fitness. The genital papilla is a small, fleshy tube behind the anus in some fishes, from which the sperm or eggs are released; the sex of a fish often can be determined by the shape of its papilla.

Anatomy

Testes

Most male fish have two testes of similar size. In the case of sharks, the testes on the right side is usually larger. The primitive jawless fish have only a single testis, located in the midline of the body, although even this forms from the fusion of paired structures in the embryo.

Under a tough membranous shell, the tunica albuginea, the testis of some teleost fish, contains very fine coiled tubes called seminiferous tubules. The tubules are lined with a layer of cells (germ cells) that from puberty into old age, develop into sperm cells (also known as spermatozoa or male gametes). The developing sperm travel through the seminiferous tubules to the rete testis located in the mediastinum testis, to the efferent ducts, and then to the epididymis where newly created sperm cells mature. The sperm move into the vas deferens, and are eventually expelled through the urethra and out of the urethral orifice through muscular contractions.

However, most fish do not possess seminiferous tubules. Instead, the sperm are produced in spherical structures called *sperm ampullae*. These are seasonal structures, releasing their contents during the breeding season, and then being reabsorbed by the body. Before the next breeding season, new sperm ampullae begin to form and ripen. The ampullae are otherwise essentially identical to the seminiferous tubules in higher vertebrates, including the same range of cell types.

In terms of spermatogonia distribution, the structure of teleosts testes has two types: in the most common, spermatogonia occur all along the seminiferous tubules, while in Atherinomorph fish they are confined to the distal portion of these structures. Fish can present cystic or semi-cystic spermatogenesis in relation to the release phase of germ cells in cysts to the seminiferous tubules lumen.

Ovaries

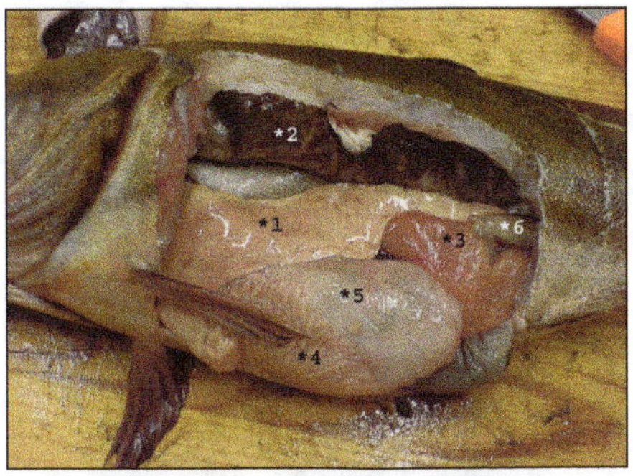

Ovary of a female Atlantic cod.

Many of the features found in ovaries are common to all vertebrates, including the presence of follicular cells and tunica albuginea There may be hundreds or even millions of fertile eggs present in the ovary of a fish at any given time. Fresh eggs may be developing from the germinal epithelium throughout life. Corpora lutea are found only in mammals, and in some elasmobranch fish; in other species, the remnants of the follicle are quickly resorbed by the ovary.] The ovary of teleosts is often contains a hollow, lymph-filled space which opens into the oviduct, and into which the eggs are shed. Most normal female fish have two ovaries. In some elasmobranchs, only the right ovary develops fully. In the primitive jawless fish, and some teleosts, there is only one ovary, formed by the fusion of the paired organs in the embryo.

Fish ovaries may be of three types: gymnovarian, secondary gymnovarian or cystovarian. In the first type, the oocytes are released directly into the coelomic cavity and then enter the ostium, then through the oviduct and are eliminated. Secondary gymnovarian ovaries shed ova into the coelom from which they go directly into the oviduct. In the third type, the oocytes are conveyed to the exterior through the oviduct. Gymnovaries are the primitive condition found in lungfish, sturgeon, and bowfin. Cystovaries characterize most teleosts, where the ovary lumen has continuity with the oviduct. Secondary gymnovaries are found in salmonids and a few other teleosts.

Eggs

Salmon eggs in different stages of development. In some only a few cells grow on top of the yolk, in the lower right the blood vessels surround the yolk and in the upper left the black eyes are visible.

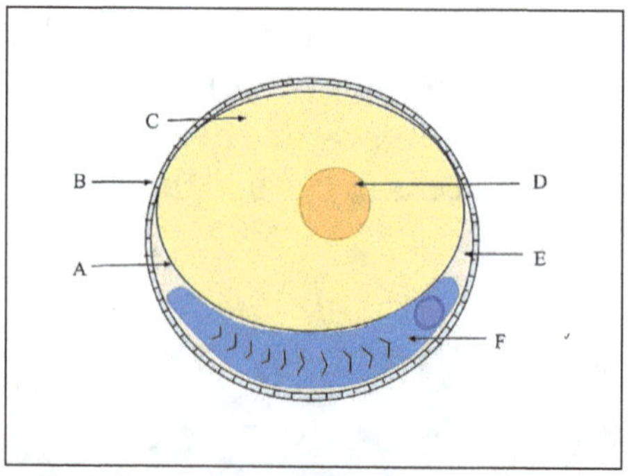

Diagram of a fish egg: A. vitelline membrane B. chorion C. yolk D. oil globule E. perivitelline space F. embryo.

Salmon Fry hatching. The larva has grown around the remains of the yolk and the remains of the soft, transparent egg are discarded.

The eggs of fish and amphibians are jellylike. Cartilagenous fish (sharks, skates, rays, chimaeras) eggs are fertilized internally and exhibit a wide variety of both internal and external embryonic development. Most fish species spawn eggs that are fertilized externally, typically with the male inseminating the eggs after the female lays them. These eggs do not have a shell and would dry out in the air. Even air-breathing amphibians lay their eggs in water, or in protective foam as with the Coast foam-nest treefrog.

Intromittent Organs

This male mosquitofish has a gonopodium, an anal fin which functions as an intromittent organ.

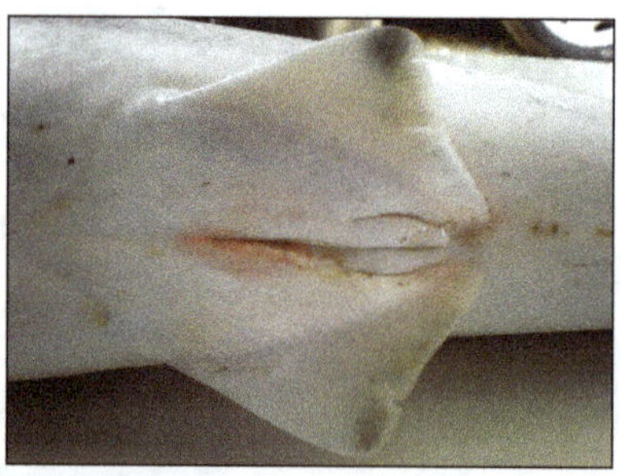

This young male spinner shark has claspers, a modification to the pelvic fins which also function as intromittent organs.

Male cartilaginous fishes (sharks and rays), as well as the males of some live-bearing ray finned fishes, have fins that have been modified to function as intromittent organs, reproductive appendages which allow internal fertilization. In ray finned fish they are called *gonopodiums* or *andropodiums*, and in cartilaginous fish they are called *claspers*.

Gonopodia are found on the males of some species in the Anablepidae and Poeciliidae families. They are anal fins that have been modified to function as movable intromittent organs and are used to impregnate females with milt during mating. The third, fourth and fifth rays of the male's anal fin are formed into a tube-like structure in which the sperm of the fish is ejected. When ready for mating, the gonopodium becomes erect and points forward towards the female. The male shortly inserts the organ into the sex opening of the female, with hook-like adaptations that allow the fish to grip onto the female to ensure impregnation. If a female remains stationary and her partner contacts her vent with his gonopodium, she is fertilized. The sperm is preserved in the female's oviduct. This allows females to fertilize themselves at any time without further assistance from males. In some species, the gonopodium may be half the total body length. Occasionally the fin is too long to be used, as in the "lyretail" breeds of *Xiphophorus helleri*. Hormone treated females may develop gonopodia. These are useless for breeding. Similar organs with similar characteristics are found in other fishes, for example the *andropodium* in the *Hemirhamphodon* or in the Goodeidae.

Claspers are found on the males of cartilaginous fishes. They are the posterior part of the pelvic fins that have also been modified to function as intromittent organs, and are used to channel semen into the female's cloaca during copulation. The act of mating in sharks usually includes raising one of the claspers to allow water into a siphon through a specific orifice. The clasper is then inserted into the cloaca, where it opens like an umbrella to anchor its position. The siphon then begins to contract expelling water and sperm.

Physiology

Oogonia development in teleosts fish varies according to the group, and the determination of oogenesis dynamics allows the understanding of maturation and fertilisation processes. Changes in the nucleus, ooplasm, and the surrounding layers characterize the oocyte maturation process.

Postovulatory follicles are structures formed after oocyte release; they do not have endocrine function, present a wide irregular lumen, and are rapidly reabsorbed in a process involving the apoptosis of follicular cells. A degenerative process called follicular atresia reabsorbs vitellogenic oocytes not spawned. This process can also occur, but less frequently, in oocytes in other development stages.

Some fish are hermaphrodites, having both testes and ovaries either at different phases in their life cycle or, as in hamlets, have them simultaneously.

Reproductive Strategies

In fish, fertilisation of eggs can be either external or internal. In many species of fish, fins have been modified to allow Internal fertilisation. Similarly, development of the embryo can be external or internal, although some species show a change between the two at various stages of embryo development. Thierry Lodé described reproductive strategies in terms of the development of the zygote and the interrelationship with the parents; there are five classifications - ovuliparity, oviparity, ovo-viviparity, histotrophic viviparity and hemotrophic viviparity.

Ovuliparity

Ovuliparity means the female lays unfertilised eggs (ova), which must then be externally fertilised. Examples of ovuliparous fish include salmon, goldfish, cichlids, tuna and eels. In the majority of these species, fertilisation takes place outside the mother's body, with the male and female fish shedding their gametes into the surrounding water.

Oviparity

Oviparity is where fertilisation occurs internally and so the female sheds zygotes (or newly developing embryos) into the water, often with important outer tissues added. Over 97% of all known fish are oviparous. In oviparous fish, internal fertilisation requires the male to use some sort of intromittent organ to deliver sperm into the genital opening of the female. Examples include the oviparous sharks, such as the horn shark, and oviparous rays, such as skates. In these cases, the male is equipped with a pair of modified pelvic fins known as claspers.

Marine fish can produce high numbers of eggs which are often released into the open water column. The eggs have an average diameter of 1 millimetre (0.039 in). The eggs are generally surrounded by the extraembryonic membranes but do not develop a shell, hard or soft, around these membranes. Some fish have thick, leathery coats, especially if they must withstand physical force or desiccation. These type of eggs can also be very small and fragile.

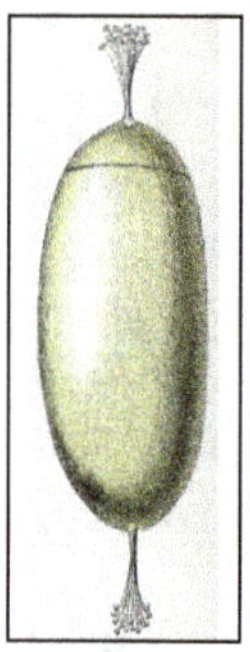
Egg of lamprey.

Egg of catshark (mermaids' purse).

Egg of bullhead shark.

Egg of chimaera.

The newly hatched young of oviparous fish are called larvae. They are usually poorly formed, carry a large yolk sac (for nourishment) and are very different in appearance from juvenile and adult specimens. The larval period in oviparous fish is relatively short (usually only several weeks), and larvae rapidly grow and change appearance and structure (a process termed metamorphosis) to become juveniles. During this transition larvae must switch from their yolk sac to feeding on zooplankton prey, a process which depends on typically inadequate zooplankton density, starving many larvae.

Ovoviviparit

In ovoviviparous fish the eggs develop inside the mother's body after internal fertilisation but receive little or no nourishment directly from the mother, depending instead on a food reserve inside the egg, the yolk Each embryo develops in its own egg. Familiar examples of ovoviviparous fish include guppies, angel sharks, and coelacanths.

Viviparity

There are two types of viviparity, differentiated by how the offspring gain their nutrients.

Histotrophic (tissue eating) viviparity means embryos develop in the female's oviducts but obtain nutrients by consuming other tissues, such as ova (oophagy) or zygotes. This has been observed primarily among sharks such as the shortfin mako and porbeagle, but is known for a few bony fish as well such as the halfbeak *Nomorhamphus ebrardtii*. An unusual mode of vivipary is adelphophagy or intrauterine cannibalism, in which the largest embryos eat weaker, smaller unborn siblings. This is most commonly found among sharks such as the grey nurse shark, but has also been reported for *Nomorhamphus ebrardtii*.

Hemotrophic (blood eating) viviparity means embryos develop in the female's (or male's) oviduct and nutrients are provided directly by the parent, typically via a structure similar to, or analogous to the placenta seen in mammals. Examples of hemotrophic fish include the surfperches, splitfins, lemon shark, seahorses and pipefish.

Aquarists commonly refer to ovoviviparous and viviparous fish as livebearers.

Hermaphroditism

Female groupers change their sex to male if no male is available.

An anemone fish couple guarding their anemone. If the female dies, a juvenile male moves in, and the resident male changes sex.

Hermaphroditism occurs when a given individual in a species possesses both male and female reproductive organs, or can alternative between possessing first one, and then the other. Hermaphroditism is common in invertebrates but rare in vertebrates. It can be contrasted with gonochorism, where each individual in a species is either male or female, and remains that way throughout their lives. Most fish are gonochorists, but hermaphroditism is known to occur in 14 families of teleost fishes.

Female anglerfish, Haplophryne mollis, with atrophied males attached.

Parthenogenesis was first described among vertebrates in the Amazon molly.

Usually hermaphrodites are *sequential*, meaning they can switch sex, usually from female to male (protogyny). This can happen if a dominant male is removed from a group of females. The largest female in the harem can switch sex over a few days and replace the dominant male. This is found amongst coral reef fishes such as groupers, parrotfishes and wrasses. It is less common for a male to switch to a female (protandry). As an example, most wrasses are protogynous hermaphrodites within a haremic mating system. Hermaphroditism allows for complex mating systems. Wrasses exhibit three different mating systems: polygynous, lek-like, and promiscuous mating systems. Group spawning and pair spawning occur within mating systems. The type of spawning that occurs depends on male body size. Labroids typically exhibit broadcast spawning, releasing high amounts of planktonic eggs, which are broadcast by tidal currents; adult wrasses have no interaction with offspring. Wrasse of a particular subgroup of the family Labridae, Labrini, do not exhibit broadcast spawning.

Less commonly hermaphrodites can be *synchronous*, meaning they simultaneously possess both ovaries and testicles and can function as either sex at any one time. Black hamlets "take turns releasing sperm and eggs during spawning. Because such egg trading is advantageous to both individuals, hamlets are typically monogamous for short periods of time–an unusual situation in fishes." The sex of many fishes is not fixed, but can change with physical and social changes to the environment where the fish lives.

Particularly among fishes, hermaphroditism can pay off in situations where one sex is more likely to survive and reproduce, perhaps because it is larger. Anemone fishes are sequential hermaphrodites which are born as males, and become females only when they are mature. Anemone fishes live together monogamously in an anemone, protected by the anemone stings. The males do not have to compete with other males, and female anemone fish are typically larger. When a female dies a juvenile (male) anemone fish moves in, and "the resident male then turns into a female and reproductive advantages of the large female–small male combination continue". In other fishes sex changes are reversible. For example, if some gobies are grouped by sex (male or female), some will switch sex.

The mangrove rivulus *Kryptolebias marmoratus* produces both eggs and sperm by meiosis and routinely reproduces by self-fertilization. Each individual hermaphrodite normally fertilizes itself when an egg and sperm that it has produced by an internal organ unite inside the fish's body. In nature, this mode of reproduction can yield highly homozygous lines composed of individuals so genetically uniform as to be, in effect, identical to one another. The capacity for selfing in these fishes has apparently persisted for at least several hundred thousand years.

Although inbreeding, especially in the extreme form of self-fertilization, is ordinarily regarded as detrimental because it leads to expression of deleterious recessive alleles, self-fertilization does provide the benefit of *fertilization assurance* (reproductive assurance) at each generation.

Sexual Parasitism

Sexual parasitism is a mode of sexual reproduction, unique to anglerfish, in which the males of a species are much smaller than the females, and rely on the females for food and protection from predators. The males give nothing back except the sperm which the females need in order to produce the next generation.

Some anglerfish, like those of the deep sea ceratioid group, employ this unusual mating method. Because individuals are very thinly distributed, encounters are also very rare. Therefore, finding a mate is problematic. When scientists first started capturing ceratioid anglerfish, they noticed that all the specimens were female. These individuals were a few centimetres in size and almost all of them had what appeared to be parasites attached to them. It turned out that these "parasites" were highly reduced male ceratioid anglerfish. This indicates the anglerfish use a polyandrous mating system.

The methods by which the anglerfish locate mates are variable. Some species have minute eyes unfit for identifying females, while others have underdeveloped nostrils, making it unlikely that they effectively find females using olfaction. When a male finds a female, he bites into her skin, and releases an enzyme that digests the skin of his mouth and her body, fusing the pair down to the blood-vessel level. The male becomes dependent on the female host for survival by receiving nutrients via their now-shared circulatory system, and provides sperm to the female in return. After fusing, males increase in volume and become much larger relative to free-living males of the species. They live and remain reproductively functional as long as the female stays alive, and can take part in multiple spawnings. This extreme sexual dimorphism ensures that when the female is ready to spawn she has a mate immediately available. Multiple males can be incorporated into a single individual female with up to eight males in some species, though some taxa appear to have a one male per female rule.

One explanation for the evolution of sexual parasitism is that the relative low density of females in deep-sea environments leaves little opportunity for mate choice among anglerfish. Females remain large to accommodate fecundity, as is evidenced by their large ovaries and eggs. Males would be expected to shrink to reduce metabolic costs in resource-poor environments and would develop highly specialized female-finding abilities. If a male manages to find a female parasitic attachment, then it is ultimately more likely to improve lifetime fitness relative to free living, particularly when the prospect of finding future mates is poor. An additional advantage to parasitism is that the male's sperm can be used in multiple fertilizations, as he stays always available to the female for mating. Higher densities of male-female encounters might correlate with species that demonstrate facultative parasitism or simply use a more traditional temporary contact mating.

Parthenogenesis

Parthenogenesis is a form of asexual reproduction in which growth and development of embryos occur without fertilization. In animals, parthenogenesis means development of an embryo from an unfertilized egg cell. The first all-female (unisexual) reproduction in vertebrates was described in the *Amazon molly* in 1932. Since then at least 50 species of unisexual vertebrate have been described, including at least 20 fish, 25 lizards, a single snake species, frogs, and salamanders. As with all types of asexual reproduction, there are both costs (low genetic diversity and therefore susceptibility to adverse mutations that might occur) and benefits (reproduction without the need for a male) associated with parthenogenesis.

Parthenogenesis in sharks has been confirmed in the bonnethead and zebra shark. Other, usually sexual species, may occasionally reproduce parthenogenetically, and the hammerhead and blacktip sharks are recent additions to the known list of facultative parthenogenetic vertebrates.

A special case of parthenogenesis is gynogenesis. In this type of reproduction, offspring are produced by the same mechanism as in parthenogenesis, however, the egg is stimulated to develop simply by the *presence* of sperm - the sperm cells do not contribute any genetic material to the offspring. Because gynogenetic species are all female, activation of their eggs requires mating with males of a closely related species for the needed stimulus. The Amazon molly, (pictured), reproduces by gynogenesis.

Others

The elkhorn sculpin (*Alcichthys elongatus*) is a marine teleost with a unique reproductive mode called "internal gametic association". Sperm are introduced into the ovary by copulation and then enter the micropylar canal of ovulated eggs in the ovarian cavity. However, actual sperm-egg fusion does not occur until the eggs have been released into sea water.

Inbreeding

Inbreeding Depression

The effect of inbreeding on reproductive behavior was studied in the poeciliid fish *Heterandria formosa*. One generation of full-sib mating was found to decrease reproductive performance and likely reproductive success of male progeny. Other traits that displayed inbreeding depression were offspring viability and maturation time of both males and females.

Exposure of zebra fish to a chemical environmental agent, analogous to that caused by anthropogenic pollution, amplified the effects of inbreeding on key reproductive traits. Embryo viability was significantly reduced in inbred exposed fish and there was a tendency for inbred males to sire fewer offspring.

The behaviors of juvenile Coho salmon with either low or medium inbreeding were compared in paired contests. Fish with low inbreeding showed almost twice the aggressive pursuit in defending territory than fish with medium inbreeding, and furthermore had a higher specific growth rate. A significant effect of inbreeding depression on juvenile survival was also found, but only in high-density competitive environments, suggesting that intra-specific competition can magnify the deleterious effects of inbreeding.

Inbreeding Avoidance

Inbreeding ordinarily has negative fitness consequences (inbreeding depression), and as a result species have evolved mechanisms to avoid inbreeding. Numerous inbreeding avoidance mechanisms operating prior to mating have been described. However, inbreeding avoidance mechanisms that operate subsequent to copulation are less well known. In guppies, a post-copulatory mechanism of inbreeding avoidance occurs based on competition between sperm of rival males for achieving fertilisation. In competitions between sperm from an unrelated male and from a full sibling male, a significant bias in paternity towards the unrelated male was observed.

Inbreeding depression is considered to be due largely to the expression of homozygous deleterious recessive mutations. Outcrossing between unrelated individuals results in the beneficial masking of deleterious recessive mutations in progeny.

Examples:

Female goldfish spawn (discharge) eggs into the water, encouraged by male goldfish who simultaneously discharge sperm which externally fertilizes the eggs.

Within two or three days, the vulnerable goldfish eggs hatch into larvae, and rapidly develo into fry Goldfish.

Goldfish, like all cyprinids, are egg-layers. They usually start breeding after a significant temperature change, often in spring. Males chase females, prompting them to release their eggs by bumping and nudging them. As the female goldfish spawns her eggs, the male goldfish stays close behind fertilizing them. Their eggs are adhesive and attach to aquatic vegetation. The eggs hatch within 48 to 72 hours. Within a week or so, the fry begins to assume its final shape, although a year may pass before they develop a mature goldfish colour; until then they are a metallic brown like their wild ancestors. In their first weeks of life, the fry grow quickly—an adaptation born of the high risk of getting devoured by the adult goldfish.

Carp

A member of the Cyprinidae, carp spawn in times between April and August, largely dependent upon the climate and conditions they live in. Oxygen levels of the water, availability of food, size of each fish, age, number of times the fish has spawned before and water temperature are all factors known to effect when and how many eggs each carp will spawn at any one time.

Siamese Fighting Fish

Prior to spawning, male Siamese fighting fish build bubble nests of varying sizes at the surface of the water. When a male becomes interested in a female, he will flare his gills, twist his body, and spread his fins. The female darkens in colour and curves her body back and forth. The act of spawning takes place in a "nuptial embrace" where the male wraps his body around the female, each embrace resulting in the release of 10-40 eggs until the female is exhausted of eggs. The male, from his side, releases milt into the water and fertilization takes place externally. During and after spawning, the male uses his mouth to retrieve sinking eggs and deposit them in the bubble nest (during mating the female sometimes assists her partner, but more often she will simply devour all the eggs that she manages to catch). Once the female has released all of her eggs, she is chased away from the male's territory, as it is likely that she'll eat the eggs due to hunger. The eggs then remain in the male's care. He keeps them in the bubble nest, making sure none fall to the bottom and repairing the nest as needed. Incubation lasts for 24–36 hours, and the newly hatched larvae remain in the nest for the next 2–3 days, until their yolk sacs are fully absorbed. Afterwards the fry leave the nest and the free-swimming stage begins.

Siamese fighting fish build bubble nests of varying sizes.

A pair of Siamese fighting fish spawning under their bubble nest.

One-day-old Siamese fighting fish larvae in a bubble nest - their yolk sacs have not yet been absorbed.

A 15-day-old free-swimming fry of a Siamese fighting fish.

Spawning and Sexual Strategies

Spawn is the eggs and sperm released or deposited into water by aquatic animals. As a verb, to spawn refers to the process of releasing the eggs and sperm, and the act of both sexes is called spawning. Most aquatic animals, except for aquatic mammals and reptiles, reproduce through the process of spawning.

The spawn (eggs) of a clownfish. The black spots are the developing eyes.

Spawn consists of the reproductive cells (gametes) of many aquatic animals, some of which will become fertilized and produce offspring. The process of spawning typically involves females releasing ova (unfertilized eggs) into the water, often in large quantities, while males simultaneously or sequentially release spermatozoa (milt) to fertilize the eggs.

Most fish reproduce by spawning, as do most other aquatic animals, including crustaceans such as crabs and shrimps, molluscs such as oysters and squid, echinoderms such as sea urchins and sea cucumbers, amphibians such as frogs and newts, aquatic insects such as mayflies and mosquitoes and corals, which are actually small aquatic animals—not plants. Fungi, such as mushrooms, are also said to "spawn" a white, fibrous matter that forms the matrix from which they grow.

There are many variations in the way spawning occurs, depending on sexual differences in anatomy, how the sexes relate to each other, where and how the spawn is released and whether or how the spawn is subsequently guarded.

Pacific salmon are semelparous or "big bang" spawners, which means they die shortly after spawning.

The pickled and dehydrated roe of mullet.

Marine animals, and particularly bony fish, commonly reproduce by *broadcast spawning*. This is an external method of reproduction where the female releases many unfertilised eggs into the water. At the same time, a male or many males release a lot of sperm into the water which fertilises some of these eggs. The eggs contain a drop of nutrient oil to sustain the embryo as it develops inside the egg case. The oil also provides buoyancy, so the eggs float and drift with the current. The strategy for survival of broadcast spawning is to disperse the fertilised eggs, preferably away from the coast into the relative safety of the open ocean. There the larvae develop as they consume their fat stores, and eventually hatch from the egg capsule into miniature versions of their parents. To survive, they must then become miniature predators themselves, feeding on plankton. Fish eventually encounter others of their own kind (conspecifics), where they form aggregations and learn to school.

Internally, the sexes of most marine animals can be determined by looking at the gonads. For example, male testes of spawning fish are smooth and white and account for up to 12% of the mass of the fish, while female ovaries are granular and orange or yellow, accounting for up to 70% of the fish's mass. Male lampreys, hagfish and salmon discharge their sperm into the body cavity where it is expelled through pores in the abdomen. Male sharks and rays can pass sperm along a duct into a seminal vesicle, where they store it for a while before it is expelled, while teleosts usually employ separate sperm ducts. Externally, many marine animals, even when spawning, show little sexual dimorphism (difference in body shape or size) or little difference in colouration. Where species are dimorphic, such as sharks or guppies, the males often have penis-like intromittent organs in the form of a modified fin.

A species is semelparous if its individuals spawn only once in their lifetime, and iteroparous if its individuals spawn more than once. The term semelparity comes from the *semel*, once, and *pario*, to beget, while iteroparity comes from *itero*, to repeat, and *pario*, to beget.

Semelparity is sometimes called "big bang" reproduction, since the single reproductive event of semelparous organisms is usually large and fatal to the spawners. The classic example of a semelparous animal is the Pacific salmon,which lives for many years in the ocean before swimming to the freshwater stream of its birth, spawning, and then dying. Other spawning animals which are semelparous include mayflies, squid, octopus, smelt, capelin and some amphibians.Semelparity is often associated with r-strategists. However, most fish and other spawning animals are iteroparous.

When the internal ovaries or egg masses of fish and certain marine animals are ripe for spawning they are called roe. Roe from certain species, such as shrimp, scallop, crab and sea urchins, are sought as human delicacies in many parts of the world. *Caviar* is a name for the processed, salted roe of non-fertilized sturgeon. The term *soft roe* or *white roe* denotes fish milt. Lobster roe is called *coral* because it turns bright red when cooked. Roe (reproductive organs) are usually eaten either raw or briefly cooked.

"The reproductive behaviour of fishes is remarkably diversified: they may be oviparous (lay eggs), ovoviparous (retain the eggs in the body until they hatch), or viviparous (have a direct tissue connection with the developing embryos and give birth to live young). All cartilaginous fishes—the elasmobranches (e.g., sharks, rays, and skates)—employ internal fertilization and usually lay large, heavy-shelled eggs or give birth to live young. The most characteristic features of the more primitive bony fishes is the assemblage of polyandrous (many males) breeding aggregations in open water and the absence of parental care".

There are two main reproduction methods in fish. The first method is by laying eggs and the second by *live-bearing* (producing their young alive).

In the first method, the female fish lays eggs either on the sea floor or on the leaves of an aquatic plant. A male fish fertilizes the eggs, and both then work together to protect the eggs/babies from danger until they can defend themselves.

In the second method, the male fish uses its anal fin to transmit sperm into the female fish and fertilize the fish eggs. Later, the female gives live birth to her fry.

Sexual Strategies

Basic Strategies

The four basic mating systems		
	Single female	Multiple females
Single male	Monogamy	Polygyny
Multiple males	Polyandry	Polygynandry

Cutthroat trout are monogamous pair spawners.

The anglerfish Haplophryne mollis is polyan-drous. This female is trailing the atrophied remains of males she has encountered

Monogamy occurs when one male mates with one female exclusively. This is also called *pair spawning*.Most fish are not monogamous, and when they are, they often alternate with non-monogamous behaviours. Monogamy can occur when feeding and breeding grounds are small, when it is difficult for fish to find partners, or when both sexes look after the young. Many tropical cichlids, which rear their young together in locations where they must fiercely defend against competitors and predators are monogamous. “In some pipefishes and seahorses, development of eggs takes a long time before the female can place them in the brood pouch of a male, where they are fertilized. While the male is pregnant, the female starts a new batch of eggs, which are ready at about the same time that the male gives birth to the young from the previous mating. This close timing of development promotes monogamy, especially if the likelihood of encountering another potential mate is low”.

Polygyny occurs when one male gets exclusive mating rights with multiple females. In polygyny a large conspicuous male usually defends females from other males or defends a breeding site. The females choose large males that are successfully defending prime breeding sites which the females find attractive. For example, sculpin males defend “caves” underneath rocks which are suitable for the incubation of embryos.

Another way males get to mate with several females is through the use of leks. Leks are places where many fish come together, and the males display to each another. Based on these displays, each female then selects the male they want to be their mate. For example, among the cichlid *Cyrtocara eucinostomus* in Lake Malawi, up to 50,000 large and colourful males display together on a lek four kilometres long. The females, which are mouth brooders, choose which male they want to fertilize their eggs.

Polyandry occurs when one female gets exclusive mating rights with multiple males. This happens among fish like clownfish that change their sex. It can also happen when males do the brooding but can cannot handle all the eggs the female produce, such as with some pipefish.

The males in some deep sea anglerfishes are much smaller than the females. When they find a female they bite into her skin, releasing an enzyme that digests the skin of their mouth and her body and fusing the pair down to the blood-vessel level. The male then slowly atrophies, losing first his digestive organs, then his brain, heart, and eyes, ending as nothing more than a pair of gonads, which release sperm in response to hormones in the female's bloodstream indicating egg release. This ensures that, when the female is ready to spawn, she has a mate immediately available. A single anglerfish female can "mate" with many males in this manner.

Polygynandry occurs when multiple males mate indiscriminately with multiple females. This mutual promiscuity is the approach most commonly used by spawning animals, and is perhaps the "original fish mating system." Common examples are forage fish, such as herrings, which form huge mating shoals in shallow water. The water becomes milky with sperm and the bottom is draped with millions of fertilized eggs.

Cuckoldry

Small male bluegill sunfishes cuckold large males by adopting *sneaker* or *satellite* strategies.

Female groupers change their sex to male if no male is available.

Alternate male strategies which allow small males to engage in cuckoldry can develop in species where spawning is dominated by large and aggressive males. Cuckoldry is a variant of polyandry, and can occur with *sneak spawners* (sometimes called *streak spawners*). A sneak spawner is a male that rushes in to join the spawning rush of a spawning pair. A spawning rush occurs when a fish makes a burst of speed, usually on a near vertical incline, releasing gametes at the apex, followed by a rapid return to the lake or sea floor or fish aggregation. Sneaking males do not take part in courtship. In salmon and trout, for example, *jack males* are common. These are small silvery males that migrate upstream along with the standard, large, hook-nosed males and that spawn by sneaking into a redd (spawning nest) to release sperm simultaneously with a mated pair. This behaviour is an evolutionarily stable strategy for reproduction, because it is favoured by natural selection just like the "standard" strategy of large males.

An anemone fish couple guarding their anemone. If the female dies, a juvenile male moves in, and the resident male changes sex.

Cuckoldry occurs in many fish species, including dragonets, parrotfishes and wrasses on tropical reefs and the bluegill sunfish in fresh water. Sneaker males that become too large to hide effectively become *satellite males*. With bluegill sunfish, satellite males mimic the behaviour and colouration of the females. They hover over a nest containing a pair of courting sunfish, and gradually descend to reach the pair just as they spawn. Males may need to be 6 or 7 years old to function capably as parental males, but may be able to function as sneaker or satellite males when they are as young as 2 or 3 years old. The smaller satellite and sneaker males may get mauled by the more powerful parental males, but they spawn when they are younger and they do not put energy into parental care.

Hermaphroditism

Hermaphroditism occurs when a given individual in a species possesses both male and female reproductive organs, or can alternate between possessing first one, and then the other. Hermaphroditism is common in invertebrates but rare in vertebrates. It can be contrasted with gonochorism, where each individual in a species is either male or female, and remains that way throughout their lives. Most fish are gonochorists, but hermaphroditism is known to occur in 14 families of teleost fishes.

Usually hermaphrodites are *sequential*, meaning they can switch sex, usually from female to male (protogyny). This can happen if a dominant male is removed from a group of females. The largest female in the harem can switch sex over a few days and replace the dominant male. This is found amongst coral reef fishes such as groupers, parrotfishes and wrasses. It is less common for a male to switch to a female (protandry). As an example, most wrasses are protogynous hermaphrodites within a haremic mating system. Hermaphroditism allows for complex mating systems. Wrasses exhibit three different mating systems: polygynous, lek-like, and promiscuous mating systems. Group spawning and pair spawning occur within mating systems. The type of spawning that occurs depends on male body size. Labroids typically exhibit broadcast spawning, releasing high amounts of planktonic eggs, which are broadcast by tidal currents; adult wrasses have no interaction with offspring. Wrasse of a particular subgroup of the family Labridae, Labrini, do not exhibit broadcast spawning.

Less commonly hermaphrodites can be *synchronous*, meaning they simultaneously possess both ovaries and testicles and can function as either sex at any one time. Black hamlets "take turns releasing sperm and eggs during spawning. Because such egg trading is advantageous to both individuals, hamlets are typically monogamous for short periods of time–an unusual situation in fishes." The sex of many fishes is not fixed, but can change with physical and social changes to the environment where the fish lives.

Particularly among fishes, hermaphroditism can pay off in situations where one sex is more likely to survive and reproduce, perhaps because it is larger. Anemone fishes are sequential hermaphrodites which are born as males, and become females only when they are mature. Anemone fishes live together monogamously in an anemone, protected by the anemone stings. The males do not have to compete with other males, and female anemone fish are typically larger. When a female dies a juvenile (male) anemone fish moves in, and "the resident male then turns into a female and reproductive advantages of the large female–small male combination continue".In other fishes sex changes are reversible. For example, if some gobies are grouped by sex (male or female), some will switch sex.

Unisexuality

Unisexuality occurs when a species is all-male or all-female. Unisexuality occurs in some fish species, and can take complex forms. *Squalius alburnoides*, a minnow found in several river basins in Portugal and Spain, appears to be an all-male species. The existence of this species illustrates the potential complexity of mating systems in fish. The species originated as a hybrid between two species, and is diploid, but not hermaphroditic. It can have triploid and tetraploid forms, including all-female forms that reproduce mainly through hybridogenesis.

It is rare to find true parthenogenesis in fishes, where females produce female offspring with no input from males. All-female species include the Texas silverside, *Menidia clarkhubbs*ias well as a complex of Mexican mollies. Parthenogenesis has been recently observed in hammerhead sharks and blacktip sharks. It is also known to occur in crayfishand amphibians.

Spawning Strategies

This section is patterned after a classification of the spawning behaviours of fish by Balon into reproductive guilds. This classification is based on how the eggs are fertilized (internal or external spawners), where the eggs are deposited (pelagic or benthic spawners), and whether and how the parents look after the eggs after spawning (bearers, guarders and nonguarders).

Nonguarders

Nonguarders do not protect their eggs and offspring after spawning.

Open Substrate Spawners

Nonguarders: Open substrate spawners:

- Pelagic spawners.
- Benthic spawners:
  - Spawners on coarse bottoms:
    - Pelagic free embryo and larvae.
    - Benthic free embryo and larvae.
  - Spawners on plants:
    - Obligatory.
    - Nonobligatory.
  - Spawners on fine substrates.
- Terrestrial spawners.

Pike usually spawn on vegetation flooded by high water. Their eggs adhere to the stems of plants.

Nonguarders: Brood hiders.

- Benthic spawners.
- Crevice spawners.
- Spawners on invertebrates.
- Beach spawners.

Bitterlings transfer responsibility for the care of their young to mussels.
This male bitterling is exhibiting spawning colours.

Open substrate spawners scatter their eggs in the environment. They usually spawn in shoals without complex courtship rituals, and males outnumber females.

Broadcast spawners: Release their gametes (sperm and eggs) into open water for external fertilisation. There is no subsequent parental care. About 75% of coral species are broadcasters, the majority of which are hermatypic, or reef-building corals.

- Pelagic spawners: A type of broadcast spawners, spawn in the open sea, mostly near the surface. They are usually pelagic fish such as tuna and sardines. Some demersal fish leave the bottom to spawn pelagically, particularly coral reef fish such as parrotfish and wrasses. Pelagic spawning means water currents widely disperse the young. The eggs, embryos and larvae of pelagic spawners contain oil globules or have a high water content. As a result, they are buoyant and are widely dispersed by currents. The downside is that mortality is high, because they can be eaten so easily by pelagic predators or they can drift into unsuitable areas. Females compensate by spawning large numbers of eggs and extending their spawning periods. Pelagic spawners that live in or around coral reefs can spawn a small number of eggs almost daily over a period of months. These fishes have complex breeding behaviours including sex changes, harems, leks and territoriality.
- Benthic spawners: Deposit their spawn on or near the bottom of the sea (or lake). They are usually demersal fish such as cod and flatfish. These species typically spawn without ceremony; they do not engage in elaborate courtship rituals. Each female is usually followed by several males who fertilize the eggs as they are released. Various strategies ensure the eggs and embryos remain in place, and do not drift with the current. The eggs can adhere to other eggs or to whatever they are deposited on, or the eggs can be laid in long strings which are wrapped around plants or rocks. Some eggs take on water after they are released, so they can be dropped into cracks where they swell and wedge themselves in place.
    - Egg scatterers: Scatter adhesive or non-adhesive eggs to fall to the substrate, into plants, or float to the surface. These species do not look after their brood and even eat their own eggs. These are often schooling fish which spawn in groups or pairs, often laying a large number of small eggs. The fry hatch quickly.
    - Egg depositers: Deposit eggs on a substrate (tank glass, wood, rocks, plants). Egg depositors usually lay fewer eggs than egg-scatterers, although the eggs are larger. Egg depositors fall into two groups: those that care for their eggs, and those that do not. Among egg depositors that care for their eggs are cichlids and some catfish. Egg depositors that care for their young can be divided into two groups: cavity spawners and open spawners.
    - Cavity spawners: Lay eggs in a cave or cavity. These fish form pairs and have advanced brood care where the eggs are defended and cleaned. The eggs take a few days to hatch, and the fry are often guarded by the parents. Various catfish, Cyprinidae, and killifish make up the majority. Cavity spawners can be contrasted with open (shelter) spawners, which lay their eggs on an open surface.

Brood Hiders

Brood hiders hide their eggs but do not give parental care after they have hidden them. Brood hiders are mostly benthic spawners that bury the fertilized eggs. For example, among salmon and trout the female digs a nest with her tail in gravel. These nests are called redds. The female then lays her eggs while the male fertilizes them, while both fish defend the redd if necessary from other members of the same species. Then the female buries the nest, and the nest site is abandoned. In North America, some minnows build nests out of piles of stones rather than dig holes. The minnow males have tubercles on their head and body which they use to help them defend the nest site.

- Egg buriers - can inhabit waters that dry up at some time of the year. An example are annual killifish which lay their eggs in mud. The parents mature quickly and lay their eggs before dying when the water dries up. The eggs remain in a dormant stage until rains stimulate hatching.

Bitterlings have a remarkable reproduction strategy where parents transfer responsibility for the care of their young to mussels. The female extends her ovipositor into the mantle cavity of the mussel and deposits her eggs between the gill filaments. The male then ejects his sperm into the mussel's inhalant water current and fertilization takes place within the gills of the host. The same female may use a number of mussels, and she deposits only one or two yellow, oval eggs into each. Early developmental stages are protected from predation within the body of the mussel. After 3 to 4 weeks larvae swim away from the host to continue life on their own.

Guarders

Guarders: Substrate spawners.

- Rock tenders.
- Plant tenders.
- Terrestrial tenders.
- Pelagic tenders.

Damselfish are substrate spawners. This one keeps her spawn in a gastropod shell.

Guarders: Nest spawners.

- Rock and gravel nesters.
- Sand nesters.
- Plant-material nesters.

    - Gluemakers.
    - Nongluemakers.
- Bubble nesters.
- Hole nesters.
- Misc-materials nesters.
- Anemone nesters.

The stickleback glues plant material to make its nest.

Guarders protect their eggs and offspring after spawning by practicing *parental care* (also called *brood care*). Parental care is an "investment by parents in offspring that increases the offspring's chances of surviving (and hence reproducing). In fish, parental care can take a variety of forms including guarding, nest building, fanning, splashing, removal of dead eggs, retrieval of straying fry, external egg carrying, egg burying, moving eggs or young, ectodermal feeding, oral brooding, internal gestation, brood-pouch egg carrying, etc".

Territorial behaviour is generally necessary for guarders, and the embryos are almost always guarded by males (apart from cichlids). There is a need to be territorial because looking after embryos usually includes defending the site where they are being looked after. It also often means there is competition for the best egg-laying sites. Elaborate courtship behaviour is usual among guarders.

Guarding males keep the embryos safe from predators, keep oxygen levels high by fanning water currents, and keep the area free from dead embryos and debris. They protect the embryos until they hatch, and often look after the larval stages as well. The time spent guarding can range from a few days to several months.

Substrate Spawners

Some guarders build nests (*nest spawners*) and some do not (*substrate spawners*), though the difference between the two groups can be small. Substrate spawners clean off a suitable area of surface suitable for egg laying, and look after the area, but they do not actively build a nest.

Baby paradise fish just hatched, gathered under the surface of a bubble nest.

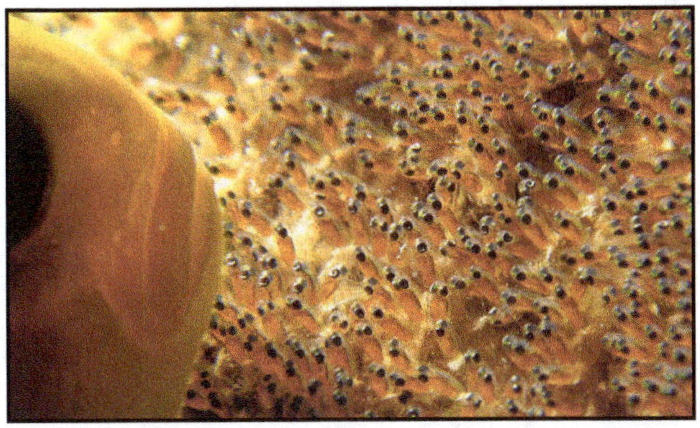

Anemone fish nest in an anemone. Here a male is protecting spawn produced by his partner.

Bearers

Bearers: External

- Transfer brooders.
- Auxiliary brooders.
- Mouth brooders.
- Gill-chamber brooders.
- Pouch brooders.

Male seahorses are pouch brooders.

A female cichlid mouthbrooding fry which can be seen looking out her mouth.

Bearers are fish that carry their embryos (and sometimes their young) around with them, either externally or internally.

External Bearers

Mouth brooders - carry eggs or larvae in their mouth. Mouth brooders can be ovophiles or larvophiles. Ovophile or egg-loving mouth-brooders lay their eggs in a pit, which are sucked up into the mouth of the female. The small number of large eggs hatch in the mother's mouth, and the fry remain there for a period of time. Fertilization often occurs with the help of egg-spots, which are colorful spots on the anal fin of the male. When the female sees these spots, she tries to pick up the egg-spots, but instead gets sperm that fertilizes the eggs in her mouth. Many cichlids and some labyrinth fish are ovophile mouth-brooders. Larvophile or larvae-loving mouth-brooders lay their eggs on a substrate and guard them until the eggs hatch. After hatching, the female picks up the fry and keeps them in her mouth. When the fry can fend for themselves, they are released. Some eartheaters are larvophile mouthbrooders.

Internal Bearers

Facultative Internal Bearers

The beginning of the evolutionary process of livebearing starts with *facultative* (optional) internal bearing. The process occurs in several species of oviparous (egg-laying) killifishes which spawn in the normal way on the substrate, but in the process accidentally fertilize eggs which the female retains and does not spawn. These eggs are spawned later, usually without allowing much time for embryonic development.

- Bearers: InternalFacultative internal bearers.
- Obligate internal bearers.
- Livebearers.

Guppies are livebearers. This one has been pregnant for about 26 days.

Obligate Internal Bearers

The next step in the evolution of livebearing is *obligate* (by necessity) internal bearing, where the female retains all the embryos. "The only source of nutrition for these embryos, however, is the egg yolk, as in externally spawned eggs. This situation, also referred to as ovoviviparity, is characteristic of marine rock fishes and the Lake Baikal sculpins. This strategy allows these fish to have fecundities approaching those of pelagic fish with external fertilization, but it also enables them to protect the young during their most vulnerable stage of development. By contrast, sharks and rays using this strategy produce a relatively small number of embryos and retain them for a few weeks to 16 months or longer. The shorter times spans are characteristic of species that eventually deposit their embryos in the environment, surrounded by a horny capsule; whereas the longer periods are characteristic of sharks that retain the embryos until they are ready to emerge as actively swimming young".

Viviparous Fish

However, some fish do not fit these categories. The livebearing largespring gambusia (*Gambusia geiseri*) was thought to be ovoviviparous until it was shown in 2001 that the embryos received nutrients from the mother.

Spawning Grounds

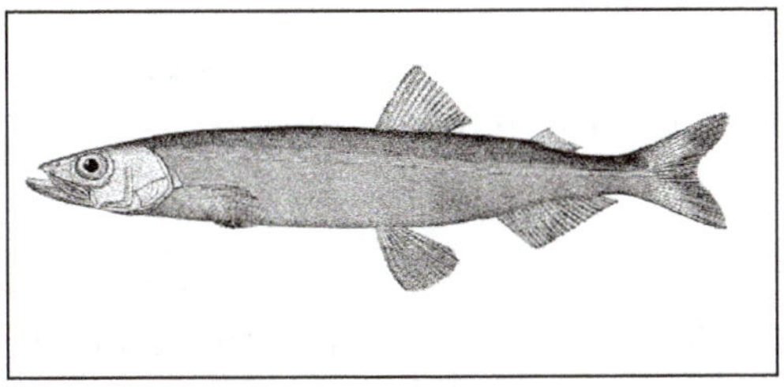

Capelin migrate huge distances to their spawning grounds.

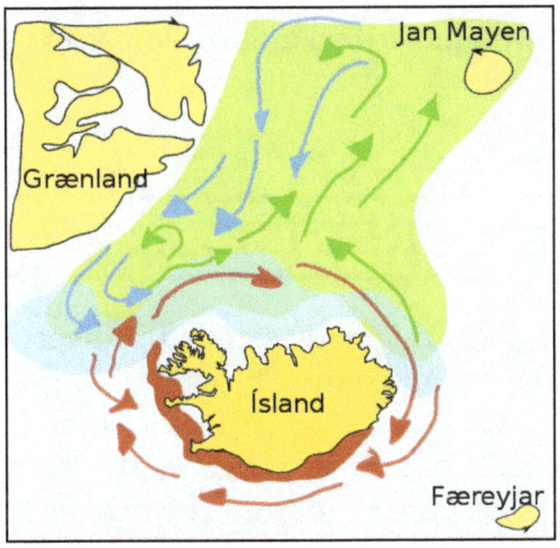

Migration of capelin around Iceland. Capelin on the way to feeding grounds is green, capelin on the way back is blue, the breeding grounds are red.

Spawning grounds are the areas of water where aquatic animals spawn, or produce their eggs. After spawning, the spawn may or may not drift to new grounds which become their nursery grounds. Many species undertake migrations each year, and sometimes great migrations, to reach their spawning grounds. For example, lakes and river watersheds can be major spawning grounds for anadromous fish such as salmon. These days, it is often necessary to construct fish ladders and other bypass systems so salmon can navigate their way past hydroelectric dams or other obstructions such as weirs on their way to spawning grounds. Coastal fish often use mangroves and estuaries as spawning grounds, while reef fish can find adjacent seagrass meadows that make good spawning grounds. Short-finned eels can travel anything up to three or four thousand kilometres to their spawning ground in deep water somewhere in the Coral Sea.

Forage fish often make great migrations between their spawning, feeding and nursery grounds. Schools of a particular stock usually travel in a triangle between these grounds. For example, one stock of herrings have their spawning ground in southern Norway, their feeding ground in Iceland, and their nursery ground in northern Norway. Wide triangular journeys such as these may be important because forage fish, when feeding, cannot distinguish their own offspring.

Capelin are a forage fish of the smelt family found in the Atlantic and Arctic oceans. In summer, they graze on dense swarms of plankton at the edge of the ice shelf. Larger capelin also eat krill and other crustaceans. The capelin move inshore in large schools to spawn and migrate in spring and summer to feed in plankton rich areas between Iceland, Greenland, and Jan Mayen. The migration is affected by ocean currents. Around Iceland maturing capelin make large northward feeding migrations in spring and summer. The return migration takes place in September to November. The spawning migration starts north of Iceland in December or January.

The diagram on the right shows the main spawning grounds and larval drift routes. Capelin on the way to feeding grounds is coloured green, capelin on the way back is blue, and the breeding grounds are red.

Referred to as "the greatest shoal on earth", the sardine run occurs when millions of sardines migrate from their spawning grounds south of the southern tip of Africa northward along the Eastern Cape coastline. Chinook salmon make the longest freshwater migration of any salmon, over 3,000 kilometres (1,900 mi) up the Yukon River to spawning grounds upstream of Whitehorse, Yukon. Some green sea turtles swim more than 2,600 kilometres (1,600 mi) to reach their spawning grounds.

REPRODUCTION IN REPTILES

Although anatomy and physiology varies greatly among the diverse class Reptilia, the basic structure of the reproductive tract is fairly consistent. The paired ovaries and testes, which range in color from yellow to grayish-pink, are located dorsomedially within the coelom although their exact location is species-specific. The right gonad sits cranial to the left, particularly in snakes. Females possess a right and left oviduct, but no true uterus. The oviduct empties directly into the cloaca through a genital papillae.

Secondary Sex Characteristics

The very first step of any signalment, determining your patient's gender, can come to a screeching halt with many reptiles. Observation of secondary sex characteristics can prove invaluable in determining gender.

The Jackson's or three-horned chameleon (Chamaeleo jacksoni xantholophus) is an example of sexual dimorphism. Only the male possesses the three rostral horns shown here.

Common Secondary Sex Characteristics in Reptiles

Lizards	• Males often have a bigger, more robust appearance overall and a larger head. • The male Jackson's chameleon has three rostral horns. • Femoral pores are larger, more prominent in iguanas and geckos. • The presence of the hemipenes can create a dramatic bulge at the tail base of some male lizards. • Many male iguanid lizards posses enlarged postanal scales.
Snakes	• Spurs are generally larger in male boids.
Turtles and tortoises	• The plastron or bottom shell is often concave in males. • The tail is usually larger and longer in males. • The iris of the male Eastern box turtle is frequently red or bright orange. Females possess a brown or light colored iris.
The external spurs are remnants of the boid's vestigial pelvic girdle. The spurs are located just lateral to the vent or the opening of the cloaca.	

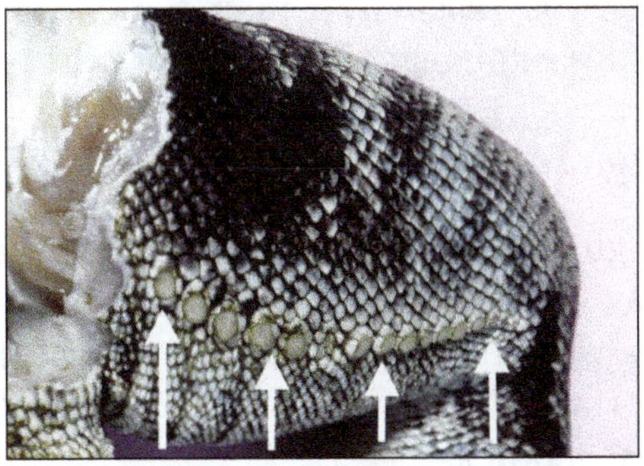

Femoral pores (arrows) are larger and more prominent in some male lizards such as this green iguana (Iguana iguana) shown here.

In many instances, clinical techniques (Box) must be performed to determine a patient's sex. Cloacal probing is the most common technique performed in snakes.

Clinical Sexing Techniques in the Reptile

- Cloacal probing.
- Hydrostatic eversion of hemipenes.
- Manual eversion of hemipenes.
- Ultrasound.
- Radiographs.
- Endoscopy (or surgery).

Oviparity or Egg Laying

All chelonians (turtles and tortoises), pythons, most colubrids, monitors, and most iguanas lay eggs. Eggs can often be detected during visual examination in many species of snakes, although larger species like pythons require palpation. Viviparous or live bearing reptiles include many skinks, chameleons, boas, all rattlesnakes, and most vipers.

Egg Production

The process of egg production in reptiles is very similar o that of birds.

- Rising estrogen levels promote vitellogenesis or yolk production within the liver
- Yolk is then transported via the blood stream to the reproductive tract where it is deposited in oocytes. During this time, lab results in female reptiles frequently reveals high levels of triglycerides, hypercholesterolemia, and hypercalcemia.
- Triggers for ovulation are poorly understood in the reptile, but when ovulation > does occur, oocytes enter the oviduct and become fertilized. (Females of some species can store sperm for an extended period and produce fertile eggs after being separated from males for a long time).
- As oocytes move through the oviduct, albumen is added and the shell is layed down.

Depending on the species, the egg can normally remain within the oviduct anywhere from 9 days to 6 months, however most reptiles retain eggs for 1 to 2 months.

Many females can Lay Eggs in the Absence of a Male

Many lizard species, including iguanas, geckos, and some snakes such as the corn snake can begin folliculogenesis and even lay eggs without the presence of a mate. Although these eggs are frequently infertile, this is not always the case since some species can store sperm for extended periods of time. This fact is important to keep in mind, since it means egg-related problems cannot be ruled out in many "single", pet reptiles.

Thinly Shelled Eggs

The thinly shelled eggs produced by many normal reptiles may surprise anyone familiar with the hard, strong avian eggshell. All snakes, most lizards, and some turtles produce thinly shelled eggs that are normally pliant. The eggs of tortoises, many geckos, and crocodilians are more rigid.

Phallus or Hemipenes

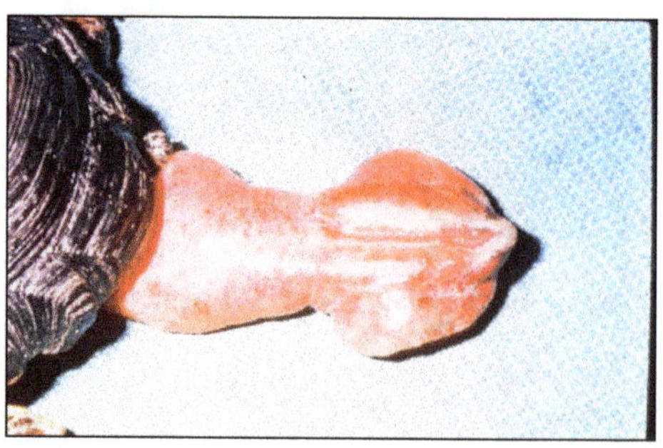

Normal extended phallus in a box turtle (Genus Terrapene).

The chelonian phallus sits in a groove on the cloacal floor. When engorged, the phallus extends cranioventrally.

Male snakes and lizards posses paired copulatory organs called hemipenes that normally range in color from pink to black. Hemipenes sit within pouches at the base of the tail just caudal to the cloaca. They are connected to the testes by the ductus deferens, and each hemipenis is functionally complete. During copulation, only one hemipenis everts into the female's cloaca. Sperm travels along a groove on the outside of the everted organ.

Hemipenes of a Western diamondback rattlesnake (Crotalus atrox).

Hemipenes in a green iguana (Iguana iguana) cadaver.
Hemipenes are generally pink in the live specimen.

Sexual Maturity

Sexual maturity is determined primarily by size. The time it takes to reach a particular size can vary dramatically with husbandry, therefore sexual maturity is determined primarily by size as opposed to age in reptiles. Denardo (2006) has stated that under "optimal conditions":

- Snakes usually reach sexual maturity at 2-3 years.
- Small lizards at 1-2 years.
- Large lizards at 3-4 years.
- Chelonians at 5-7 years.

Breeding Season

Most reptiles possess a distinct breeding season. Although some species reproduce throughout the year, most reptiles have a distinct breeding season. Although rainfall can be an important trigger for equatorial species, a change in temperature serves as the most important environmental stimulus in temperate zone species, with reproductive activity beginning after a period of seasonal cooling. Tropical boids are a notable exception to this rule of thumb since they tend to breed during cooler periods of the year. In the United States, reproductive activity in the green iguana (Iguana iguana) usually occurs between October and April.

The Normal Gravid Lizard

The gravid lizard frequently produces anywhere from 20 to 40 eggs. As ovarian follicles enlarge, they eventually fill the entire coelom compressing the gastrointestinal tract and frequently resulting in anorexia. Prominent coelomic distension may be seen and individual eggs may even become visible against the body wall.

Iguana owners frequently describe loss of appetite 1 to 4 weeks before egg laying. Weight loss and some loss of body condition (reduced muscle mass over the pelvis, shoulders and limbs) can also be detected as eggs continue to mature. A behavioral change is also common. The lizard first becomes hyperactive, scratching and digging in various substrates as she intensely searches for an appropriate nesting site. Eventually she will become less active, and will spend more time basking. Aggression can also be observed.

Possible Findings in the Gravid Lizard

- Anorexia.
- Restlessness, hyperactivity at first.
- Reduced activity later.
- Aggression.
- Coelomic distension.
- Weight loss.
- Loss of muscle mass.
- Female chameleons often become darker in color.

Differences in the clinical appearance of a normal gravid lizard and one suffering from follicular stasis or dystocia can be subtle. Frequently the most obvious difference between a gravid lizard and one in dystocia is her attitude. The gravid female is alert, aware, and active (as possible) while the female in follicular stasis or dystocia may be lethargic, persistently anorectic, and possibly emaciated.

REPRODUCTION IN AMPHIBIANS

Male Dendropsophus microcephalus calling.

Sexual selection in amphibians involves sexual selection processes in amphibians, including frogs, salamanders and newts. Prolonged breeders, the majority of frog species, have breeding seasons at regular intervals where male-male competition occurs with males arriving at the waters edge first in large number and producing a wide range of vocalizations, with variations in depth of calls the speed of calls and other complex behaviours to attract mates. The fittest males will have the deepest croaks and the best territories, with females known to make their mate choices at least partly based on the males depth of croaking. This has led to sexual dimorphism, with females being larger than males in 90% of species, males in 10% and males fighting for groups of females.

There is a direct competition between males to win the attention of the females in salamanders and newts, with elaborate courtship displays to keep the females attention long enough to get her interested in choosing him to mate with. Some species store sperm through long breeding seasons, as the extra time may allow for interactions with rival sperm.

Selection in Salamanders

Smooth newt.

Italian crested newt.

Red-backed salamander.

Important aspects of courtship displays include male alert posture, approach and pursuit, the marking of the substrate trail, and tail undulations. Once a male has the attention of the female, he may release pheromones, which are used for individual recognition, mate recognition, and mate choice decisions, and generally increase female receptivity towards the male's spermatophore and improve his chances of reproductive success with her. In the red-legged salamander *(Plethodon shermani)*, males directly apply pheromones onto the female's nares. In the aquatic smooth newt (*Lissotriton vulgaris*), the pheromones are delivered indirectly; the male wafts them towards the female with his tail. In some species, it has been shown that the size a larger, more elaborate male's tail has played a role in the female's decision to mate. In the newt species of the Italian crested newt (*Triturus carnifex*), males that possessed large mid-dorsal body crests were more favorable to the females looking to mate.

Female choice is the final step in the males attempt to reproduce; if the female does not find the male's traits desirable, reproduction will not be successful. As it is in most species, body size is an important and most common desired trait amongst females. Body size is indicative of fitness and genetic quality; and because it is easily observable, it can be an indicator for how a female will pick her mate. A larger body size positively correlates to higher paternity; thus larger males are more successful in siring more offspring than smaller males. Male salamanders also prefer larger females because they produce more ova than smaller females.

In sexual coercion, a form of male sexual intimidation which includes forced copulations, repeated mating attempts with the same female, and punishing resisting females. males forcibly mate with females; an example is the red-backed salamander (*Plethodon cinereus*). It has also been proposed as a third form of sexual selection, along with intersexual selection and intrasexual selection, even though prevailing scientific opinion is that sexual coercion is a manifestation of sexual conflict, stemming out of the natural discord between males and females. In some species of salamanders, females that are territorial and aggressive also intimidate males. *Plethodon cinereus* females aggressively punish males who are exhibiting social polygyny. Males have been shown to punish females that partake in mating polyandry, forcing them into mating monogamously.

Populations of salamanders tend to be heavily male-biased, which results in a higher number of males vying for access to the lesser number of females. Males will use different tactics in an effort to keep other males away from the female of interest and ensure their own paternity and reproductive success, such as sexual defense, which is used to steal other males spermatophore, which may have already deposited into the female, and sexual interference, when a male pushes the female away from another courting male, such as sperm capping, whereby a competing male lays his spermatophore directly on top of those previously deposited by another male. Females also participate in sexual interference: a rival female places herself between the courting male and the female he is courting, and attempts to engage the male in their own courtship dance, a tactic for avoiding sperm competition. The male removes the sperm from a female who had mated with a previous male: many species of salamanders will remove the spermatophore from the previous male and insert his own; thus, ensuring his paternity.

Sperm storage is a method by which females can control the sperm that fertilizes their eggs. Female salamanders and newts, most notably *Triturus vulgaris*, possess sperm-storage glands called spermathecae in the walls of their cloacae. Sperm storage has evolved independently to control reproduction in response to the environment. Females have the ability to store multiple spermatophores until they are needed for fertilization. Sperm storage leads to the post-copulatory sexual selection process of cryptic female choice, which occurs when a female removes the deposited sperm or favors one set of spermatophore over another, during or after copulation. Spermatophores are composed of a cap containing the spermatozoa and a gelatinous support arrangement which fastens the spermatophore to a substrate. Spermatophores are very central to reproductive success; therefore, they are handled carefully. Mating success has a direct correlation to favorable phenotypic traits such as tail loss, adult body size, and age.

Vocal signaling in frogs

Gray tree frog.

Glass frog..

The vocalizations that male frogs use to call for females are crucial for reproductive success in frogs. Initially, male frogs will produce advertisement calls, considered a pre-mating isolating mechanism because females use them to isolate conspecific males, as cited in which convey information about species, sexual readiness, and size, with females moving towards the call they find most attractive. In *Hyla versicolor*, females favor long calls rather than short callswhile in *Dendropsophus microcephalus* they are attracted to short and high rate calling.

Northern cricket frog.

Boophis madagascariensis.

If a female shows interest male frogs will switch to courtship calls, some of which are simply an increase in rate, as cited by Wells and Schwartz, which can contain fifty percent more notes than advertisement calls, and can become more complex when a male is at a specific oviposition site as cited by the glass frog (*Hyalinobatrachium fleischmanni*) from Panama gives long calls when a female frog is first seen and will switch to multiple staccato chirps once a female indicates interest.

Since anurans are highly sensitive to motion, females use conspicuous movements to communicate interest. A female frog may also send vocal signals of her own to males, motivating copulation and competitive behaviors. Female Iberian midwife toads (*Alytes cisternasii*) use courtship calls to distinguish themselves from other male competitors. The African common platanna (*Xenopus laevis*) uses duet calling when males are not visible, allowing females to distinguish territorial males from satellite males. The Emei music frog (*Babina daunchina*) female calling stimulates male movement towards sound or phonotaxis.

Calling is an energetically expensive activity, requiring high levels of aerobic respiration. Selection favors the most efficient call production, a good quality call that does not require high levels of energy. Allowing a male frog to be able to call continuously for females. *Dendropsophus microcephalus*, a small tree frog, will produce 100 notes per minute for several hours, requiring up to 25 times more oxygen. Some frogs produce less energetic calls and remain in chorus for longer periods of time with the ability to remain active longer being better for mating success than a higher energy vocal display.

The morphological and biochemical processes of call production are influenced by sexual selection. Male frogs have developed vocal sacs that enable louder calls, with the larger tympanum increasing the radiation of the vocal signal. Males also have larger trunk muscles with highly aerobic muscle fibers, many mitochondria, and fat storage.

The sensory exploitation hypothesis states that females have natural preexisting preferences for sexual signals that have been favored by sexual selection, with males producing calls that are known to be favorable for females. A recent study of advertisement calls from 72 species of anurans has indicated that most call features in males show phylogenetic correlation, indicating that species-specific evolution of calls might be constrained by a lack of genetic variation or because female preferences for call features have diverged to only a limited extent. Male anurans will often alter the rate of calling in order to intimidate other males though females tend to favor advertisement and courtship calls over aggressive ones, indicating that aggressive vocalizations are a form of male- male competition.

Advertisement and aggressive calls are usually similar in frequency but produced at different rates. *Geocrinia victoriana*, an eastern Australian smooth frog, has a long aggressive note to ward off competitors and shorter notes to attract females. In Northern cricket frogs (*Acris crepitans*) male aggressive calls serve a dual purpose of attracting females and intimidating other male callers. The Madagascar bright eyed frog (*Boophis madagascariensis*) is able to produce 28 different types of aggressive calls by modifying pulse rate and notes.

Males can intercept vocal signals that were not meant for them in a tactic known as eavesdropping. The *Hyla versicolor* male increases calling rate and duration once a female is in vicinity. Eavesdroppers will intercept these increased signals and begin to call for the female in a more aggressive manner, which generally leads to escalated conflicts. Eavesdropping also allows male frogs to assess the body size of their competitors. The frequency of vocal signals is dependent on the mass of the laryngeal apparatus, and a large one will produce lower frequency calls, positively correlated to overall body size. Male frogs will use eavesdropping to decide whether or not to initiate aggressive conflicts. Male anurans are more likely to attack a frog with higher frequency calls. It can also inform frogs when the competition has become too intense, where it can reduce or stop calling altogether and move to a different location outside of the chorus. In order to save energy they attempt to call females where there is less competition and a better chance to court a female. Once a satellite male hears a change in vocalizations of other males he will begin courtship calling in an attempt to steal a female, with some males now invests more energy into a competitive courtship call which may sway the female's interest. Satellite males may contribute 20% of sexual reproduction in a populations.

Polygynandry in Frogs

Allobates femoralis.

Bibron's toadlet.

Anuran mating systems are promiscuous: females and males breed with multiple mates and females exhibit sequential polyandry when one female is fertilized by more than one male at different times. Anurans are external fertilizers so there is no sperm competition. Polyandry has benefits for females such as fertility insurance, paternal care insurance and dispersal of mortality risk. Female anurans have a high risk of mating with sperm depleted males due to frequent mating and tend to select the same few males displaying fitness qualities. The requirement for successful copulation is that the male and female gametes have to be released at the same time and at the right orientation.

In anurans, 70% of polyandrous species exhibit paternal care. In *Allobates femoralis* the female's offspring survival depends on the male's carrying abilities, as they are responsible for carrying tadpoles from terrestrial to aquatic sites. If a female only mates with one male and he is a poor parent, she may lose all of her offspring. Females reduce the risk of losing all of their offspring because it is likely that some of the males they mated with make good parents. Sequential polyandry will allow females to have a multiple clutches in different nest sites, reducing the risk of abiotic factors, such as temperature, or biotic factors like predators affecting development of the offspring.

Some male behaviors that may have led to polyandry are displacement, where a male is gripped onto and mating with a female, another male may make an effort to mate as well and displace the male that was originally copulating; forced copulation, where there is a high density of aggressive males, and female abandonment, where a female but spots a better one to mate with and abandons the less favorable one. This can lead to sequential polyandry if the male leaves before the entire clutch is released and an additional male fertilizes these eggs as well. In Bibron's toadlet (*Pseudophryne bibronii*) females spread out their eggs across the nests of two to eight males, with those who exhibited higher levels of polyandrous behavior having higher mean offspring survivorship, showing that insurance against nest failure and dispersal of mortality risk is related to the evolution of polyandry.

Dimorphism and morphology

Small male and large female Fleay's barred frog (Mixophyes fleayi) in amplexus.

Male tusked frog.

Sexual dimorphism is seen as size difference between genders in salamanders[and in 90% of anuran species. Larger females is a result of selection for fecundity, including egg size or clutch size as they can store more energy and produce more offspring. Females are never more than twice the size of males.

Size is important in male-male competition, with larger males having an advantage, due to the "male combat hypothesis" which suggests that male combative behavior is positively related to male body size. Males in the tusked frog (*Adelotus brevis*) and the fanged frog (*Limnonectes kuhlii*) use fangs and tusks as weapons to defend their calling sites. Larger body size and larger weaponry leads to higher reproductive rates. However, data suggest that this male mechanism only operates in superfamily Dentrobatoidea, possibly due to alternative mating techniques that reduce selection pressure on body size. Large African bullfrog (*Pyxicephalus adspersus*) males defend territories while small intimidated males move away and become satellite males. These smaller males use eavesdropping techniques and mate with females who are in range of their satellite location.

Sexual Role Reversal in Dendrobates auratus

Poison dart frog Dendrobates auratus.

Initial observation of mating in the green poison-dart frog *Dendrobates auratus* shows convincing evidence of sex-role reversal. Small female clutch size, female-female competition, and large parental care investment by males are characteristic of sexual role reversal. To test whether this sexual role reversal was true, two hypotheses were tested; the sexual role reversal hypothesis and the parental quality hypothesis.

For the sexual role reversal hypothesis to be true, females would compete for as many males as possible, and this was not supported because male-male competition for mates was as common as female-female competition. In sexual role reversal males usually become more selective; however, in *Dendrobates auratus* females were more selective when choosing males, supporting the parental quality hypothesis, which predicted that females would compete for the parental care of particular males. Females were defensive of their mating partners and used courtship techniques to prevent males from mating with other females and destroyed rivals' eggs. These observations support the parental quality hypothesis. Summers weakens the sexual role reversal hypothesis and strengthens the case parental quality selection.

REPRODUCTION IN AVES

Bird reproduction starts the same way as in mammals by the joining of an egg or ovum with a sperm cell in the oviduct. This fertilised ovum then forms the nucleus of the egg and the formation of the yolk, whites and shell can begin. The sperm from the male is transferred to the female when they touch cloaca, an act referred to as the cloacal kiss.

Egg Formation

The first animals to develop the use of eggs with shells were the reptiles, in order that the egg would not dry out when not laid in water. With birds evolving from reptiles they have also made use of this incredible adaptation although they have taken it even further developing the egg into an amazing life support system for the growing embryo.

Most animals have two ovaries and oviducts, the passage for the egg to travel to the uterus. Birds however only have one and as the fertilised ovum travels down it a layer of albumen is applied around it, followed by the shell membranes. The shell itself is formed in the uterus where, depending on the movement of the egg, pigmentation forms patterns in the shell material.

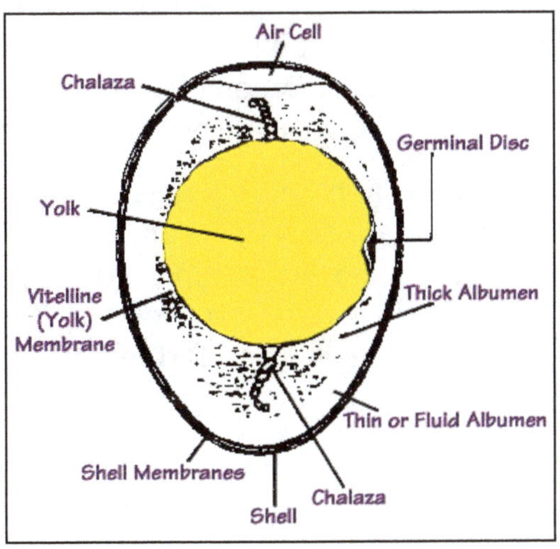

Egg Structure

An egg comprises several main elements, the yolk, the albumen, the membranes and the shell. The yolk is a fat rich food store with a small white spot, known as the germinal spot, and it is from here that the embryo begins to develop.

The whites of the egg, the albumen, supplies the embryo with water, provides some shock protection and helps gases to move in and out of the egg during development.

The membranes are three in number, one around the yolk known as the vitelline membrane, and two around the albumen, the inner and outer shell membranes. They provide a barrier against bacterial penetration and the air sac or air cell forms between the two shell membranes.

The shell is comprised of protein fibres and calcium carbonate or chalk merged into a complex mesh like structure and despite its smooth appearance is permeable to gases. It allows oxygen through to the growing embryo and waste gases and moisture to escape the egg. As the embryo develops the air sac increases in size and allows the mature chick its first breath.

There are also two twisted strands in the albumen attached to the yolk, called the chalaza, which hold the yolk in place and suspend the growing embryo in safety.

The yolk and albumen provide much for the developing embryo and all it needs from the outside world is oxygen and warmth.

The shape of most eggs is, obviously, egg shaped, but there are those that are adapted to suit particular circumstances. As an example cliff nesting birds have more pointed eggs, which makes them roll in a tight circle so they don't fall off the cliff if disturbed.

Laying Schedule

The timing and number of clutches of eggs produced by different species varies dependent upon circumstances. The Blue Tit, for instance, feed their young mainly on insect larvae and so tend to produce one large clutch in early summer when this food source is more abundant. Conversely the Blackbird will lay two or three smaller clutches as they have a more abundant, steady food source and therefore have longer to raise multiple clutches.

Both of these situations however still share the same incubation behaviour. The female will lay one egg per day until the clutch is complete and it is only at this point that incubation starts. Even though the eggs may have different ages, this method of incubation results in all the young hatching around the same time and all have an equal chance at feeding time and for survival.

Some species such as birds of prey as well as others lay eggs two or three days apart and start incubation as soon as the egg is laid. This means that the young hatch at different intervals and when food is scarce the youngest hatchlings will not be able to compete for food and will not survive. In hard times such as this these fatalities will even become food for its erstwhile brothers and sisters and although this seems a bit gruesome after all the effort invested in breeding it helps prevent all the chicks being lost in harsh conditions.

Camouflage

The colour of the eggs varies with different species and nesting habits. Birds that build their nests in holes tend to have light coloured eggs, white or blue, that are easier to locate in dark conditions and so avoid accidental damage. The hidden nature of the nesting site provides the camouflage. Ground laying birds, such as plovers, produce highly camouflaged eggs to prevent. Although Cuckoos already produce variably coloured and patterned eggs they do have the ability to mimic the appearance of other species eggs.

Incubation

In most cases it is the female parent that incubates the eggs and to help transfer heat effectively she will develop a brood patch as the breeding season approaches. This brood patch has an area of skin with densely packed blood vessels and hardly any feathers, which produces more heat, due to the blood vessels, and this is transferred more easily to the egg due to the lack of feathers. Eggs are also turned periodically to ensure an even distribution of warmth. In species where the males also incubate eggs they will also develop a brood patch but they spend much less time at this than females. In all cases, however, the brood patch will disappear at the end of the breeding season.

Hatching

It is possible that a developed chick may be able to communicate with its parent a day or two before hatching by answering its parents call with some vocal sounds. It will then start to use the hard tip of its bill, called the egg-tooth, to break out of the egg by laying on its back and pushing the egg-tooth into the shell above it. Once it has weakened the shell enough it will try to push the two halves of the eggshell apart. This may take a few days to accomplish and after a few days of hatching, the young lose the egg-tooth.

Nestlings

Hatchlings fall into two main categories, nidicolous and nidifugous. Nidicolous hatchlings are practically helpless, blind and naked, are wholly dependent on their parents and do not leave the nest until fledged. They tend to belong to species with the shorter incubation such as Blue Tit.

Nidifugous hatchlings are born with a covering of downy feathers and are able the leave the nest and fend for themselves. Many species of duck, Moorhen and Coot have nidifugous young.

Juveniles

Juvenile birds are those that have developed their first flight feathers and are able to leave the nest and fend for themselves. They may look very much like their parents, as in the Wren, or may be considerably different, as in the Robin.

Disturbing Birds

It is against the law in the UK to disturb wild birds or their eggs, punishable by fines and even imprisonment. There are individuals who collect eggs but this is wholly illegal, not to mention immoral and destructive.

Parental Care

In birds, 90% to 95% of species are monogamous, meaning the male and female remain together for breeding for a few years or until one mate dies. Birds of all types, from parrots to eagles and falcons, are monogamous. Usually, the parents take turns incubating the eggs. Birds usually incubate their eggs after the last one has been laid. In polygamous species, where there is more than one mate, one parent does all of the incubating. The wild turkey is an example of a polygamous bird.

The length and type of parental care varies widely amongst different species of birds. At one extreme, in a group of birds called the magapodes (which are chicken-like birds), parental care ends at hatching. In this case, the newly-hatched chick digs itself out of the nest mound without parental help and can take care of itself right away. These birds are called precocial. Other precocial birds include the domestic chicken and many species of ducks and geese. At the other extreme, many seabirds care for their young for extended periods of time. For example, the chicks of the Great Frigatebird receive intensive parental care for six months, or until they are ready to fly, and then take an additional 14 months of being fed by the parents. These birds are the opposite of precocial birds and are called altricial.

In most animals, male parental care is rare. But it is very common in birds. Often both parents share tasks such as defense of territory and nest site, incubation, and the feeding of chicks. Since birds often take great care of their young, some birds have evolved a behavior calledbrood parasitism. This happens when a bird leaves her eggs in another bird's nest. The host bird often accepts and raises the parasite bird's eggs.

REPRODUCTION IN MAMMALS

Goat kids will stay with their mother until they are weaned.

Most mammals are viviparous, giving birth to live young. However, the five species of monotreme, the platypuses and the echidnas, lay eggs. The monotremes have a sex determination system different from that of most other mammals. In particular, the sex chromosomes of a platypus are more like those of a chicken than those of a therian mammal.

The mammary glands of mammals are specialized to produce milk, a liquid used by newborns as their primary source of nutrition. The monotremes branched early from other mammals and do not have the teats seen in most mammals, but they do have mammary glands. The young lick the milk from a mammary patch on the mother's belly.

Viviparous mammals are in the subclass Theria; those living today are in the Marsupialia and Placentalia infraclasses. A marsupial has a short gestation period, typically shorter than its estrous cycle, and gives birth to an underdeveloped (altricial) newborn that then undergoes further development; in many species, this takes place within a pouch-like sac, the marsupium, located in the front of the mother's abdomen. Some placentals, e.g. guinea pig, give birth to fully developed (precocial) young, usually after long gestation periods, while some others, e.g. mouse, give birth to underdeveloped young.

Reproduction in Placental Mammals

Male Placental Mammals

The mammalian male reproductive system contains two main divisions, the penis and the testicles, the latter of which is where sperm are produced. In humans, both of these organs are outside the abdominal cavity, but they can be primarily housed within the abdomen in other animals. For instance, a dog's penis is covered by a penile sheath except when mating. Having the testicles outside the abdomen best facilitates temperature regulation of the sperm, which require specific temperatures to survive. The external location may also cause a reduction in the heat-induced contribution to the spontaneous mutation rate in male germinal tissue. Sperm are the smaller of the two gametes and are generally very short-lived, requiring males to produce them continuously from the time of sexual maturity until death. The produced sperm are stored in the epididymis until ejaculation. The sperm cells are motile and they swim using tail-like flagella to propel themselves towards the ovum. The sperm follows temperature gradients (thermotaxis) and chemical gradients (chemotaxis) to locate the ovum.

Female Placentals

The mammalian female reproductive system likewise contains two main divisions: the vagina and uterus, which act as the receptacle for the sperm, and the ovaries, which produce the female's ova. All of these parts are always internal. The vagina is attached to the uterus through the cervix, while the uterus is attached to the ovaries via the Fallopian tubes. At certain intervals, the ovaries release an ovum, which passes through the fallopian tube into the uterus.

If, in this transit, it meets with sperm, the egg selects sperm with which to merge; this is termed fertilization. The fertilization usually occurs in the oviducts, but can happen in the uterus itself. The zygote then implants itself in the wall of the uterus, where it begins the processes of embryogenesis and morphogenesis. When developed enough to survive outside the womb, the cervix dilates and contractions of the uterus propel the fetus through the birth canal, which is the vagina.

The ova, which are the female sex cells, are much larger than the sperm and are normally formed within the ovaries of the fetus before its birth. They are mostly fixed in location within the ovary until their transit to the uterus, and contain nutrients for the later zygote and embryo. Over a regular interval, in response to hormonal signals, a process of oogenesis matures one ovum which is released and sent down the Fallopian tube. If not fertilized, this egg is released through menstruation in humans and other great apes, and reabsorbed in other mammals in the estrus cycle.

Gestation

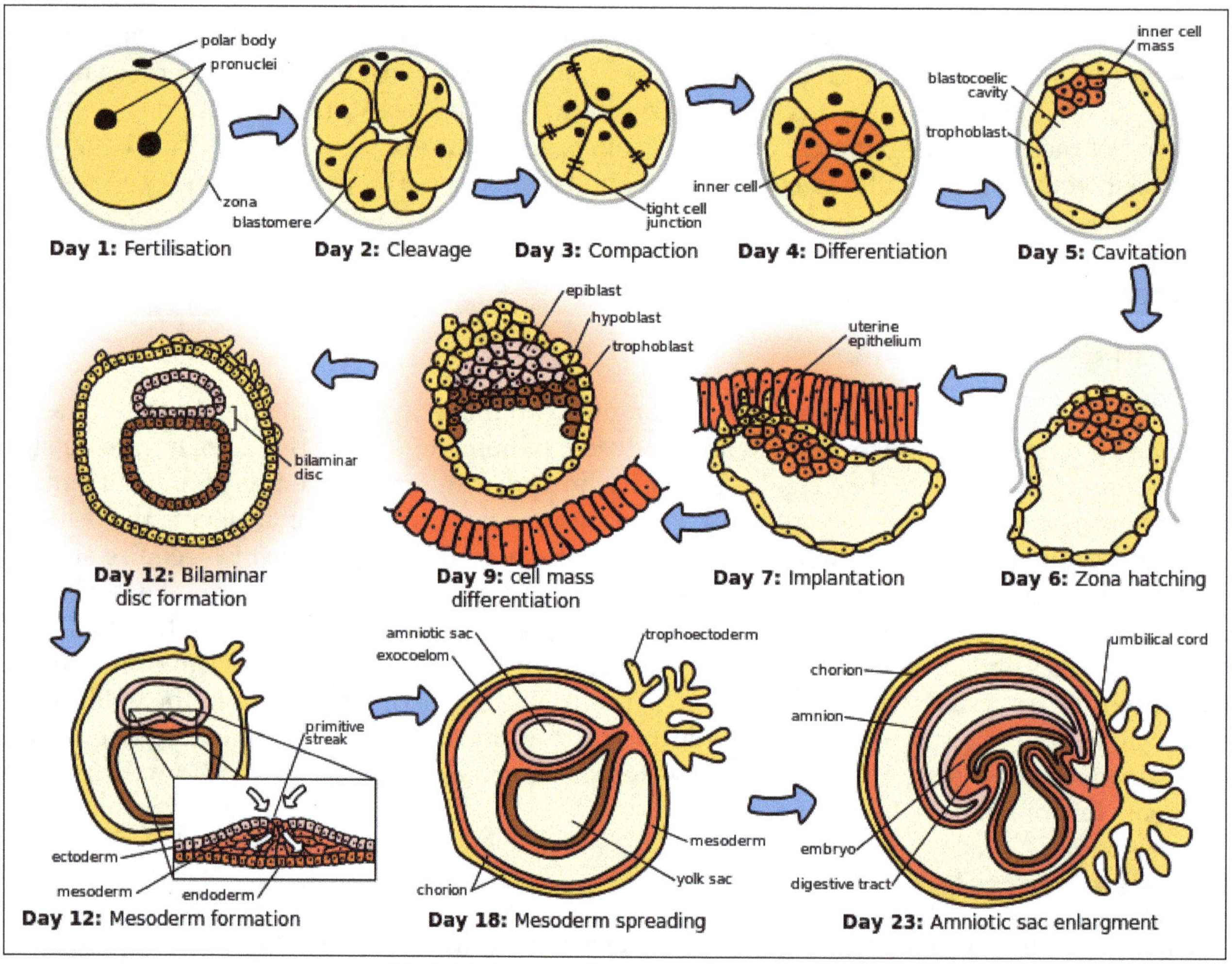

The initial stages of human embryogenesis.

Gestation, called *pregnancy* in humans, is the period of time during which the fetus develops, dividing via mitosis inside the female. During this time, the fetus receives all of its nutrition and oxygenated blood from the female, filtered through the placenta, which is attached to the fetus' abdomen via an umbilical cord. This drain of nutrients can be quite taxing on the female, who is required to ingest slightly higher levels of calories. In addition, certain vitamins and other nutrients are required in greater quantities than normal, often creating abnormal eating habits. The length of gestation, called the gestation period, varies greatly from species to species; it is 40 weeks in humans, 56–60 in giraffes and 16 days in hamsters.

Birth

Once the fetus is sufficiently developing, chemical signals start the process of birth, which begins with contractions of the uterus and the dilation of the cervix. The fetus then descends to the cervix, where it is pushed out into the vagina, and eventually out of the female. The newborn, which is called an infant in humans, should typically begin respiration on its own shortly after birth. Not long after, the placenta is passed as well.

Reproduction in Monotremes

The key anatomical difference between monotremes and other mammals gives them their name; *monotreme* means "single opening", referring to the single duct (the cloaca) for their urinary, defecatory, and reproductive systems. Monotremes, like reptiles, have a single cloaca; marsupials also have a separate genital tract; whereas most placental mammal females have separate openings for reproduction (the vagina), urination (the urethra), and defecation (the anus). In monotremes, only semen passes through the penis; urine is excreted through the cloaca. The monotreme penis is similar to that of turtles, and is covered by a preputial sac.

Monotreme eggs are retained for some time within the mother and receive nutrients directly from her, and they generally hatch within 10 days after laying, much shorter than the incubation period of sauropsid eggs. Newborn monotremes are larval and fetus-like, much like newborn marsupials (and perhaps all non-placental mammals[)), and like them have relatively well-developed forelimbs that enable them to crawl around. In fact, given that monotremes lack nipples, their puggles crawl about more frequently than marsupial joeys in search of milk; this difference raises questions about the supposed developmental restrictions on marsupial forelimbs.

Monotremes lactate from their mammary glands via openings in their skin, rather than through nipples. All five extant species show prolonged parental care of infants, with low rates of reproduction and relatively long life-spans.

Monotremes are also noteworthy in their zygotic development. Most mammal zygotes go through holoblastic cleavage, meaning that after fertilization, the ovum splits into multiple, divisible daughter cells. In contrast, the zygotes of monotremes, like those of birds and reptiles, undergo meroblastic (partial) division. This means the cells at the yolk's edge have cytoplasm continuous with that of the egg, which allows the yolk and embryo to exchange waste and nutrients with the surrounding cytoplasm.

PROCESS OF GAMETOGENESIS IN ANIMALS

The series of changes through which the germinal epithelial cells pass to produce germ cells or gametes is called gametogenesis. The development of eggs is oogenesis and that of the sperm is spermatogenesis. The phases of development in both the gametes are essentially similar.

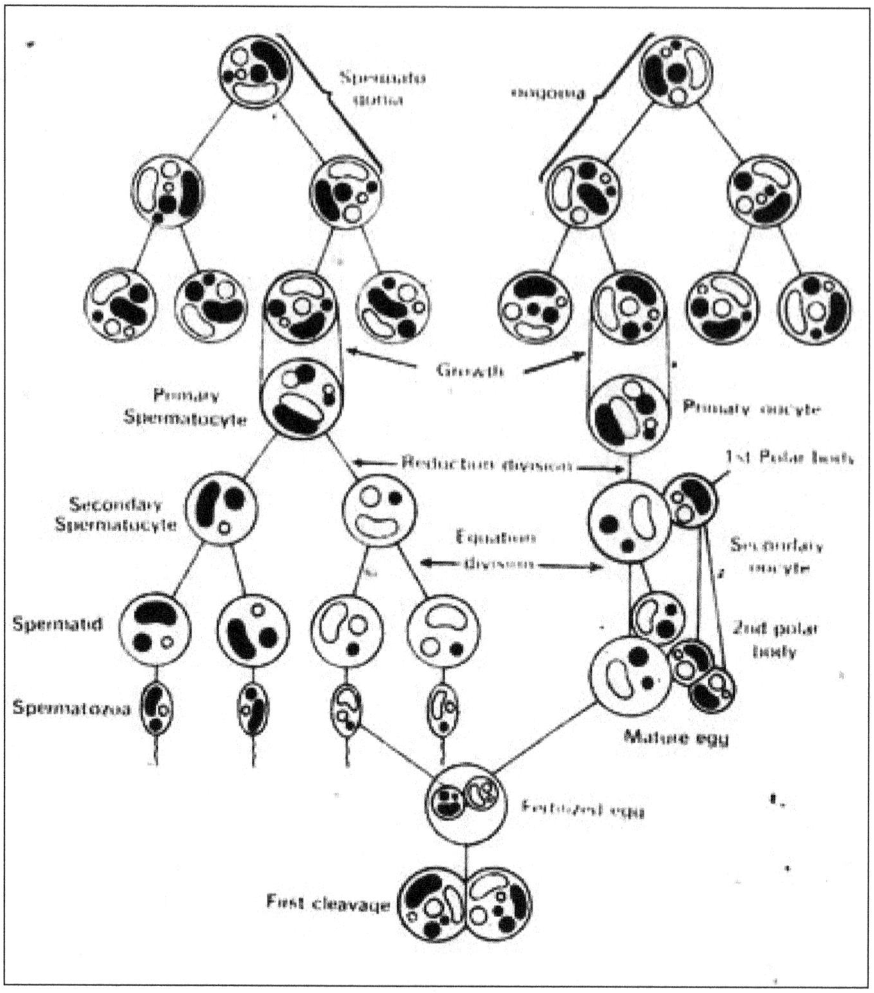

Gametogenesis.

Spermatogenesis

- It occurs in the testes. The walls of the seminiferous tubules of vertebrates contain primordial germ cells.
- The primordial germ cells begin to enlarge with increased metabolism and undergo mitotic division, and produce spermatogonia.
- The spermatogonium divides meiotically (first meiosis) resulting in two spermatocytes.
- Due to second meiotic division (mitosis) of spermatocyte four haploid-spermatids are formed.
- Spermatids metamorphose to spermatozoa. This is spermiogenesis.
    - In this process spermatid increases in size; centriole divides into two; distal centriole forms the main axis of the tail the axial filament is surrounded by a fibre coat; mitochondria form a spiral sheath; golgi complex forms acroblast and give rise to acrosome.

- The mature spermatozoon has three distinct zones head, trunk and tail. It can move actively in fluid.

Oogenesis

- It occurs in the ovary. The phases of division in multiplication are similar to those in spermatogenesis.
- The primary germ cells of the ovary undergo several mitotic divisions to produce oogonia.
- The oogonium gives rise to primary oocytes mitotically.
- A primary oocyte undergoes first meiotic division and produces one large secondary oocyte and one small first polar body or polocyte.
- The oocyte and the polocyte divide mitotically (second meiotic division). The oocyte produces one large ootid and one small second polar body and the polocyte two polar bodies.
- In oogenesis one haploid ootid and three polar bodies are formed. On maturity, the ootid is transformed into ovum.

Fertilization

Fertilization is complete fusion of sperm and ovum except the chromosomes, resulting in a single diploid cell, the Zygote. The union of the nuclei of male and female gametes is karyogamy and that of their cytoplasm isplasmogamy.

The steps in fertilization are

Approach of the Spermatozoon to the Egg

- By chemotactic movement the sperm swims towards the egg.
- The jelly coat of the egg produces fertilizin and the sperms produce antifertilizin.
- The fertilizin and the antifertilizin combine to form an initial bond to facilitate penetration of spermatozoon into the egg.

Penetration of the sperm

- The bound or agglutinated spermatozoon produces an enzyme, lysine, which dissolves egg membranes in the local area and clear the path for spermatozoon to reach the egg surface.
- The acrosomal filament is pushed through the jelly and vitelline membrane, and touches the surface membrane of the egg cytoplasm.

Reaction in Egg During Sperm Penetration

- Coming in contact with the acrosomal filament the egg cytoplasm bulges forward to produce a conical projection, the fertilization cone.

- It gradually engulfs the spermatozoon and carries it inward.
- Simultaneously profound cortical reaction occurs in the cytoplasm.
- Immediately after penetration of the spermatozoon the vitelline membrane separates from the plasma membrane. The vitelline membrane thickens and forms the fertilization membrane.
- The cortical granules swell, become liquified and the fluid is released in the space between egg cytoplasm and fertilization membrane.

Behaviour of Sperm Within the Ovum

- The head and middle piece enter the egg cytoplasm but the tail is left outside.
- The egg completes its meiosis immediately and loses the centriole.
- The sperm rotates 180° and its middle part comes to the front end. The path along which the sperm moves within the egg is known as fertilization path.
- The egg pronucleus moves to the periphery and the movement of the sperm is oriented to it. This is copulation path.
- The sperm pronucleus swells up. The centrosome and the centriole in the middle piece form aster around the two pronuclei.
- The sperm centriole divides into two, move to the poles being connected by fibres.
- The nuclear membranes disappear, the chromosomes of the male and female pronuclei are arranged in the equator of the spindle.
- It is similar to metaphase plate of meiosis and immediately followed by the cleavage of the zygote.

Importance of Fertilization

- The sperm entry activates the secondary oocyte to complete its maturation division.
- Restores diploid number of chromosomes and recombines the maternal and paternal genetic materials.
- Evokes development and cleavage of egg to form a complete organism.

PROCESS OF EMBRYOGENESIS IN ANIMALS

Embryogenesis is the process by which the embryo is formed and develops. It starts with the fertilization of the ovum, egg, which, after fertilization, is then called a zygote. The zygote undergoes rapid mitotic divisions, the formation of two exact genetic replicates of the original cell, with no significant growth (a process known as cleavage) and cellular differentiation, leading to development of an embryo.

Zygote

The egg cell (and hence the fertilized egg) is always asymmetric, having an "animal pole" (future ectoderm and mesoderm), two of three primitive tissue types, and a "vegetal pole" (future endoderm), it is also covered with different protective envelopes.

The first envelope, the one which is in contact with the membrane of the egg, is made of glycoproteins and is called vitelline membrane (zona pellucida in mammals). Different taxa show different cellular and acellular envelopes outside of the vitelline membrane.

Cleavage

The zygote undergoes rapid cell divisions with no significant growth, producing a cluster of cells that is the same size as the original zygote. The different cells derived from cleavage, up to the blastula stage, are called blastomeres. Depending mostly on the amount of yolk in the egg, the cleavage can be holoblastic (total) or meroblastic (partial).

Holoblastic cleavage occurs in animals with little yolk in their eggs, such as humans and other mammals who receive nourishment as embryos from the mother, via placenta or milk. On the other hand, meroblastic cleavage occurs in animals whose eggs have more yolk; i.e., birds and reptiles. Because cleavage is impeded in the vegetal pole, there is a very uneven distribution and size of cells, being more and bigger at the animal pole of the zygote.

In holoblastic eggs the first cleavage always occurs along the vegetal-animal axis of the egg, the second cleavage is perpendicular to the first.

From here the spatial arrangement of blastomeres, due to different planes of cleavage, in various organisms, follow various patterns.

Blastula and Gastrula

After the cleavage has produced over 100 cells, the embryo is called a blastula. The blastula is usually a spherical layer of cells (the blastoderm) surrounding a fluid-filled or yolk-filled cavity (the blastocoel). Mammals at this stage form a structure called the blastocyst, characterised by an inner cell mass that is not present in the blastula. The blastocyst must not be confused with the blastula; even though they are similar in structure, their cells have different fates.

During gastrulation cells migrate to the interior of the blastula, consequently forming two (diploblastic) or three (triploblastic) germ layers. The embryo during this process is called a gastrula. The germ layers are referred to as the ectoderm, mesoderm and endoderm. In diploblastic animals only the ectoderm and the endoderm are present.

Among the different animals, different combinations of the following processes occur to place the cells in the interior of the embryo:

- Epiboly: Expansion of one cell sheet over other cells.
- Ingression: Cells move with pseudopods.
- Invagination: Forming the mouth, anus, archenteron.

- Delamination: The external cells divide, leaving the daughter cells in the cavity.
- Polar proliferation.

Other major changes during gastrulation:

- Heavy RNA transcription using embryonic genes; up to this point the RNAs used were maternal (stored in the unfertilized egg).
- Cells start major differentiation processes, losing their pluripotentiality.
    - In most animals, a blastopore is formed at the point where cells are entering the embryo. Two major groups of animals can be distinguished according to the blastopore's fate. In deuterostomes, the anus forms from the blastopore, while in protostomes it develops into the mouth.

Organogenesis

At some point after the different germ layers are defined, organogenesis begins. The first stage in vertebrates is called neurulation, where the neural plate folds forming the neural tube. Other common organs or structures which arise at this time include the heart and somites, but from now on embryogenesis follows no common pattern among the different taxa of the animal kingdom.

In most animals organogenesis along with morphogenesis will result in a larva. The hatching of the larva, which must then undergo metamorphosis, marks the end of embryonic development.

References

- Types-Of-Reproduction-In-Fishes: aquafind.com, Retrieved 03 February, 2019
- Reproductive-anatomy-physiology-in-reptiles-nine-key-facts: lafeber.com, Retrieved 25 July, 2019
- Bird-Reproduction: twootz.com, Retrieved 10 April, 2019
- Berglund A (1997) "Mating systems and sex allocation" Pages 237–265 in JJ Godon, ed. Behavioural ecology of teleost fishes. Oxford University Press. ISBN 0-19-850503-5
- Marshall Graves, Jennifer A. (2008). "Weird Animal Genomes and the Evolution of Vertebrate Sex and Sex Chromosomes" (PDF). Annual Review of Genetics. 42: 568–586. doi:10.1146/annurev.genet.42.110807.091714. PMID 18983263
- Process-of-gametogenesis-in-animals, animals-3249: notesonzoology.com, Retrieved 18 August, 2019
- Julian Lombardi (6 December 2012). Comparative Vertebrate Reproduction. Springer Science & Business Media. ISBN 978-1-4615-4937-6

Adaptation in Vertebrates

CHAPTER 5

The evolution of the living organism to better suit its environment and habitat is called adaptation. Adaptation in vertebrates includes structural and functional adaptations of pisces and mammals, behavior and morphological adaptations of reptiles, flight adaptation in aves, amphibians' adaptation to water balance and cold survival. This chapter discusses in detail these adaptation theories related to vertebrates.

STRUCTURAL AND FUNCTIONAL ADAPTATIONS OF FISHES

Locomotion in Water

To the human eye, some fishes appear capable of swimming at extremely high speeds. But our judgment is unconsciously tempered by our own experience that water is a highly resistant medium through which to move. Most fishes, such as a trout or a minnow, can swim maximally about 10 body lengths per second, obviously an impressive performance by human standards. Yet when these speeds are translated into kilometers per hour it means that a 30 cm (1 foot) trout can swim only about 10.4 km (6.5 miles) per hour. As a general rule, the larger the fish the faster it can swim.

Trunk musculature of a teleost fish, partly dissected to show internal arrangement of the muscle bands(myomeres). The myomeres are folded into a complex, nested grouping, an arrangement that favors stronger and more controlled swimming.

The propulsive mechanism of a fish is its trunk and tail musculature. The axial, locomotory musculature is composed of zigzag bands, called myomeres. Muscle fibers in each myomere are relatively short and connect the tough connective tissue partitions that separate each myomere from the next. On the surface the myomeres take the shape of a Wlying on its side but internally the bands are complexly folded and nested so that the pull of each myomere extends over several vertebrae.

This arrangement produces more power and finer control of movement since many myomeres are involved in bending a given segment of the body.

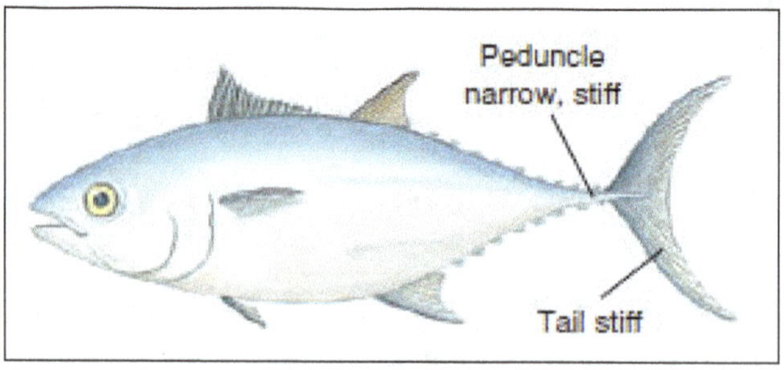

Bluefin tuna, showing adaptations for fast swimming. Powerful trunk muscles pull on the slender tail stalk. Since the body does not bend, all of the thrust comes from beats of the stiff sickle-shaped tail.

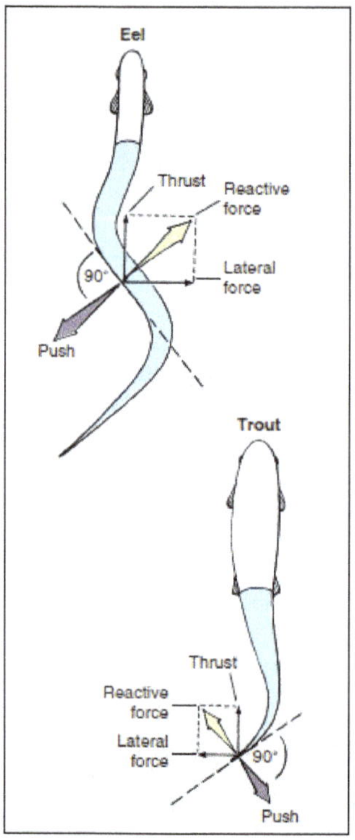

Movements of swimming fishes, showing the forces developed by an eel-shaped and spindleshaped fish.

Understanding how fishes swim can be approached by studying the motion of a very flexible fish such as an eel. The movement is serpentine, not unlike that of a snake, with waves of contraction moving backward along the body by alternate contraction of the myomeres on either side. The anterior end of the body bends less than the posterior end, so that each undulation increases in amplitude as it travels along the body. While undulations move backward, the bending of the body pushes laterally against the water, producing a reactive force that is directed forward, but at an angle. It can be analyzed as having two components: thrust, which is used to overcome drag and propels the fish forward, and lateral force, which tends to make the fish's head "yaw," or deviate from the course in the same direction as the tail. This side-to-side head movement is very obvious in a swimming eel or shark, but many fishes have a large, rigid head with enough surface resistance to minimize yaw.

The movement of an eel is reasonably efficient at low speed, but its body shape generates too much frictional drag for rapid swimming. Fishes that swim rapidly, such as trout, are less flexible and limit body undulations mostly to the caudal region. Muscle force generated in the large anterior muscle mass is transferred through tendons to the relatively nonmuscular caudal peduncle and tail where thrust is generated. This form of swimming reaches its highest development in the tunas, whose bodies do not flex at all. Virtually all thrust is derived from powerful beats of the tail fin . Many fast oceanic fishes such as marlin, swordfish, amberjacks, and wahoo have sweptback tail fins shaped much like a sickle. Such fins are the aquatic counterpart of the high-aspect ratio wings of the swiftest birds.

Swimming is the most economical form of animal locomotion, largely because aquatic animals are almost perfectly supported by their medium and need expend little energy to overcome the force of gravity. If we compare the energy cost per kilogram of body weight of traveling 1 km by different forms of locomotion, we find swimming costs only 0.39 kcal (salmon) as compared with 1.45 kcal for flying (gull) and 5.43 for walking (ground squirrel). However, part of the unfinished business of biology is understanding how fish and aquatic mammals are able to move through the water while creating almost no turbulence. The secret lies in the way aquatic animals bend their bodies and fins (or flukes) to swim and in the friction-reducing properties of the body surface.

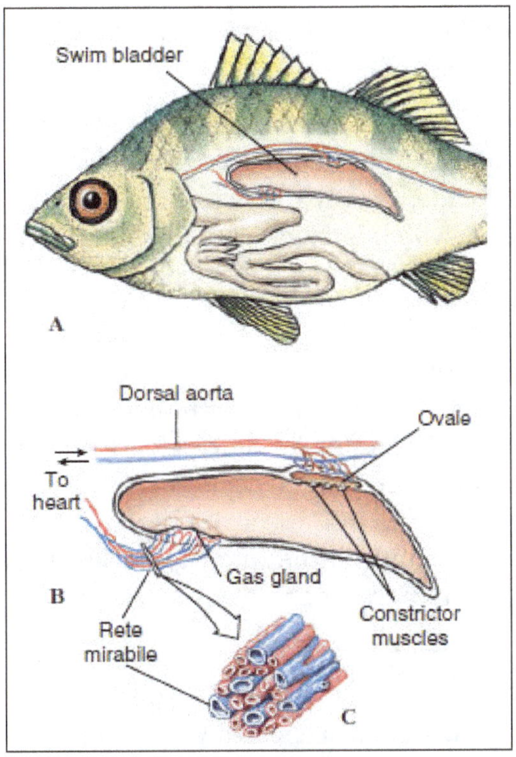

A, Swim bladder of a teleost fish. The swim bladder lies in the coelom just beneath the vertebral column. **B**, Gas is secreted into the swim bladder by the gas gland.

Gas from the blood is moved into the gas gland by the rete mirabile, a complex array of tightly-packed capillaries that act as a countercurrent multiplier to build up the oxygen concentration. The arrangement of venous and arterial capillaries in the rete is shown in C. To release gas during ascent, a muscular valve opens, allowing gas to enter the ovale from which the gas is removed by the circulation.

By adjusting the volume of gas in the swim bladder, a fish can achieve neutral buoyancy and remain suspended indefinitely at any depth with no muscular effort. There are severe technical problems, however. If the fish descends to a greater depth, the swim bladder gas is compressed so that the fish becomes heavier and tends to sink. Gas must be added to the bladder to establish a new equilibrium buoyancy. If the fish swims upward, the gas in the bladder expands, making the fish lighter. Unless gas is removed, the fish will rise with ever-increasing speed while the bladder continues to expand.

Gas may be removed from the swim bladder in one of two ways. The more primitive phystostomous (Gr., phys, bladder, stoma, mouth) fishes (trout, for example) have a pneumatic duct that connects the swim bladder to the esophagus. These fishes may simply expel air out through the pneumatic duct. More advanced teleosts exhibit the physoclistous (Gr., phys, bladder, clist, closed) condition in which the pneumatic duct is lost in adults. In physoclistous fishes, gas must be secreted into the blood from the ovale, a vascularized area. Both types of fishes require gas to be secreted into the swim bladder from the blood, although a few shallow-water-inhabiting phystostomes may gulp air to fill their swim bladder.

Gas is secreted into the swim bladder at the highly specialized gas gland. The gas gland is supplied by a remarkable network of blood capillaries, called the rete mirabile("marvelous net") that functions as a countercurrent exchange system to trap gases, especially oxygen, and prevent their loss to the circulation.

The amazing effectiveness of this device is exemplified by a fish living at a depth of 2400 m (8000 feet). To keep the bladder inflated at that depth, the gas inside (mostly oxygen, but also variable amounts of nitrogen, carbon dioxide, argon, and even some carbon monoxide) must have a pressure exceeding 240 atmospheres, which is much greater than the pressure in a fully charged steel gas cylinder. Yet the oxygen pressure in the fish's blood cannot exceed 0.2 atmosphere—equal to the oxygen pressure at the sea surface.

Physiologists who were at first baffled by the secretion mechanism now understand how it operates. In brief, the gas gland secretes lactic acid, which enters the blood, causing a localized high acidity in the rete mirabile that forces hemoglobin to release its load of oxygen. The capillaries in the rete are arranged so that the released oxygen accumulates in the rete, eventually reaching such a high pressure that the oxygen diffuses into the swim bladder. The final gas pressure attained in the swim bladder depends on the length of the rete capillaries; they are relatively short in fishes living near the surface, but are extremely long in deep-sea fishes.

Respiration

Fish gills are composed of thin filaments, each covered with a thin epidermal membrane that is folded repeatedly into platelike lamellae. These are richly supplied with blood vessels. The gills are located inside the pharyngeal cavity and are covered with a movable flap, the operculum. This arrangement provides excellent protection to the delicate gill filaments, streamlines the body, and makes possible a pumping system for moving water through the mouth, across the gills, and out the operculum. Instead of opercular flaps as in bony fishes, the elasmobranchs have a series of gill slits out of which the water flows. In both elasmobranchs and bony fishes the branchial mechanism is arranged to pump water continuously and smoothly over the gills, although to an observer it appears that fish breathing is pulsatile. The flow of water is opposite to the direction of blood flow (countercurrent flow), the best arrangement for extracting the greatest possible amount of oxygen from the water. Some bony fishes can remove as much as 85% of the oxygen from water passing over their gills. Very active fishes, such as herring and mackerel, can obtain sufficient water for their high oxygen demands only by swimming forward continuously to force water into the open mouth and across the gills. This process is called ram ventilation. Such fish will be asphyxiated if placed in an aquarium that restricts free swimming movements, even if the water is saturated with oxygen.

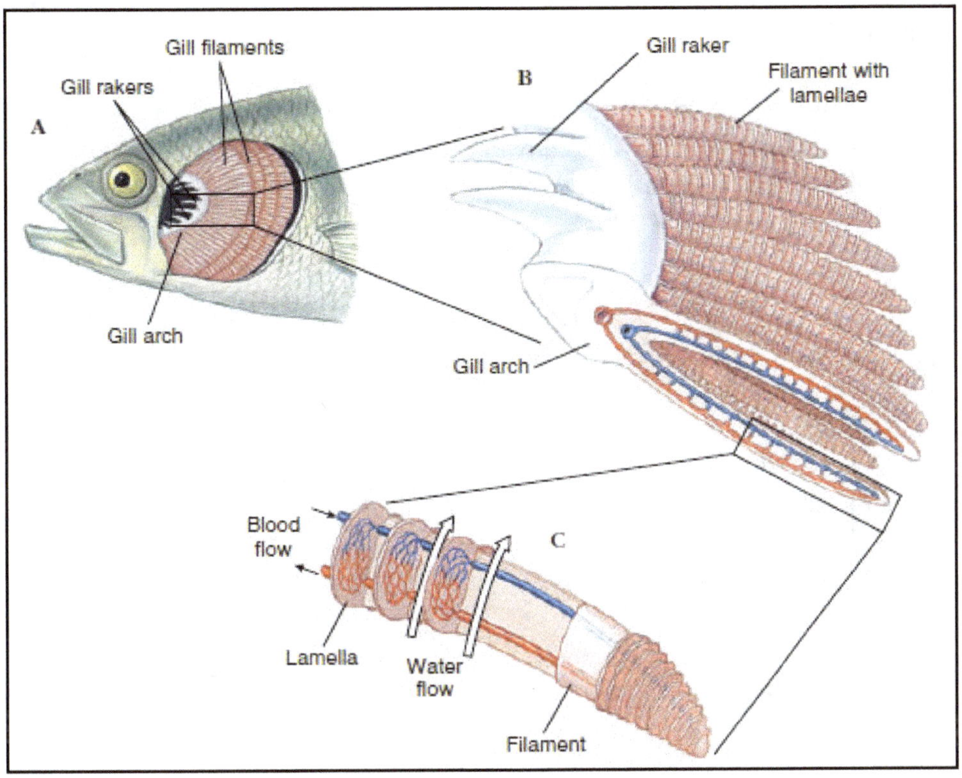

Gills of fish. Bony, protective flap covering the gills (operculum) has been removed, A, to reveal branchial chamber containing the gills. There are four gill arches on each side, each bearing numerous filaments. A portion of gill arch (B) shows gill rakers that project forward to strain out food and debris, and gill filaments that project to the rear. A single gill filament (C) is dissected to show the blood capillaries within the platelike lamellae. Direction of water flow (large arrows) is opposite the direction of blood flow.

A surprising number of fishes can live out of water for varying lengths of time by breathing air. Several devices are employed by different fishes. We already have described the lungs of the lungfishes, Polypterus, and the extinct rhipidistians. Freshwater eels often make overland excursions during rainy weather, using the skin as a major respiratory surface. The bowfin, Amia, has both gills and a lunglike swim bladder. At low temperatures it uses only its gills, but as the temperature and the fish's activity increase, it breathes mostly air with its swim bladder. The electric eel, Electrophorus (Gr. e-lektron, something bright, + phoros, to bear), has degenerate gills and must supplement gill respiration by gulping air through its vascular mouth cavity. One of the best air breathers of all is the Indian climbing perch Anabas (Gr. anabaino-, to go up), which spends most of its time on land near the water's edge, breathing air through special air chambers above muchreduced gills.

Osmotic Regulation

Fresh water is an extremely dilute medium with a salt concentration (0.001 to 0.005 gram moles per liter [M]) much below that of the blood of freshwater fishes (0.2 to 0.3 M). Water therefore tends to enter their bodies osmotically, and salt is lost by diffusion outward. Although the scaled and mucous-covered body surface is almost totally impermeable to water, water gain and salt loss do occur across thin membranes of the gills. Freshwater fishes are hyperosmotic regulators that have several defenses against these problems. First, the excess water is pumped out by the opisthonephric kidney, which is capable of forming very dilute urine. Second, special saltabsorbing cells located in the gill epithelium actively move salt ions, principally sodium and chloride, from the water to the blood. This, together with salt present in the fish's food, replaces diffusive salt loss. These mechanisms are so efficient that a freshwater fish devotes only a small part of its total energy expenditure to keeping itself in osmotic balance.

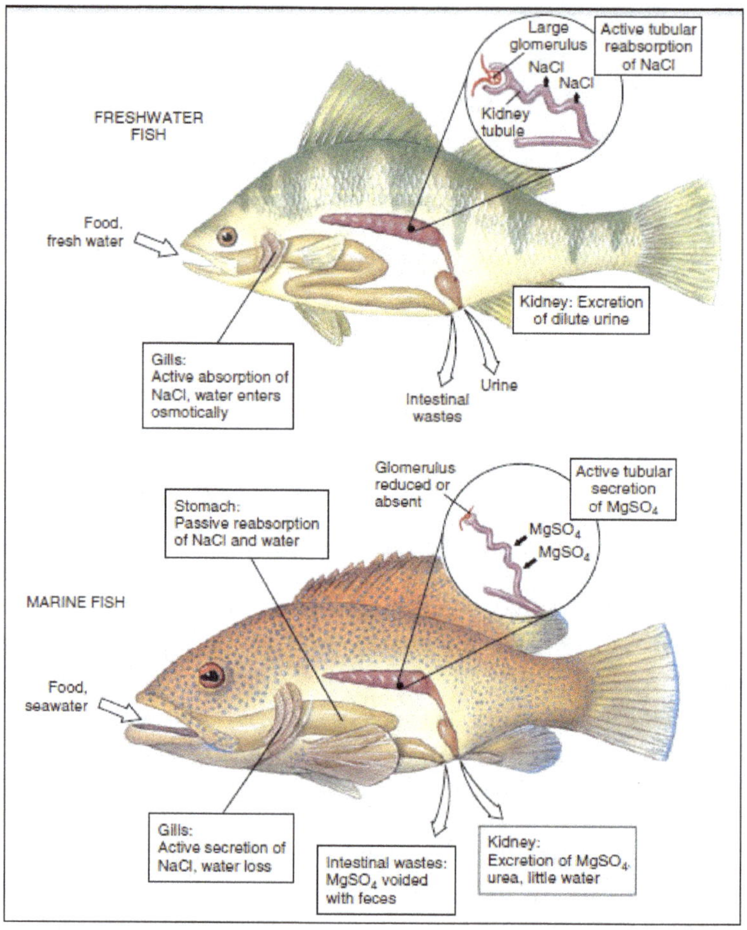

Osmotic regulation in freshwater and marine bony fishes. A freshwater fish maintains osmotic and ionic balance in its dilute environment by actively absorbing sodium chloride across the gills (somesalt is gained with food). To flushout excess that constantly enters the body, the glomerularkidney produces a dilute urine by reabsorbing sodium chloride. A marine fish must drink seawater to replace water lost osmotically to its salty environment. Sodium chloride and water are absorbed from the stomach. Excess sodium chloride is actively transported outward by the gills. Divalent sea salts, mostly magnesium sulfate, are eliminated with feces and secreted by the tubular kidney.

Marine bony fishes are hypoosmotic regulators that encounter a completely different set of problems. Having a much lower blood salt concentration (0.3 to 0.4 M) than the seawater around them (about 1 M), they tend to lose water and gain salt. The marine teleost fish quite literally risks drying out, much like a desert mammal deprived of water. Again, marine bony fishes, like their freshwater counterparts, have evolved an appropriate set of defenses. To compensate for water loss, the marine teleost drinks seawater. Although this behavior obviously brings needed water into the body, it is unfortunately accompanied by a great deal of unneeded salt. Unwanted salt is disposed in two ways: (1) the major sea salt ions (sodium, chloride, and potassium) are carried by the blood to the gills where they are secreted outward by special salt-secretory cells; and (2) the remaining ions, mostly the divalent ions (magnesium, sulfate, and calcium), are left in the intestine and voided with the feces. However, a small but significant fraction of these residual divalent salts in the intestine, some 10% to 40% of the total, penetrates the intestinal mucosa and enters the bloodstream. These ions are excreted by the kidney. Unlike the freshwater fish kidney, which forms its urine by the usual filtrationresorption sequence typical of most vertebrate kidneys, the marine fish's kidney excretes divalent ions by tubular secretion. Since very little if any filtrate is formed, the glomeruli have lost their importance and disappeared altogether in some marine teleosts. The pipefishes, and the goosefish shown in Figure, are examples of "aglomerular" marine fishes.

Feeding Behavior

For any fish, feeding is one of the main concerns of day-to-day living. Although many a luckless angler would swear otherwise, the fact is that a fish devotes more time and energy to eating, or searching for food to eat, than to anything else. Throughout the long evolution of fishes, there has been unrelenting selective pressure for those adaptations that enable a fish to come out on the better end of the eat-or-beeaten contest. Certainly the most farreaching single event was the evolution of jaws. Their possessors were freed from a largely passive filterfeeding existence and could adopt a predatory mode of life. Improved means of capturing larger prey demanded stronger muscles, more agile movement, better balance, and improved special senses. More than any other aspect of its life habit, feeding behavior shapes the fish.

Wolf eel, Anarrhichthys ocellatus, feeding on a sea cucumber it has captured and pulled to the opening of its den.

Goosefish Lophius piscatorius awaits its meal. Above its head swings a modified dorsal fin spine ending in a fleshy tentacle that contracts and expands in a convincing wormlike manner. When a fish approaches the alluring bait, the huge mouth opens suddenly, creating a strong inward current that sweeps the prey inside. In a split second all is over.

Most fishes are carnivores that prey on a myriad of animal foods from zooplankton and insect larvae to large vertebrates. Some deep-sea fishes are capable of eating victims nearly twice their own size—an adaptation for life in a world where meals are necessarily infrequent. Most advanced ray-finned fishes cannot masticate their food as we can because doing so would block the current of water across the gills. Some, however, such as the wolf eel, have molarlike teeth in the jaws for crushing their prey, which may include hard-bodied crustaceans. Others that do grind their food use powerful pharyngeal teeth in the throat. Most carnivorous fish almost invariably swallow their prey whole, using sharppointed teeth in the jaws and on the roof of the mouth to seize their prey. The incompressibility of water makes the task even easier for many largemouthed predators. When the mouth is suddenly opened, a negative pressure is created that sweeps the victim inside.

A second group of fishes are herbivores that eat plants and algae. Although plant eaters are relatively few in number, they are crucial intermediates in the food chain, especially in freshwater rivers, lakes, and ponds that contain very little plankton.

Suspension-feeders that crop the abundant microorganisms of the sea form a third and diverse group of fishes ranging from fish larvae to basking sharks. However, the most characteristic group of plankton feeders are herringlike fishes (menhaden, herring, anchovies, capelin, pilchards, and others), mostly pelagic (open-sea dwellers) fishes that travel in large schools. Both phytoplankton and the smaller zooplankton are strained from the water with the sievelike gill rakers. Because plankton feeders are the most abundant of all marine fishes, they are important food for numerous larger but less abundant carnivores. Many freshwater fishes also depend on plankton for food.

A fourth group of fishes contains omnivores that feed on both plant and animal food. Finally there are scavengers that feed on organic debris (detritus) and parasites that suck the body fluids of other fishes.

Digestion in most fishes follows the vertebrate plan. Except in several fishes that lack stomachs altogether, the food proceeds from stomach to tubular intestine, which tends to be short in carnivores but may be extremely long and coiled in herbivorous forms. In the herbivorous grass carp, for example, the intestine may be nine times the body length, an adaptation for the lengthy digestion required for plant carbohydrates. In carnivores, some protein digestion may be initiated in the acid medium of the stomach, but the principal function of the stomach is to store the often large and infrequent meals while awaiting their reception by the intestine.

Digestion and absorption proceed simultaneously in the intestine. A curious feature of ray-finned fishes, especially the teleosts, is the presence of numerous pyloric ceca found in no other vertebrate group. Their primary function appears to be fat absorption, although all classes of digestive enzymes (protein-, carbohydrate-, and fat-splitting) are secreted there.

Migration Eel

For centuries naturalists had been puzzled about the life history of the freshwater eel Anguilla (an-gwil´la) (L. eel), a common and commercially important species of coastal streams of the North Atlantic. Eels are catadromous (Gr. kata, down, + dromos, running), meaning that they spend most of their lives in fresh water but migrate to the sea to spawn. Each fall, large numbers of eels were seen swimming down the rivers toward the sea, but no adults ever returned. Each spring countless numbers of young eels, called "elvers", each about the size of a wooden matchstick, appeared in the coastal rivers and began swimming upstream. Beyond the assumption that eels must spawn somewhere at sea, location of their breeding grounds was completely unknown.

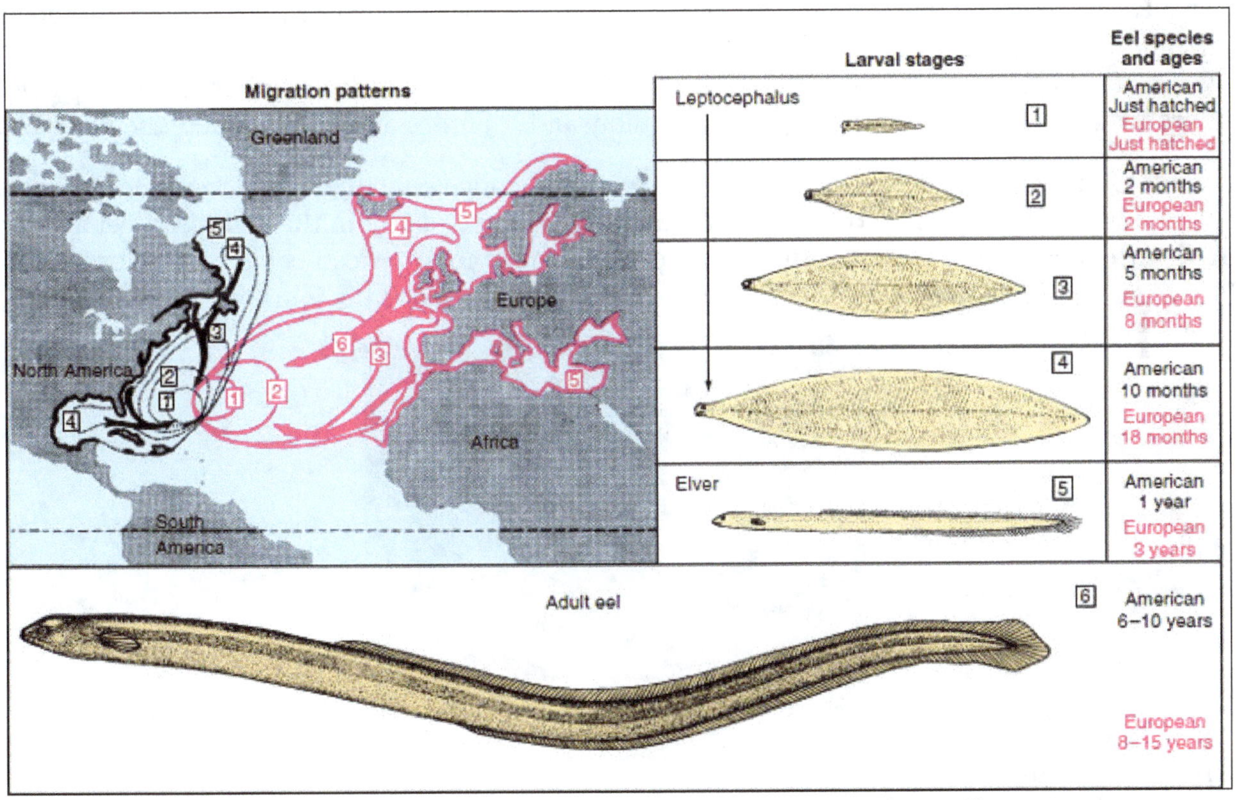

Life histories of the European eel, Anguilla anguilla, and American eel, Anguilla rostrata.

Migration patterns of European species are shown in pink. Migration patterns of American species are shown in black. Boxed numbers refer to stages of development. Note that the American eel completes its larval metamorphosis and sea journey in one year. It requires nearly three years for the European eel to complete its much longer journey.

The first clue was provided by two Italian scientists, Grassi and Calandruccio, who in 1896 reported that elvers were not larval eels but rather were relatively advanced juveniles. True larval eels, they discovered, were tiny, leafshaped, completely transparent creatures that bore absolutely no resemblance to an eel. They had been called leptocephali (Gr. leptos, slender, + kephale, head) by early naturalists, who never suspected their true identity. In 1905 Johann Schmidt, supported by the Danish government, began a systematic study of eel biology that he continued until his death in 1933. With cooperation of captains of commercial vessels plying the Atlantic, thousands of leptocephali were caught in different areas of the Atlantic with the plankton nets Schmidt supplied them. By noting where larvae in different stages of development were captured, Schmidt and his colleagues eventually reconstructed the spawning migrations.

When adult eels leave the coastal rivers of Europe and North America, they swim steadily and apparently at great depth for 1 to 2 months until they reach the Sargasso Sea, a vast area of warm oceanic water southeast of Bermuda. Here, at depths of 300 m or more, the eels spawn and die. The minute larvae then begin an incredible journey back to the coastal rivers of Europe. Drifting with the Gulf Stream and preyed on constantly by numerous predators, they reach the middle of the Atlantic after 2 years. By the end of the third year they arrive in the coastal waters of Europe where the leptocephali metamorphose into elvers, with an unmistakable eel-like body form. Here the males and females part company; males remain in the brackish waters of coastal rivers and estuaries while females continue up the rivers, often traveling hundreds of miles upstream. After 8 to 15 years of growth, the females, now 1 m or more long, return to the sea to join the smaller males; both return to the ancestral breeding grounds thousands of miles away to complete the life cycle.

Schmidt found that the American eel (Anguilla rostrata) could be distinguished from the European eel (A. vulgaris) because it had fewer vertebrae— an average of 107 in the American eel as compared with an average 114 in the European species. Since the American eel is much closer to the North American coastline, it requires only about 8 months to make the journey.

Migrating Pacific sockeye salmon (Oncorhynchus nerka).

Homing Salmon

The life history of salmon is nearly as remarkable as that of the eel and certainly has received far more popular attention. Salmon are anadromous (Gr. anadromos, running upward); that is, they spend their adult lives at sea but return to fresh water to spawn. The Atlantic salmon (Salmo salar) (L. salmo, salmon, sal, salt) and the Pacific salmon (six species in the genus Oncorhynchus [on-ko-rink´us] [Gr. onkos, hook, + rhynchos, snout]) have this practice, but there are important differences among the seven species. The Atlantic salmon may make repeated upstream spawning runs. The six Pacific salmon species (king, sockeye, silver, humpback, chum, and Japanese masu) each make a single spawning run, after which they die.

The virtually infallible homing instinct of the Pacific species is legendary: after migrating downstream as a smolt, a sockeye salmon ranges many hundreds of miles over the Pacific for nearly 4 years, grows to 2 to 5 kg in weight, and then returns almost unerringly to spawn in the headwaters of its parent stream. Some straying does occur and is an important means of increasing gene flow and populating new streams.

Experiments by A. D. Hasler and others have shown that homing salmon are guided upstream by the characteristic odor of their parent stream. When the salmon finally reach the spawning beds of their parents (where they themselves were hatched), they spawn and die. The following spring, newly hatched fry transform into smolts before and during the downstream migration. At this time they are imprinted with the distinctive odor of the stream, which is apparently a mosaic of compounds released by the characteristic vegetation and soil in the watershed of the parent stream. They also seem to imprint on the odors of other streams they pass while migrating downriver and use these odors in reverse sequence as a map during the upriver migration as returning adults.

How do salmon find their way to the mouth of the coastal river from the trackless miles of the open ocean? Salmon move hundreds of miles away from the coast, much too far to be able to detect the odor of their parent stream. Experiments suggest that some migrating fish, like birds, can navigate by orienting to the position of the sun. However, migrant salmon can navigate on cloudy days and at night, indicating that sun navigation, if used at all, cannot be the salmon's only navigational cue. Fish also (again, like birds) appear able to detect and navigate to the earth's magnetic field. Finally, fishery biologists concede that salmon may not require precise navigational abilities at all, but instead may use ocean currents, temperature gradients, and food availability to reach the general coastal area where "their" river is located. From this point, they would navigate by their imprinted odor map, making correct turns at each stream junction until they reach their natal stream.

Rainbow surfperch Hypsurus caryi giving birth. All of the West Coast surfperches (family Embiotocidae) are ovoviviparous.

Reproduction and Growth

In a group as diverse as the fishes, it is no surprise to find extraordinary variations on the basic theme of sexual reproduction. Most fishes favor a simple theme: they are dioecious, with external fertilization and external developmentof the eggs and embryos (oviparity). However, as tropical fish enthusiasts are well aware, the ever-popular ovoviviparous guppies and mollies of home aquaria bear their young alive after development in the ovarian cavity of the mother, some viviparous sharks develop a kind of placental attachment through which the young are nourished during gestation.

Many marine fishes are extraordinarily profligate egg producers. Males and females come together in great schools and release vast numbers of gametes into the water to drift with the current. Large female cod may release 4 to 6 million eggs at a single spawning. Less than one in a million will survive the numerous perils of the ocean to reach reproductive maturity.

Male banded jawfish Opistognathus macrognathus orally brooding its eggs. The male retrieves the female's spawn and incubates the eggs until they hatch. During brief periods when the jawfish is feeding, the eggs are left in the burrow.

Unlike the minute, buoyant, transparent eggs of pelagic marine teleosts, those of many near-shore bottomdwelling (benthic) species are larger, typically yolky, nonbuoyant, and adhesive. Some bury their eggs, many attach them to vegetation, some deposit them in nests, and some even incubate them in their mouths. Many benthic spawners guard their eggs. Intruders expecting an easy meal of eggs may be met with a vivid and often belligerent display by the guard, which is almost always the male.

Freshwater fishes almost invariably produce nonbuoyant eggs. Those, such as perch, that provide no parental care simply scatter their myriads of eggs among weeds or along the bottom. Freshwater fishes that do provide some form of egg care, such as bullhead catfishes and some darters, produce fewer, larger eggs that enjoy a better chance for survival.

Elaborate preliminaries to mating are the rule for freshwater fishes. The female Pacific salmon, for example, performs a ritualized mating "dance" with her breeding partner after arriving at the spawning bed in a fast-flowing, gravel-bottomed stream. She then turns on her side and scoops out a nest with her tail. As the eggs are laid by the female, they are fertilized by the male. After the female covers the eggs with gravel, the exhausted fish dies and drifts downstream.

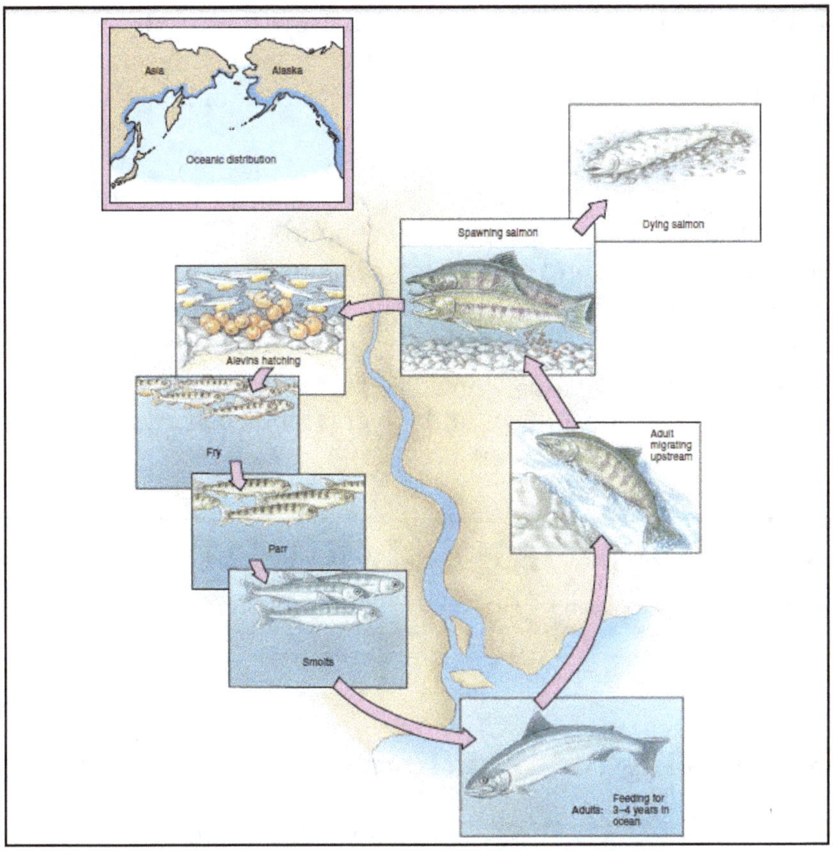

Spawning Pacific salmon and development of the eggs and young.

Soon after the egg of an oviparous species is laid and fertilized, it takes up water and the outer layer hardens. Cleavage follows, and the blastoderm forms, sitting astride a relatively enormous yolk mass. Soon the yolk mass is enclosed by the developing blastoderm, which then begins to assume a fishlike shape. The fish hatches as a larva carrying a semitransparent sac of yolk, which provides its food supply until the mouth and digestive tract have developed. The larva then begins searching for its own food. After a period of growth the larva undergoes a metamorphosis, especially dramatic in many marine species such as the freshwater eel. Body shape is refashioned, fin and color patterns change, and the animal becomes a juvenile bearing the unmistakable definitive body form of its species.

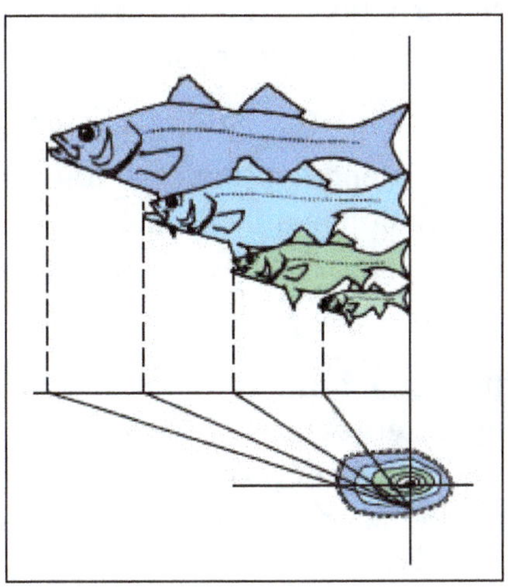

Scale growth. Fish scales disclose seasonal changes in growth rate. Growth is interrupted during winter, producing year marks (annuli). Each year's increment in scale growth is a ratio to the annual increase in body length. Otoliths (ear stones) and certain bones can also be used in some species to determine age and growth rate.

Growth is temperature dependent. Consequently, fish living in temperate regions grow rapidly in summer when temperatures are high and food is abundant but nearly stop growing in winter. Annual rings in the scales, otoliths, and other bony parts reflect this seasonal growth, a distinctive record of convenience to fishery biologists who wish to determine a fish's age. Unlike birds and mammals, which stop growing after reaching adult size, most fishes after attaining reproductive maturity continue to grow for as long as they live. This may be a selective advantage, since the larger the fish, the more gametes it produces and the greater its contribution to future generations.

BEHAVIOR AND MORPHOLOGICAL ADAPTATIONS OF REPTILES

Defensive Behaviors

Catalepsy, Death Feigning and Tonic Immobility

This behavior occurs commonly in some groups of snakes (e.g. hognose snakes, False spitting cobras) and has been described in lizards and crocodilians.

Hognose snakes (Heterodon sp.) are well known for their complex death feigning behavior. When first disturbed hognose snakes will exhibit an elaborate bluff display consisting of an exaggerated S-coil, loud hissing and false strikes. The tail is often tightly coiled and can be elevated. If grasped or further harassed, these snakes will begin to writhe violently with the mouth hung limply open and often will excrete urates and feces. Violent writhing continues for a short time and then the snakes assumes an inverted, limp posture, usually with the mouth open and the tongue hanging out. If the snakes is turned over onto its ventrum, it will immediately turn back onto its dorsum. When the threat is removed, the hognose snake slowly rights itself and crawls away. Pseudoxenodontines, closely related to the hognose snakes, and false spitting cobras (Hemachatus sp.) are also well known for their elaborate death feigning behavior that closely resembles that of the hognose snakes.

For lizards and crocodilians, chameleons in the genus Brookesia and leaf tailed geckos (Uroplatus sp) exhibit tonic immobility when threatened as does the microteiid, echinosaura horrida. Death feigning occurs in at least one teiid (Callopistis flavipunctatus), one Cordylid (Gerrhosaurus major) and three monitor species. Hatchling caiman crocodylus vocalize, bite and struggle on land but feign death if grasped under water.

The evolution of death feigning and tonic immobility may be related to removal of movement; a widespread class of stimuli among vertebrates, that elicits killing behavior by a predator. Reduction or elimination of the cues for killing might provide the potential prey animal with later opportunities for escape, especially from predators that transport whole, immobile prey to their young (e.g. raptors, canids). Alternative explanations involve the observation that many of the snakes that death feign eat frogs and toads, have enlarged adrenal glands, and at least one (Heterodon sp) exhibits bradycardia during death feigning. death feigning behavior is facilitated by typically high levels of circulating catecholamines and that it might represent a non-adaptive consequence of toad-eating. Evidence against this is that the death feigning response occurs repeatedly if a threat is rapidly presented and removed from a hognose snake and that the false spitting cobra remains alert and ready to bite while death feigning.

Squirting Blood from the Eyes and or Nostrils

A number of species of horned lizards (Phrynosoma sp) and at least one species of boid (Tropidophis sp) squirt or exude blood from the eyes and or nostrils as a defensive mechanism. In horned lizards the ability to squirt blood is derived from modifications to the cephalic circulation. Restriction of blood flow from the head results in increased vascular pressure. Subsequent contraction, of the protrusure oculi muscles, ruptures capillaries in and around the eyes which causes a thin stream of blood to be ejected up to 2 meters. This defensive mechanism seems to be an especially effective deterrent to some mammalian predators (e.g. Canids).

Tail Display

Tail displays as a defense mechanism are fairly common being seen in pipe snakes (Anilliidae), shield tailed snakes (Uropeltidae), ring necked snakes (Colubridae), burrowing pythons, sand boas and rubber boas (Boidae), coral snakes and shield nosed snakes (Elapidae) as well as a few lizard species. This behavior probably developed to divert a predators attack away from the vulnerable head to the more disposable tail.

The display varies between the different species and families but usually involves hiding the head beneath the body (coral snakes, shield tailed snakes) or within a coiled up body (sand boas, burrowing python). The tail is then waved around in the air, coiled tightly or moved in a manner that mimics a striking head (rubber boas). Many species (pipe snakes, coral snakes, ring necked snakes) have a brightly colored contrasting ventrum that is exposed when the tail is displayed. This can serve to startle the predator and give the snake a chance to escape; the tails of many of these individuals display scars from previous attacks that attest to the effectiveness of this defensive behavior. In many species, especially the coral snakes, the tail display is accompanied by writhing and cloacal discharge.

These displays can be misinterpreted as an animal in pain or possible neurologic disease.

Everting Hemipenes

Another defense mechanism that is employed by many reptiles and is especially well developed in some boids (blood python) elapids (coral snakes) and monitor species, is eversion of one or both hemipenes. When harassed, these animals will often writhe violently, empty there cloacal contents and evert one or both hemipenes. Since this behavior is often associated with manipulation of the reptile by the owner the novice owner may misinterpret this as an injury; the owner has caused the abdominal organs to prolapse. In the vast majority of cases the hempenis or hemipenes will reduce back into the tail upon cessation of the stimuli. A similar phenomena can occur during defecation.

Puffing up Body, Throat and Hissing

A portion of many defensive displays in reptiles and amphibians involve bluffs and threats. One of the more common of these behaviors is increasing the body size by inflating the lungs and puffing up the body. Subsequently, the air is rapidly expelled from the lungs in a long loud hiss. This display behavior is very common in many families of snakes and lizards. Pine snakes have developed a septum at the glottal opening (glottal keel) the serves to amplify the sound produced during hissing. The behavior can deter attack by a predator by being extremely vigorous as in many snakes and lizards or by being passive serving to increase the body size beyond the gap of many predators as in the Uromastyx sp.

For the herpetoculturist, this rapid influx and efflux of air may resemble an animal with respiratory disease. Inflation of the throat region in bearded dragons, chameleons, monitors and boid and pine snakes can look like excess fluid build up.

As an adjunct to the above, many reptiles especially monitors and rat snakes (Asian rat snakes) will flatten the caudal area of the head and the cranial are of the neck to increase size in a threat display. This display also has the effect of uncovering bright colors of the skin that are normally concealed by the scales. These bright colors can startle an attacker and give the animal time to escape. Many of the Asian rat snakes and monitors will also hold the head in a bent position and move the tongue slowly and stiffly during the display. This can be misinterpreted as respiratory disease or injury to the muscles or nerves of the head and neck.

Chameleon Color Change-dropping from Branch

Chameleons are for the most part strictly diurnal animals that cease all movement as soon as the sun goes down or the cage lights are turned out. Most species will sleep at the ends of thin branches and their color becomes blanched almost to a white. If they are disturbed in the wild or captivity at night they drop as if dead from the branch. This is an effective escape behavior that allows the chameleon to avoid predation but the uninformed owner may think the color change is abnormal or that the fall from the branch indicates a sick animal.

Chameleons are also noted for and are especially adept at color change due to an elaborate system of chromatophores in the dermis. If two unfamiliar chameleons, two male chameleons or a non-receptive female is placed with a courting male a violent color change can occur. This may be interpreted as illness. These changes are behaviorally and hormonally induced and illness secondary to stress can ensue if husbandry changes are not instituted.

Fragile Skin in Geckos

Most geckos species are nocturnal, have special scansorial foot pads for adhesion to walls and ceilings and rely on crypsis or rapid locomotion for escape from predators such as snakes, owls and bats. Tail autonomy is almost ubiquitous. A number of species like the common house gecko (Aristelliger hechti), Gehyra mutilata, Pachydactylus manquensis, Phelsuma sp, Teratoscincus scincus and Ctenotus lenista have evolved very fragile skin. These species will loose large portions of epidermis even when handled gently. In species that have been studied such as the bronze gecko (Ailuronyx seychellensis) of the Seychelles and the Madagascan genus Geckolepis the skin on the animals back is weakened by gaps in the fibrous structure of the skin. When predators grasp the lizard most of the skin, except for a thin layer that remains to protect underlying tissues, tears away. This is probably an escape mechanism similar in function to tail loss and does not indicate a nutritional, endocrine or other similar disease.

Balling up

Another defense mechanism similar to hiding the head under a loop of coils, is hiding the head in the center of a ball of coils. A number of boids employ this method of defense. The most well known of these boid species is the ball python (Python regius) of West Africa. A number of other species including the African burrowing python (Calabaria rehinhardii) and the rosy boas (Lichanura sp.) exhibit this behavior.

When harassed these boids roll into a tight ball with the vulnerable head in the middle of the coils. Protection from trauma is the obvious reason for this behavior. Field observations of Lichanura trivirgata support a less obvious reason for the behavior. On two separate occasions when these small boas were encountered under rocks, the animals balled up and rolled out of reach of the observer into nearby crevices. This locomotor use of the balling behavior may be used to escape predators, like ringtails, that live in the area and eat snakes. A similar behavior is exhibited by the armadillo girdled tailed lizard (Cordylus cataphractus) of Africa. When threatened in the open this lizard will grasp the heavily armored tail in its mouth and form a circle to shield its more vulnerable under parts from attack.

Tail Vibrating

The vibrating of the tail in a disturbed rattlesnake is an almost universally known defensive behavior. Many other venomous and non-venomous snakes also employ tail vibrating as part of a defensive repertoire. Tail vibration as a behavior is usually seen as part of an overall defensive strategy that involves striking, writhing and emptying of the cloacal contents. The behavior is characterized by a rabid tail movement or vibration that if done in dry leaves can produce a buzz that is audible to the human ear.

This behavior is restricted to and widespread among snakes and is exhibited by many commonly kept colubrids such as king snakes, bull and pine snakes, rat snakes, racers and coachwhips. It is also a common behavior of many species of vipers. In the bushmaster (Lachesis muta) a large Central and South American viper sound production is enhanced by the peculiar shape and arrangement of the terminal scales of the tail. While the behavior would most likely not be misinterpreted as disease a veterinarian may get a call because a person believes the snake to be a rattlesnake.

Tail Loss/Autonomy

This defense mechanism is a part of broad category of defenses designed to divert an attack away from a vulnerable part of the body and allow escape. While autonomy is common in invertebrates it occurs, in vertebrates, only in some salamanders, several rodents and the majority of lizards. At least some species in all families except the Agamidae, Chamaeleontidae, Helodermatidae, Lanthonodidae, Xenosauridae and Varanidae exhibit total or partial tail autonomy when grasped by a predator. After the tail is shed it will continue to wiggle often vigorously distracting the attacker and allowing the lizard an opportunity to escape.

In most species that exhibit this behavior anatomic modifications are present in the tail that serve to direct where fractures occur. The ruptures occur at so called fracture planes that are located within a single vertebral body not between vertebral bodies. This division in the vertebrae corresponds to a similar division between two muscle segments and this zone of weakness is maintained even to the level of the skin. Blood loss is minimal in autonomy and for those lizards that escape an attack the tail will regrow over a period of months. The lost vertebrae are replaced by a cartilaginous rod and the muscles and scales that regrow are irregular in shape and may differ somewhat in color.

Tail loss while not a sign of disease has costs associated with it. The tail is often a common site for fat deposition and some lizards use the tail in grasping or social interactions. An owner should be informed that males may not breed without tails and females may not have sufficient energy stores to produce viable young during the period between tail loss and regrowth.

Salt Glands

Accessory salt glands are present in all orders of reptiles including many commonly kept as pets such as green iguanas and Uromastyx sp. These glands function to rid the animals of excess salts accumulated, during feeding (marine iguanas), as a result of the environment (sea turtles), or as a way of conserving water (iguanas, desert species). Clinically, the salts are sneezed out of the nares often in a fine spray and white deposits are seen around the nares or on the glass of the caging. The owner may report this as respiratory or fungal disease to the practitioner.

The salt water crocodile (Crocodylus porosus) also has fairly extensive salt glands located in the tongue. These glands serve the same function as those glands located in the nares and can be seen as a discoloration of the tongue that may be misinterpreted as pathology.

The reptilian kidney cannot cope with excess salts absorbed from sea water. Sea snakes and file snakes possess a modified salivary gland located beneath the tongue. Excess salt is excreted into the tongue sheath and when the snake protrudes the tongue the salt is extruded into the sea. The homolopsines (South American water snakes) possess a similar gland located in the front of the roof of the mouth. Sea turtle salt glands are modified tear glands and excess salts are excreted from the corners of the eyes.

Thermoregulation

Reptiles do not have the ability to internally regulate their body temperature. Normally, the body temperature of an exposed reptile quickly approaches that of the environment. Because of this a number of strategies are employed to maintain the body temperature at a desired level. When too cool, reptiles bask in the sun and then shuttle between a cool shaded area and an open sunny one. Many nocturnal species will expose portions of their body to the sunlight or hide under bark or rocks and absorb heat by transfer from these objects to themselves (thigmothermy). Some reptiles may also limit activity to certain times (morning and evening) thus avoiding extremes of temperature. Length of basking is controlled hormonally via the pineal gland also known as the third eye.

A few of the behaviors associated with thermoregulation could be misinterpreted as disease, trauma or pathology. Many reptiles will begin morning basking by exposing only the head from a crevice or burrow. Heat uptake in this situation is enhanced by hormonally controlled darkening of the skin through the dispersal of melanin. Seeing this color change in a pet animal could be interpreted as disease or trauma. Other strategies used are to move all or part of the body off the ground to avoid picking up heat radiated from the surface. As an example, a lacertid lizard that lives in the Namib Desert of Africa raises a foreleg and the opposite hind leg simultaneously, balancing on the other two legs. This allows the skin of the elevated feet to cool. The lizard then alternates the lifting and lowering of the diagonal pairs of legs as long as it is exposed to the hot sand. This behavior could be interpreted as pain or neurologic disease. The pineal or third eye itself may be mistaken for a defective scale or an injury to the head. Lastly, many tortoises and some monitor species sun themselves with the head and neck fully extended and the forelimbs arranged limply. They can and do look as if they are dead.

Although not strictly associated with thermoregulation, Asian water dragons (Physignathous concincinus) will commonly rest or sleep underwater. While this behavior is normal owners may believe their pet has drowned or is drowning.

Reproductive Behavior

Squamate Reproduction and Courtship

During the breeding season many reptiles undergo a color or body shape change. Broad headed skinks develop an orange head and iguanas can have an overall orange or reddish appearance to the body especially the head, dewlap and forelimbs. These changes are associated with reproductive hormones and do not represent the reddening of the skin seen in septicemia. Many male lizards also have a hormonally induced increase in the size and activity of there femoral pore glands during the breeding season. The femoral pores of green iguanas increase greatly in size and exude a waxy material containing pheromones which serve to attract female iguanas. These pores can look like abscesses to the owner.

Part of the behavioral repertoire in many iguanids includes social displays such as head bobbing, push-ups, dewlap extension and aggression (biting, tail lashing). Head bobbing is employed by both male and female iguanas with males generally displaying this behavior more frequently and with greater variety. Females bob in a jerky, erratic motion. They most commonly bob when irritated by a male iguana or when warning another iguana or owner away from food or a basking area.

Male iguanas have several different types of head bobs. All the different head bobs are executed in fluid smooth motions. The bobs are generally up and down in motion, can be fast or slow and may include some side to side movement. The shudder bob, is a warning and is characterized by a rapid vibration of the head in an up-down-sideways motion. At the completion of the movement, the head is held up for a few seconds and then a very deliberate up and down bob is performed. Subordinate males tend to bob more like females.

Head bobbing in an aggressive territorial male is accompanied by a full, rigid extension of the dewlap and extended legs lifting the body high off the ground. The body is presented laterally to object in question and the lungs are inflated. A very characteristic stilted, stiff gait is employed by the animal. This behavior and posture serves to increase the size of the silhouette presented by the iguana to the offending object.

Another behavior associated with aggression is the open mouth treat display (gaping). The mouth is opened very wide exposing the bright pink oral mucosa. The tongue will often be visible and in an extended, arched configuration and a low guttural hiss may accompany the display. These are strong warning and threat signs that may lead to full bites and tail lashing if not heeded. A modification of the gaping threat is employed by some iguanas when involved in aggressive or dominance encounters. These animals will sit with the body tensed and the mouth slightly opened. An iguana in this posture may bite at the slightest provocation.

It inconvenient to split aggressive behavior in iguanas into two categories: defensive and offensive aggression. Defensive aggression is most often associated with a threat such as another iguana or human attempting to displace the animal from a basking site, perch or food; grabs the iguana or invades the animals perceived territory. The behaviors most commonly associated with defensive aggression include the puffed-up, sideways stance, extended dewlap, open mouth threat and tail lashing; if provoked further they may bite.

Offensive aggression is associated with the breeding season or in the establishment of dominance hierarchies and usually involves sexually mature males. These iguanas will often stalk the human females in the household biting them in an attempt to breed. Many of these attacks are directed at the neck and face because male iguanas bite and hold onto the neck of the female to achieve intromission. These male iguanas will often view the human male in the household as competition and will perform head bobs and tail lashing at an increased rate towards that person. Since the person does not leave the iguanas "territory" these displays will often escalate into biting attacks. Although not the whole story, increases in aggression are generally associated with increases in testosterone. This breeding season aggression usually lasts for3-4 months and can occur anytime of the year but generally is from November-April.

Suggested ways to deal with breeding season aggression have been many and varied. Certain colors of cloths and time of month (menses/ovulation) for females can trigger attacks. Owners should be aware of what they are wearing and doing during attacks so they can become aware of a pattern if one exists. Adopting an aggressive stance or attitude can sometimes help to deter attacks. Reducing the photoperiod to 10 hours of light per day will sometimes reduce the frequency and vigorousness of attacks. Changing the room and or the cage the iguana lives in sometimes will help to reduce aggression as can providing a surrogate iguana, such as a towel a stuffed toy or a sock, to mate with has worked well for some iguana owners.

Castration can also be a helpful adjunct to one or more of the above mentioned suggestions for reducing aggression in mature male green iguanas. In one recent study, 16 mature male green iguanas were randomly placed into one of three experimental groups. One group of 5 animals was castrated before the onset of breeding season aggression, a second group of 5 animals was castrated during the breeding season when exhibiting aggressive behavior and a third group of 6 animals served as the control group; these animals were sham operated but were not castrated. There was a statistically significant reduction in aggressive behavior between the group castrated prior to breeding season but not between the group castrated during breeding season and the control, non-castrated group.

The courtship behavior of many squamates involves the male grasping and biting the female in the neck and upper back region. This behavior can range from mild, causing the female no harm, to vigorous with the creation of open bleeding wounds. In snakes the males will often grasp the female by the neck with an open mouth. This behavior can look like fighting or in the case of some snake species can appear as if one animal is tryingto the eat the other.

Incubation Behavior in Pythons

Egg brooding or shivering thermogenisis is a phenomenon well known to herpetologists and herpetoculturists. This behavior is characterized by a female python coiled tightly around a clutch of eggs and exhibiting rhythmic muscular contractions (shivering) of the body which produce sufficient heat to maintain a relatively constant temperature in the egg mass over a fairly wide ambient temperature range. This behavior has been reported for (species 289, table) and may be universal in pythons.

In studies with captive pythons, rhythmic contractions begin when the temperature of the egg mass falls below the desired temperature (approx 30°C). At 30°C no contractions occur; the female simply remains coiled around the egg mass. As the ambient temperature falls the female python begins to shiver and the number of contractions increases as the temperature drops up to a maximum of approximately 30 contractions per minute. Monitoring of the temperature of the egg mass revealed a constant differential of 5-7 °C (9-15 °F) between the egg mass and the surrounding environment and maintenance of the egg mass at 30°C.

This behavior can resemble a parasitic infection (mites causing irritation) a nutritional disease (hypocalcemia) or a neuromuscular disease especially as the female python will often continue to shiver even if the egg mass is removed. To stop the shivering behavior the animal should be moved to a clean cage and the old cage thoroughly cleaned and the substrate changed to remove all scent of the egg mass.

This behavior is energetically expensive to the females; who may loose up to half of their body weight during incubation and may require 3 years to regain enough body condition to breed again. The advantages to the species are that the eggs develop rapidly and the species who demonstrate egg brooding can reproduce successfully in cooler climates.

Turtle Reproduction and Courtship

The social interactions of turtles and tortoises associated with reproduction and courtship usually involve head-bobbing, males butting and biting the females to immobilize them, and then mounting the females shell from the rear. These behaviors are often accompanied by bellowing or whistling vocalizations.

In order to achieve intromission the male must incline his body towards the vertical, to varying degrees in different species. The extreme is seen in the box turtles, who may incline beyond the vertical.

This reproductive and courtship behaviors maybe seen as fighting and the vocalizations as pain or injury by the owners. Seeing a box turtle inclined backwards may look as if he is stuck in the females shell.

Geckos Licking Eyes

The vast majority of gecko species lack eyelids and instead have a evolved a spectacle analogous to that of snakes. A common behavior exhibited by most species is to clean the spectacle by licking it with the tongue. This behavior is normal but rule outs could include ulcers, retained spectacle or any kind of irritation.

Arm Waving in Bearded Dragons

Arm waving is seen in very young bearded dragons and can be present within a few days of hatching. It seems to serve both as an intraspecies signal, for identification as a bearded dragon, and as a submissive behavior. This behavior persists commonly in adult females and in subordinate males during aggressive encounters. This behavior could be misinterpreted as a painful or injured limb.

Caudal Luring

A behavior to attract and capture food using the tail as a lure is a behavior most common among vipers but has been suggested in boids (Madagascan ground boa, boa constrictor) and elapids. In most cases the tip of the tail is a bright color contrasting with that of the rest of body and in some instances is modified into the shape of an insect some (Bothrops bilineatus) are swollen and pink.

The tail is waved in a rhythmic fashion and will often imitate the movement of an invertebrate. Lizards, mice, frogs and toads are attracted to within striking distance and captured and eaten. This behavior occurs in both juvenile and adult animals of various species.

Stargazing in Boids and Crotalids

An interesting behavior that has been noted in captive boids and some crotalids is the adoption of a posture where the neck is bent in such a manner as to cause the head to be tilted upwards (stargazing). A similar posture is seen in snakes suffering from parasitic, viral bacterial or fungal CNS disease. The difference between the two postures is that the animals without disease revert to a normal posture when disturbed or manipulated and the ones with disease do not.

This behavior may be a feeding posture modified for captivity. Many animals are fed from top opening cages. Even with those cages that are front opening the prey items are often introduced from above. The snakes may have adapted to looking up for food in their search for prey.

Morphological Adaptations

A number of normal structures that are morphological adaptations have evolved in the chelonia that could be seen by an owner or practitioner as a disease process or traumatic injury.

Mud and musk turtles (Kinosternidae), box turtles (Terrepene, Emydidae) and hinge backed tortoises (Kinixys sp) have evolved a hinge or hinges in the plastron or carapace so that the animal can close itself in for protection from predators and equally important from lose of moisture. The hinges vary in number and location. In most species of mud turtles two hinges are present, on cranial and one caudal, that allow the turtles to completely close itself within the shell. Musk turtles have one cranially located hinge that serves to protect the front limbs and the head. Box turtles also have one hinge in the plastron but it is located further back than in musk turtles. This location allows the front and back halves of the plastron to close; completely sealing the animal within the shell. Hinge backed tortoises have a hinge located on either side of the carapace which affords protection of the hindquarters. Other species that have developed one or more hinges include African mud turtles, Asian box turtles and Madagascan and Egyptian tortoises.

A lesser degree of movement is found among certain species of pond and river turtles. Asian leaf turtles and neotropical wood turtles have a partial hinging of the plastron with ligamentous rather than bony connections between the plastron and the carapace. This gives flexibility to the plastron but the shell cannot be closed. This adaptation appears to have evolved to allow these species to lay their exceptionally large eggs which could not otherwise fit through the shell opening. In a number of these species (cane turtle, spiny turtle) only mature females develop the kinetic plastron.

Certain large headed, aggressive species such as snapping turtles, big headed turtles and Mexican musk turtles also have movable plastrons due to ligamentous rather than bony connections between the plastron and the carapace. In these species the flexibility of the plastron allows the large head to be retracted within the shell. With the jaws open and the head protected by the shell a formidable and nearly impregnable defensive is provided.

Similarly, the pancake tortoise of East Africa (Malachochersus tornieri) has a flat, soft shell as an adult. This condition is due to permanent open spaces or fontanels between the bony plates of the carapace and the plastron. These fontanels increase in size as the tortoises grow. These species live in rock outcroppings and are good climbers. When threatened they wedge themselves in rock crevices and are very difficult to extract. All of the above are normal structural modifications and should not be interpreted as traumatic injury or as a result of nutritional or other metabolic disease.

Many species of side necked turtles (genus) can exude fluid from the plastron, generally in the area of the bridge. It is unknown if this fluid is for defense or a pheromone for courtship and reproduction. This should not be mistaken for fluid loss secondary to septicemia or trauma.

A morphological adaptation for attracting and capturing prey is possessed by the alligator snapping turtle (Malacochersus tornieri). This bifurcate modification of the tongue is wormlike in appearance and can be made to wiggle by contractions of the muscular base that it is attached to. This appendage can look like an oral parasite to the laymen.

Morphological Adaptations

The tongues of many lizards are very colorful or of two different colors. Blue tongued skinks have a bright blue tongue that they expose when harassed. Iguanas have a two toned tongue; the tip being darker than the body. These colors do not represent cyanosis in the skink or an infection and necrosis in the iguana.

Many gecko species have bilateral structures, located in the ventro-lateral cervical region, that are involved in calcium metabolism and storage. These structures are called endolymphatic glands or chalk glands. During the breeding season these glands will often hypertrophy and fill with calcium in preparation for egg laying. These structures are very prominent on radiographs and may be misinterpreted as an abnormal mass.

Male gharials, a slender snouted crocodilian from India, develop a large bulbous growth on the tip of snout during the breeding season. This growth is a normal structure and does not represent a granuloma or tumor.

In snakes the lower jaw is loosely connected by a ligament instead of a solid bony connection as is seen in most vertebrates. During feeding the two halves of the mandible can move independently and stretch out of shape to aid in prehension and swallowing of food. After feeding the snake manipulates the jaws by yawning to reposition the mandible back into normal position; this can take few minutes to accomplish. This should not be mistaken for a fracture of the mandibular symphysis.

FLIGHT ADAPTATION IN AVES

Over the aeons, birds have evolved not only wings but many other adaptations that help them to fly. Birds have a strong, but a lightweight framework of bones. This is achieved by the fusion and elimination of some bones while hollowing the remaining. Some bones of the pelvic girdle and vertebrae are fused together. Generally, there are two types of flight adaptations in birds:

- Morphological Adaptations.
- Anatomical Adaptations.

Morphological Adaptations

Body Contour

The birds have a spindle-shaped body to offer less air resistance during flight. This helps the birds to conserve energy and become more efficient at flying.

Compact Body

The body of a bird is compact, dorsally strong and ventrally heavy to maintain equilibrium in the air. Their wings are attached on the thorax, the light organs like lungs and sacs are positioned high, the heavy muscles placed centrally are other features that help in flight.

Body Covered With Feathers

The feathers are smooth, directed backwards, and closely fitting which make the body streamlined and reduce friction during flight. It lightens the body weight and protects it from the effect of environmental temperature. They also have a wide surface area for striking the air.

Feathers add to the body buoyancy. It insulates the body and prevents any loss of heat from the body. This helps the birds to bear low temperatures at higher altitudes.

Forelimbs Modified into Wings

The forelimbs are modified into wings which is the only organ of flight. These consist of a framework of bones, muscles, nerves, feathers, and blood vessels.

The wings have a large surface area. They also support the bird in the air. The wings have a thick strong leading edge with a concave lower surface and a convex upper surface. This helps in increasing the air pressure below and reducing the air pressure above. Thus the bird can fly upward and forward during flight.

Mobile Neck and Head

The birds have a long and flexible neck which helps in the movement of head important for various functions. They possess a horny beak which helps them to pick the grains and insects while feeding.

Bipedal Locomotion

The anterior part of the body of a bird helps in taking off during flight. The anterior part of the body also helps birds to land. The hindlimbs help in the locomotion on land. They can support the entire body weight of a bird.

Perching

When a bird sits on the branch of a tree, its toes wrap around the twig. This is known as perching. The muscles are so well-developed that a bird can sleep in that position without falling.

Short Tail

The tail bears long feathers that spread like a fan and function as a rudder during flight. They also help in balancing, lifting, and steering while flying and perching.

Anatomical Adaptations

Flight Muscles

The well-developed muscles control the action of the flight muscles. It weighs about 1/6th of the entire bird. The flight muscles are striated. The muscles on the wings are large. Other muscles help the above muscles in functioning.

Light and Rigid Endoskeleton

The birds have a very stout and light skeleton. The bones are hollow, filled with air sacs. They are provided with a secondary plastering to increase their rigidity. The bones are fused and lack bone marrow.

The birds lack teeth. The thoracic vertebrae are fused except for the last one. This plays an essential role in the action of wings striking the air.

Digestive System

The birds have a very high rate of metabolism. Therefore, food digests rapidly. The length of the rectum is reduced because of the minimum undigested waste. They have no gall bladder which reduces the weight of the bird.

Respiratory System

The respiratory system of birds is designed in such a manner that the food is oxidised rapidly and a large amount of energy is liberated. Since the metabolism rate is higher, a large number of oxygen molecules are required by the body. For this, the lungs are provided which occupy the entire space between the internal organs.

Circulatory System

Rapid supply of oxygen is required by the blood due to rapid metabolism rate in birds. Therefore, birds require an efficient circulatory system. Birds have a four-chambered heart that performs double circulation. This prevents the mixing of oxygenated and deoxygenated blood. Also, the birds contain a large amount of haemoglobin in their red blood cells which helps in the quick aeration of body tissues.

Warm Blooded

The temperature of the body of a bird remains high and does not change with the change in the environment. This facilitates the birds to fly at very high altitudes.

Excretory System

The nitrogenous waste is converted to less toxic organic compounds such as uric acid, and urates. They have no urinary bladder. The uriniferous tubules efficiently absorb water.

Structural and Functional Adaptations of Mammals

Integument and its Derivatives

Mammalian skin and especially its modifications distinguish mammals as a group. As the interface between an animal and its environment, the skin is strongly molded by the animal's way of life. In general the skin is thicker in mammals than in other classes of vertebrates, although as in all vertebrates it is made up of epidermis and dermis. Among mammals the dermis becomes much thicker than the epidermis. The epidermis is thinner where it is well protected by hair, but in places that are subject to much contact and use, such as the palms or soles, its outer layers become thick and cornified with keratin.

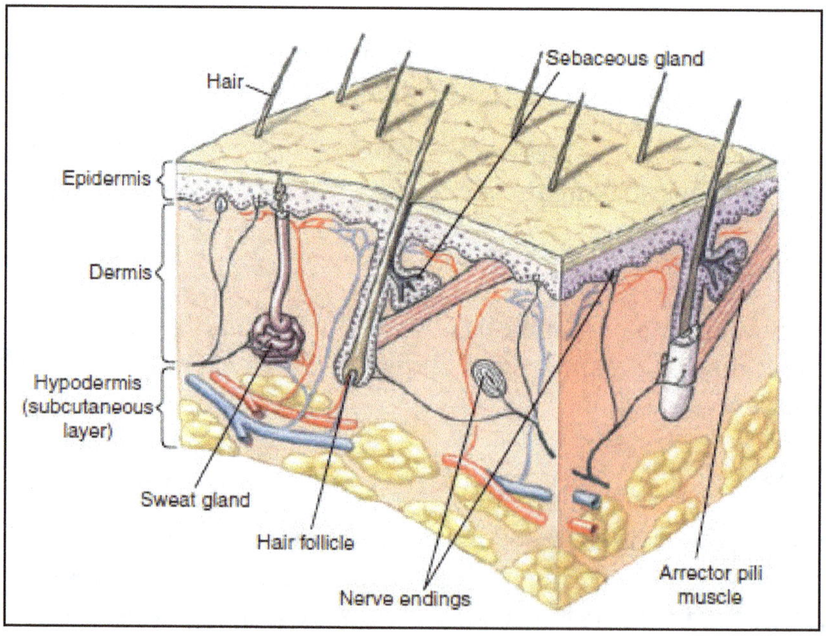

Structure of human skin (epidermis and dermis) and hypodermis, showing hair and glands.

Hair

Hair is especially characteristic of mammals, although humans are not very hairy creatures and, in whales, hair is reduced to only a few sensory bristles on the snout. A hair grows from a hair follicle that, although an epidermal structure, is sunk into the dermis of the skin. The hair grows continuously by rapid proliferation of cells in the follicle. As the hair shaft is pushed upward, new cells are carried away from their source of nourishment and die, turning into the same dense type of fibrous protein, called keratin, that constitutes nails, claws, hooves, and feathers.

Mammals characteristically have two kinds of hair forming the pelage (fur coat): (1) dense and soft under hair for insulation and (2) coarse and longer guard hair for protection against wear and to provide coloration. Underhair traps a layer of insulating air. In aquatic mammals, such as fur seals, otters, and beavers, it is so dense that it is almost impossible to wet. In water, guard hairs become wet and mat down, forming a protective blanket over the underhair.

American beaver, Castor canadensis, gnawing on an aspen tree. This second largest rodent (the South American capybara is larger) has a heavy waterproof pelage consisting of long, tough guard hairs overlying the thick, silky underhair so valued in the fur trade. Order Rodentia, family Castoridae.

(a)

(b)

Snowshoe, or varying, hare, Lepus americanus in A, brown summer coat and, in B, white winter coat. In winter, extra hair growth on the hind feet broadens the animal's support in snow. Snowshoe hares are common residents of the taiga and are an important prey for lynxes, foxes, and other carnivores. Population fluctuations of hares and their predators are closely related.Order Lagomorpha family Leporidae.

When a hair reaches a certain length, it stops growing. Normally it remains in the follicle until a new growth starts, whereupon it falls out. In most mammals there are periodic molts of the entire coat. In humans, hair is shed and replaced throughout life (although balding males confirm that replacement is not assured!).

In the simplest cases, such as foxes and seals, the coat is shed once every summer. Most mammals have two annual molts, one in the spring and one in the fall. Summer coats are always much thinner than winter coats and in some mammals it may be a different color. Several northern mustelid carnivores, for example, weasels, have white winter coats and brown-colored summer coats. It was once believed that the white inner pelage of arctic animals conserves body heat by reducing radiation loss; in fact, dark and white pelages radiate heat equally well. Winter white pelage of arctic animals is simply camouflage in a land of snow. The varying hare of North America has three annual molts: the white winter coat is replaced by a brownish gray summer coat, and this is replaced in autumn by a grayer coat, which is soon shed to reveal the winter white coat beneath. White fur of arctic mammals in winter (leukemism) is not to be confused with albinism, caused by a recessive gene that blocks pigment formation. Albinos have red eyes and pinkish skin, whereas arctic animals in their winter coats have dark eyes and often dark-colored ear tips, noses, and tail tips.

Outside the Arctic, most mammals wear somber colors that are protective. Often the species is marked with "saltand- pepper" coloration or a disruptive pattern that helps make it inconspicuous in its natural surroundings. Examples are the spots of leopards and fawns and the stripes of tigers. Skunks advertise their presence with conspicuous warning coloration.

The hair of mammals has become modified to serve many purposes. Bristles of hogs, spines of porcupines and their kin, and vibrissae on the snouts of most mammals are examples. Vibrissae, commonly called "whiskers," are really sensory hairs that provide a tactile sense to many mammals. The slightest movement of a vibrissa generates impulses in sensory nerve endings that travel to special sensory areas in the brain. Vibrissae are especially long in nocturnal and burrowing animals.

Dogs are frequent victims of the porcupine's impressive quills. Unless removed (usually by a veterinarian) quills will continue to work their way deeper in the flesh causing great distress and may lead to the victim's death.

Porcupines, hedgehogs, echidnas, and a few other mammals have developed an effective and dangerous spiny armor. When cornered, the common North American porcupine turns its back toward the attacker and lashes out with the barbed tail. The lightlyattached quills break off at their bases when they enter the skin and, aided by backward-pointed hooks on the tips, work deeply into tissues. Dogs are frequent victims but fishers, wolverines, and bobcats are able to flip the porcupine onto its back to expose vulnerable underparts

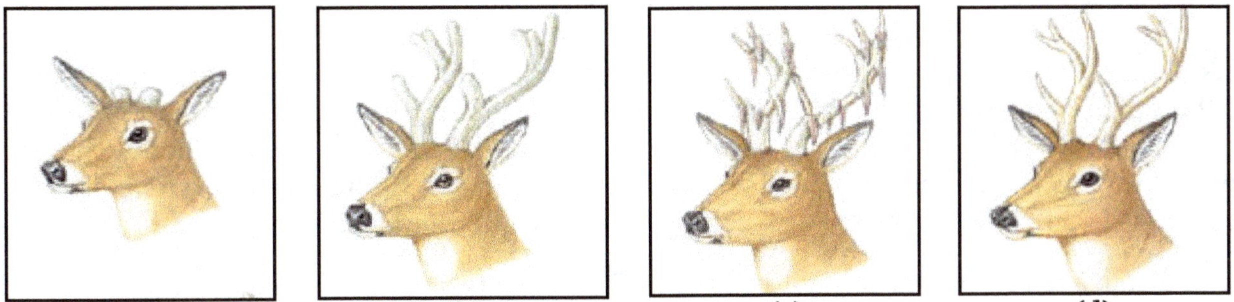

(a) (b) (c) (d)

Annual growth of buck deer antlers. A, Antlers begin growth in late spring, stimulated by pituitary gonadotropins. B, Bone grows very rapidly until halted by a rapid rise in testosterone production by the testes. C, The skin (velvet) dies and sloughs off. D, Testosterone levels peak during the fall breeding season. Antlers are shed in January as testosterone levels subside.

Horns and Antlers

Three kinds of horns or hornlike structures are found in mammals. True horns, found in ruminants (for example, sheep and cattle), are hollow sheaths of keratinized epidermis that embrace a core of bone arising from the skull. True horns are not normally shed, usually are not branched (although they may be greatly curved), grow continuously, and are found in both sexes. Horns may be absent from pronghorn antelope females but, if present, are shorter than those of the male.

Antlers of the deer family are branched and composed of solid bone when mature. During their annual spring growth, antlers develop beneath a covering of highly vascular soft skin called velvet. Except for caribou, only males of the species produce antlers. When growth of the antlers is complete just before the fall breeding season, the blood vessels constrict and the stag removes the velvet by rubbing the antlers against trees. Antlers are shed after the breeding season. New buds appear a few months later to herald the next set of antlers. For several years each new pair of antlers is larger and more elaborate than the previous set. Annual growth of antlers places a strain on the mineral metabolism, since during the growing season an older moose or elk must accumulate 50 or more pounds of calcium salts from its vegetable diet.

The rhinoceros horn is the third kind of hornlike structure. Hairlike keratinized filaments that arise from dermal papillae are cemented together to form these structures, which are not attached to the skull.

Glands

Of all vertebrates, mammals have the greatest variety of integumentary glands. Most fall into one of four classes: sweat, scent, sebaceous, and mammary. All are derivatives of the epidermis.

Sweat glands are tubular, highly coiled glands that occur over much of the body surface in most mammals. They are not present in other vertebrates. There are two kinds of sweat glands: eccrine and apocrine. Eccrine glands secrete a watery fluid that, if evaporated on the skin's surface, draws heat away from the skin and cools it. Eccrine glands occur in hairless regions, especially the foot pads, in most mammals, although in horses and most primates they are scattered over the body. They are either reduced or absent in rodents, rabbits, and whales. Apocrine glands are larger than eccrine glands and have longer and more convoluted ducts. Their secretory coil is in the dermis and extends deep into the hypodermis. They always open into a hair follicle or where a hair once was. Apocrine gland development occurs near puberty and is restricted (in the human species) to the axillae (armpits), mons pubis, breasts, prepuce, scrotum, and external auditory canals. In contrast to the watery secretions of eccrine glands, apocrine secretions are milky fluids, whitish or yellow in color, that dry on the skin to form a film. Apocrine glands are not involved in heat regulation. Their activity is correlated with certain aspects of the reproductive cycle.

Scent glands are present in nearly all mammals. Their location and functions vary greatly. They are used for communication with members of the same species, for marking territorial boundaries, for warning, or for defense. Scent-producing glands are located in orbital, metatarsal, and inter-digital regions (deer); behind the eyes and on the cheek (pica and woodchuck); penis (muskrats, beavers, and many canines); base of the tail (wolves and foxes); back of the head (dromedary); and anal region (skunks, minks, and weasels). The latter, the most odoriferous of all glands, open by ducts into the anus; their secretions can be discharged forcefully for 2 to 3 meters. During the mating season many mammals give off strong scents for attracting the opposite sex. Humans also are endowed with scent glands. However civilization has taught us to dislike our own scent, a concern that has stimulated a lucrative deodorant industry to produce an endless output of soaps and odor-masking concoctions.

Sebaceous glands are intimately associated with hair follicles, although some are free and open directly onto the surface. The cellular lining of the gland is discharged in the secretory process and must be renewed for further secretion. These gland cells become distended with a fatty accumulation, then die, and are expelled as a greasy mixture called sebum into the hair follicle. Called a "polite fat" because it does not turn rancid, it serves as a dressing to keep skin and hair pliable and glossy. Most mammals have sebaceous glands over the entire body; in humans they are most numerous in the scalp and on the face.

Mammary glands, which provide the name for mammals, are probably modified apocrine glands. Whatever their evolutionary origin, they occur on all female mammals and in a rudimentary form on all male mammals. They develop by the thickening of the epidermis to form a milk line along each side of the abdomen in the embryo. On certain parts of these lines the mammae appear while the intervening parts of the ridge disappear. Human female mammary glands begin to increase in size at puberty because of fat accumulation and reach their maximum development in approximately the twentieth year. Breasts (or mammae) undergo additional development during pregnancy. In other mammals the mammae are swollen only periodically when they are distended with milk during pregnancy and subsequent nursing of the young.

Food and Feeding

Mammals exploit an enormous variety of food sources; some mammals require highly specialized diets, whereas others are opportunistic feeders that thrive on diversified diets. Food habits and physical structure are thus inextricably linked. A mammal's adaptations for attack and defense and its specializations for finding, capturing, chewing, swallowing, and digesting food all determine a mammal's shape and habits.

Teeth, perhaps more than any other single physical characteristic, reveal the life habit of a mammal. With certain exceptions (monotremes, anteaters, certain whales), all mammals have teeth, except monotremes, anteaters, and certain whales, and their modifications are correlated with what the mammal eats.

As mammals evolved during the Mesozoic, major changes occurred in teeth and jaws. Unlike the uniform homodont dentition of the reptiles, mammalian teeth became differentiated to perform specialized functions such as cutting, seizing, gnawing, tearing, grinding, and chewing. Teeth differentiated in this manner are called heterodont. Mammalian dentition is differentiated into four types: incisors, with simple crowns and sharp edges, used mainly for snipping or biting; canines, with long conical crowns, specialized for piercing; premolars, with compressed crowns and one or two cusps, suited for shearing and slicing; and molars, with large bodies and variable cusp arrangement, used for crushing and grinding. The primitive tooth formula for most mammals, which expresses the number of each tooth type in one-half of the upper and lower jaw, was I 3/3, C 1/1, PM 4/4, M 3/3 = 44. Members of the order Insectivora (shrews), some omnivores, and carnivores come closest to this primitive pattern.

Unlike reptiles, mammals do not continuously replace their teeth throughout their lives. Most mammals grow just two sets of teeth: a temporary set, called deciduous, or milk, teeth, which is replaced by a permanent set when the skull has grown large enough to accommodate a full set. Only the incisors, canines, and premolars are deciduous; the molars are never replaced and the single permanent set must last a lifetime.

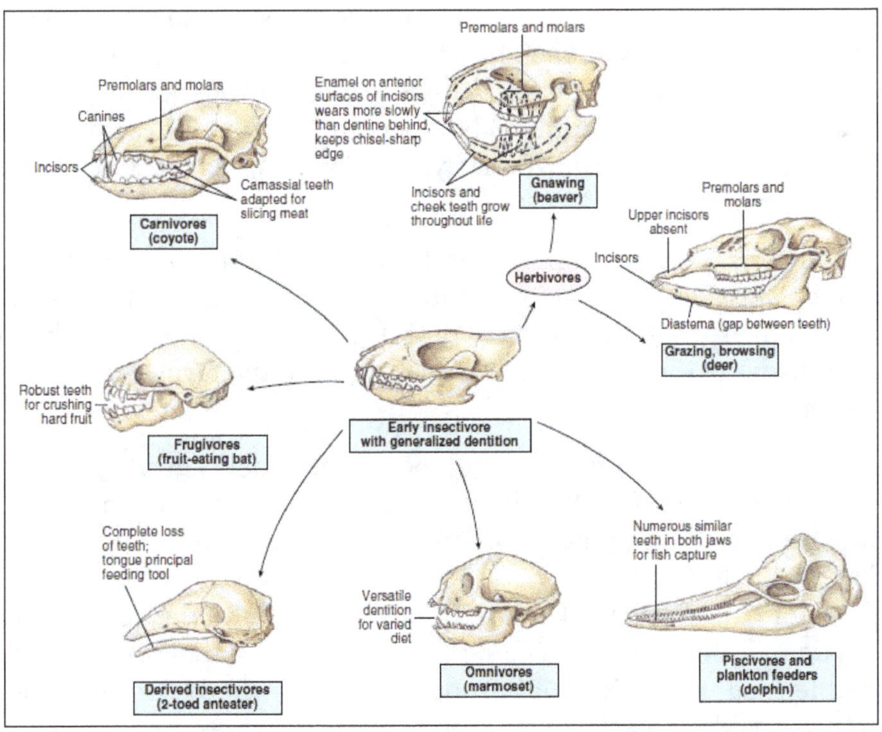

Feeding specializations of major trophic groups of eutherian mammals. Early eutherians were insectivores; all other types are descended from them.

Feeding Specializations

The feeding, or trophic, apparatus of a mammal—teeth and jaws, tongue, and alimentary canal—are adapted to its particular feeding habits. Mammals are customarily divided among four basic categories—insectivores, carnivores, omnivores, and herbivores—but many other feeding specializations have evolved in mammals, as in other living organisms, and the feeding habits of many mammals defy exact classification. The principal feeding specializations of mammals are.

Insectivores are small mammals such as shrews, moles, anteaters, and most bats. They feed on insects, as well as a variety of small invertebrates, such as worms and grubs. Since insectivores eat little fibrous vegetable matter that requires prolonged fermentation, their intestinal tract tends to be short. The insectivorous category is not a sharply distinguished one because carnivores and omnivores may include insects in their diets. Even many rodents, which are considered herbivores, may have a mixed diet of insect larvae, seeds, and fruits.

Herbivorous mammals that feed on grasses and other vegetation form two main groups: browsers and grazers, such as the ungulates (hooved mammals including horses, deer, antelope, cattle, sheep, and goats), and the gnawers, such as the rodents, and rabbits and hares. In herbivores, the canines are absent or reduced in size, whereas the molars, which are adapted for grinding, are broad and usually high-crowned. Rodents (for example, beavers) have chisel-sharp incisors that grow throughout life and must be worn away to keep pace with their continual growth.

Herbivorous mammals have a number of interesting adaptations for dealing with their fibrous diet of plant food. Cellulose, the structural carbohydrate of plants, is composed of long chains of glucose molecules, and therefore is a potentially nutritious food resource. However, the glucose molecules in cellulose are linked by a type of chemical bond that few enzymes can attack. No vertebrates synthesize cellulose-splitting enzymes. Instead, herbivorous vertebrates harbor anaerobic bacteria and protozoa in huge fermentation chambers in their gut. These microorganisms break down and metabolize cellulose, releasing a variety of fatty acids, sugars, and starches that the host animal can absorb and utilize.

Some herbivores, such as horses, zebras, rabbits, hares, elephants, and many rodents, have a gut with a spacious sidepocket, or diverticulum, called a cecum, which serves as a fermentation chamber and absorptive area. Hares, rabbits, and some rodents often eat their fecal pellets (coprophagy), giving the food a second pass through the fermenting action of the intestinal microorganisms.

Ruminants (cattle, bison, buffalo, goats, antelopes, sheep, deer, giraffes, and okapis) have a huge four-chambered stomach. As a ruminant feeds, grass passes down the esophagus to the rumen, where it is broken down by microorganisms and then formed into small balls of cud. At its leisure, the ruminant returns the cud to its mouth where the cud is deliberately chewed at length to crush the fiber. Swallowed again, the food returns to the rumen where it is digested by the cellulolytic microorganisms. The pulp passes to the reticulum, then to the omasum, where water, soluble food, and microbial products are absorbed. The remainder proceeds to the abomasum ("true" acid stomach), where proteolytic enzymes are secreted and normal digestion takes place.

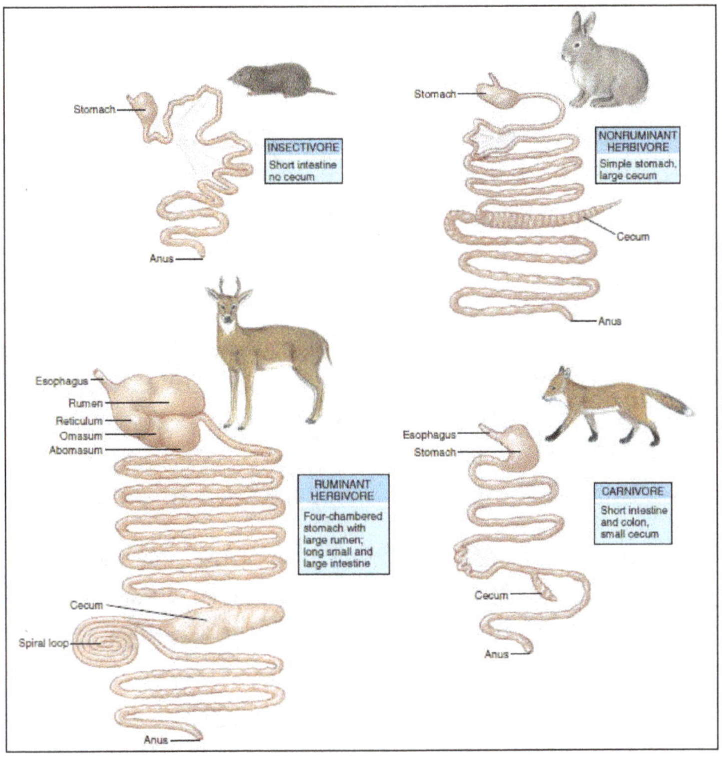

Digestive systems of mammals, showing different morphology with different diets.

Herbivores generally have large, long digestive tracts and must eat a considerable amount of plant food to survive. An African elephant weighing 6 tons must consume 135 to 150 kg (300 to 400 pounds) of rough fodder each day to obtain sufficient nourishment for life.

Carnivorous mammals feed mainly on herbivores. This group includes foxes, dogs, weasels, wolverines, fishers, cats, lions, and tigers. Carnivores are well-equipped with biting and piercing teeth and powerful clawed limbs for killing their prey. Since their protein diet is more easily digested than the woody food of herbivores, their digestive tract is shorter and the cecum small or absent. Carnivores organize their feeding into discrete meals rather than feeding continuously (as do most herbivores) and therefore have much more leisure time.

Lionesses, Panthera leo, eating a wildebeest. Lacking stamina for a long chase, lions stalk prey and then charge suddenly surprising the prey. Lions gorge themselves with their kill, then sleep and rest for periods as long as one week before eating again. Order Carnivora, family Felidae.

In general, carnivores lead more active—and by human standards more interesting—lives than do the herbivores. Since a carnivore must find and catch its prey, there is a premium on intelligence; many carnivores, such as the cats, are noted for their stealth and cunning in hunting prey. This has led to a selection of herbivores capable either of defending themselves or of detecting and escaping carnivores. Thus for herbivores, there has been a premium on keen senses, speed and agility. Some herbivores, however, survive by virtue of their sheer size (rhinos, elephants) or by defensive group behavior (for example, muskoxen).

Humans have changed the rules in the carnivore-herbivore contest. Carnivores, despite their intelligence, have suffered much from human presence and have been virtually exterminated in some areas. Small herbivores, on the other hand, with their potent reproductive ability, have consistently defeated our most ingenious efforts to banish them from our environment. The problem of rodent pests in agriculture has intensified; we have removed carnivores, which served as the herbivores' natural population control, but have not been able to devise a suitable substitute.

Omnivorous mammals—pigs, raccoons, rats, bears, and most primates, including humans—live on both plants and animals for food. Many carnivorous forms also eat fruits, berries, and grasses when hard pressed. Foxes, which usually feed on mice, small rodents, and birds, eat frozen apples, beechnuts, and corn when their normal food sources are scarce.

Eastern chipmunk, Tamias striatus, with cheek pouches stuffed with seeds to be carried to a hidden cache. It will try to store several liters of food for the winter. It hibernates but awakens periodically to eat some of its cached food. Order Rodentia, family Sciuridae.

For most mammals, searching for food and eating occupy the majority of their active life. Seasonal changes in food supplies are considerable in temperate zones. Living may be easy in the summer when food is abundant, but in winter many carnivores must range far and wide to eke out a narrow existence. Some migrate to regions where food is more abundant while others hibernate and sleep the winter months away.

Many mammals cache food stores during periods of plenty. This habit is most pronounced in rodents, such as squirrels, chipmunks, gophers, and certain mice. All tree squirrels—red, fox, and gray—collect nuts, conifer seeds, and fungi and store these in caches for winter use. Often each item is hidden in a different place (scatter hoarding) and marked by a scent to assist relocation in the future. Some of the caches of chipmunks and red squirrels can be quite large.

Body Weight and Food Consumption

The relationship between body size and metabolic rate was discussed in relation to food consumption of birds. The smaller the animal, the greater is its metabolic rate and the more it must consume relative to its body size. This happens because the metabolic rate of an animal—and therefore the amount of food it must eat to sustain this metabolic rate—varies in rough proportion to the relative surface area rather than to the body weight. Surface area is proportional to approximately 0.7 power of body weight, and the amount of food a mammal (or bird) eats also is roughly proportional to a 0.7 power of its body weight. For example, a 3 g mouse will consume per gram body weight five times more food than does a 10 kg dog and about 30 times more food than does a 50,000 kg elephant. Thus small mammals (shrews, bats, and mice) must spend much more time hunting and eating food than do large mammals. The smallest shrews weighing only 2 g may eat more than their body weight each day and will starve to death in a few hours if deprived of food. In contrast, large carnivores can remain fat and healthy with only one meal every few days. Mountain lions are known to kill an average of one deer a week, although they will kill more frequently when game is abundant.

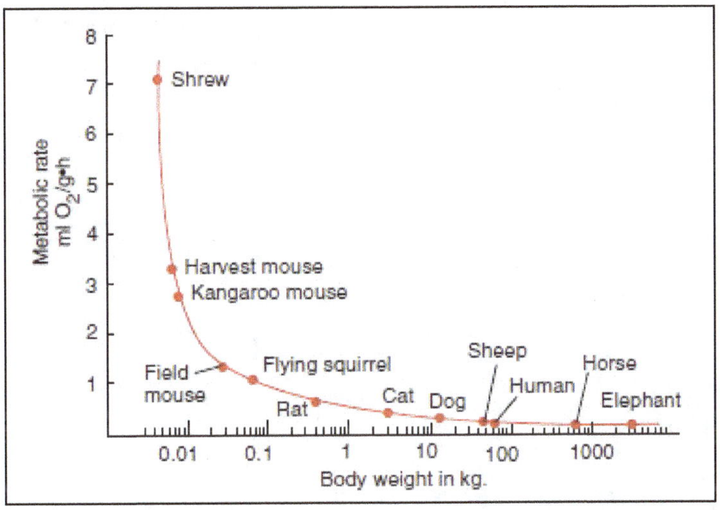

Relationship between body weight and metabolic rate for mammals. This relationship often called the "mouse-to-elephant" curve, shows that metabolic rate is intense for small mammals like shrews and mice, and declines with increasing body weight of the species.

The shorttail shrew, Blarina brevicauda, eating a grasshopper. This tiny but fierce mammal, with a prodigious appetite for insects, mice, snails, and worms, spends most of its time underground and so is seldom seen. Shrews are believed to resemble insectivorous ancestors of placental mammals. Order Insectivora, family Soricidae.

Migration

Migration is a more difficult undertaking for mammals than for birds. Not surprisingly, few mammals make regular seasonal migrations, preferring instead to center their activities in a defined and limited home range. Nevertheless, there are some striking examples of mammalian migrations. More migrators are found in North America than on any other continent.

An example is the barren-ground caribou of Canada and Alaska, which undertakes direct and purposeful mass migrations spanning 160 to 1100 km (100 to 700 miles) twice annually. From winter ranges in boreal forests (taiga), they migrate rapidly in late winter and spring to calving ranges on the barren grounds (tundra). The calves are born in mid- June. As summer progresses, caribou are increasingly harassed by warble and nostril flies that bore into their flesh, by mosquitoes that drink their blood (estimated at a liter per caribou each week during the height of the mosquito season), and by wolves that prey on their calves. They move southward in July and August, feeding little along the way. In September they reach the taiga and feed there almost continuously on low ground vegetation. Mating (rut) occurs in October.

The plains bison, before its deliberate near extinction by humans, made huge circular migrations to separate summer and winter ranges.

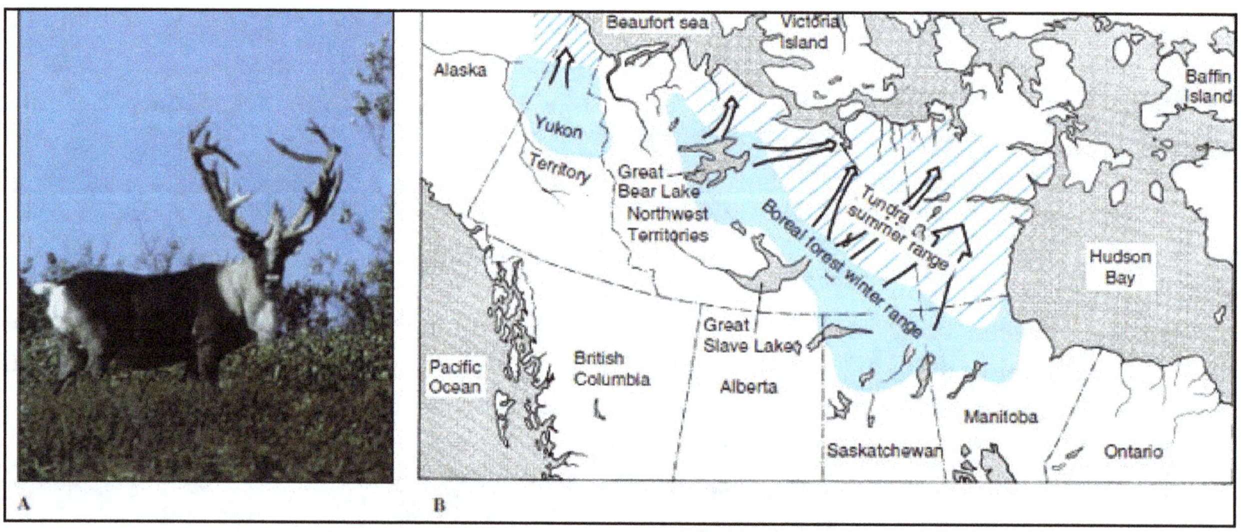

Barren-ground caribou, Rangifer tarandus, of Canada and Alaska. A, Adult male caribou in autumn pelage and antlers in velvet. B, Summer and winter ranges of some major caribou herds in Canada and Alaska (other herds not shown occur on Baffin Island and in western and central Alaska). Principal spring migration routes are indicated by arrows; routes vary considerably from year to year. The same species is known as reindeer in Europe. Order Artiodactyla, family Cervidae.

The longest mammalian migrations are made by the oceanic seals and whales. Gray whales, for example, migrate between Alaska in summer and Baja California, Mexico, in winter, an annual migration of over 18,000 km (11,250 miles). One of the most remarkable migrations is that of the fur seal, which breeds on the Pribilof Islands approximately 300 km (185 miles) off the coast of Alaska and north of the Aleutian Islands. From wintering grounds off southern California females journey as much as 2800 km (1740 miles) across open ocean, arriving in the spring at the Pribilofs where they congregate in enormous numbers. Young are born within a few hours or days after arrival of the cows. Then the bulls, having already arrived and established territories, collect harems of cows, which they guard with vigilance. After the calves have been nursed for approximately three months, cows and juveniles leave for their long migration southward. Bulls do not follow but remain in the Gulf of Alaska during the winter.

Although we might expect bats, the only winged mammals, to use their gift of flight to migrate, few of them do. Most spend winters in hibernation. Four species of American bats that migrate spend their summers in northern or western states and their winters in the southern United States or Mexico.

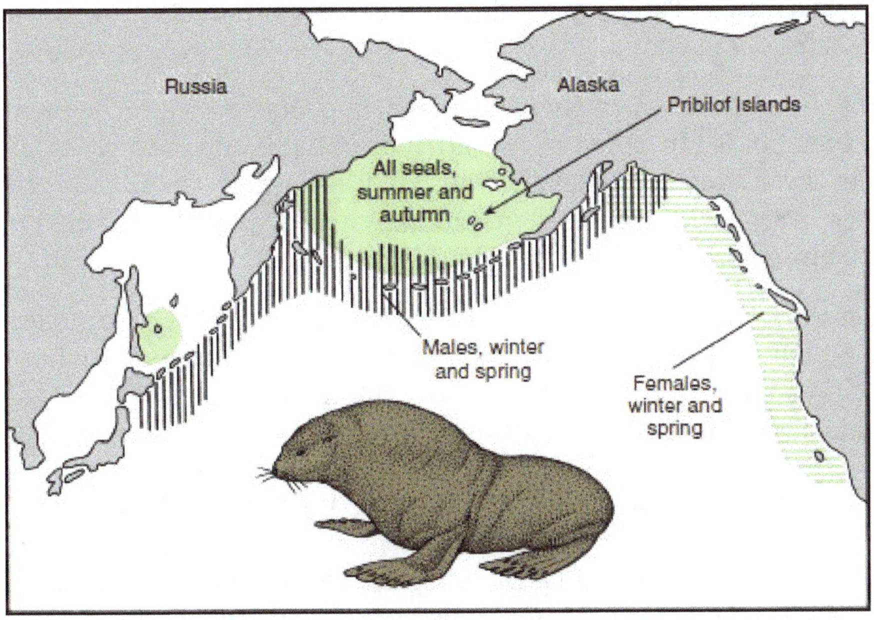

Annual migrations of the northern fur seals, showing the separate wintering grounds of males and females. Both males and females of the larger Pribilof population migrate in early summer to the Pribilof Islands, where females give birth to their pups and then mate. Order Carnivora, family Otariidae.

Eastern flying squirrel, Glaucomys sabrinus, gliding in for a landing.

Area of undersurface is nearly trebled when gliding skin is spread. Glides of 40 to 50 m are possible. Good maneuverability during flight is achieved by adjusting position of the gliding skin with special muscles. Flying squirrels are nocturnal and have superb night vision. Order Rodentia, family Sciuridae.

Flight and Echolocation

Mammals have not exploited the skies to the same extent that they have terrestrial and aquatic environments. However, many mammals scamper about in trees with amazing agility; some can glide from tree to tree and one group, the bats, is capable of full flight. Gliding and flying evolved independently in several groups of mammals, including marsupials, rodents, flying lemurs, and bats. Anyone who has watched a gibbon perform in a zoo realizes there is something akin to flight in this primate, too. Among the arboreal squirrels, all of which are nimble acrobats, by far the most efficient is the flying squirrel. These forms actually glide rather than fly, using the gliding skin that extends from the sides of the body.

Bats are nocturnal or crepuscular (active at twilight) and thus hold a niche unoccupied by most birds. Their achievement is attributed to two attributes: flight and capacity to navigate by echolocation. Together these adaptations enable bats to fly and avoid obstacles in absolute darkness, to locate and catch insects with precision, and to find their way deep into caves (a habitat largely ignored by both mammals and birds) where they sleep during the daytime hours.

Research has been concentrated on members of the family Vespertilionidae, to which most of the common North American bats belong. When in flight, bats emit short pulses 5 to 10 msec in duration in a narrow directed beam from the mouth or nose. Each pulse is frequency modulated; that is, it is highest at the beginning, up to 100,000 Hz (hertz, cycles per second), and sweeps down to perhaps 30,000 Hz at the end. Sounds of this frequency are ultrasonic to human ears which have an upper limit of about 20,000 Hz. When bats are searching for prey, they produce about 10 pulses per second. If prey is detected, the rate increases rapidly up to 200 pulses per second in the final phase of approach and capture. Pulses are spaced so the echo of each is received before the next pulse is emitted, an adaptation that prevents jamming. Since transmission-to-reception time decreases as the bat approaches an object, it can increase the pulse frequency to obtain more information about the object. Pulse length is also shortened as the bat nears the object. It is interesting that some prey of bats, certain nocturnal moths for example, have evolved ultrasonic detectors used to detect and avoid approaching bats.

External ears of bats are large, like hearing trumpets, and shaped variously in different species. Less is known about the inner ear of bats, but it obviously is capable of receiving the ultrasonic sounds emitted. Biologists believe bat navigation is so refined that the bat builds up a mental image of its surroundings from echo scanning that approaches the resolution of a visual image from eyes of diurnal animals.

For reasons not fully understood, all bats are nocturnal, even the fruiteating bats that use vision and olfaction instead of sonar to find their food. The tropics and subtropics have many nectar-feeding bats that are important pollinators for a wide variety of chiropterophilous ("batloving") plants. Flowers of these plants open at night, are white or light in color, and emit a musky, batlike odor that nectar-feeding bats find attractive.

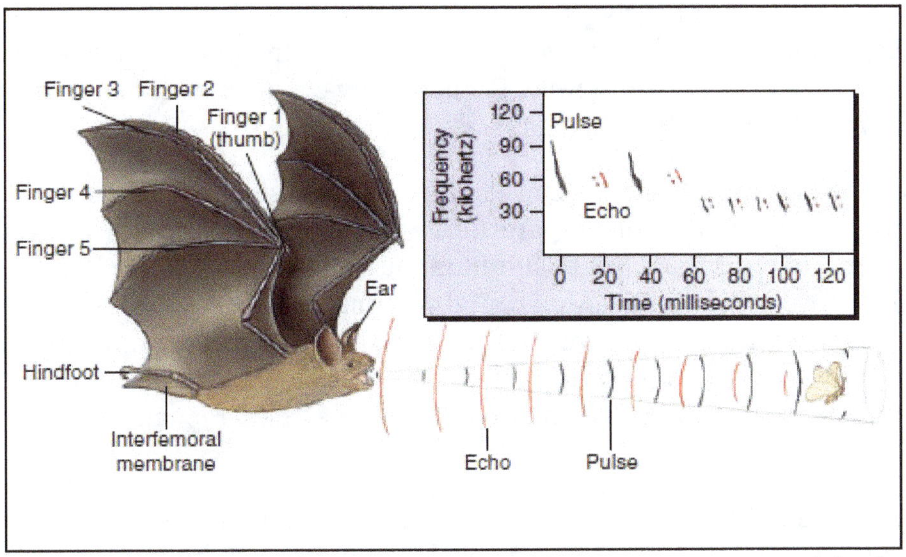

Echolocation of an insect by the little brown bat Myotis lucifugus. Frequency modulated pulses are directed in a narrow beam from the bat's mouth. As the bat nears its prey, it emits shorter, lower signals at a faster rate. Order Chiroptera, family Vespertilionidae.

The famed tropical vampire bat has razor-sharp incisors used to shave away the epidermis of its prey, exposing underlying capillaries. After infusing an anticoagulant to aid blood flow, it laps up and stores its meal in a specially modified stomach.

African lions Panthera leo mating. Lions breed at any season, although predominantly in spring andsummer.

During the short period a female is receptive, she may mate repeatedly. Three or four cubs are born after gestation of 100 days. Once the mother introduces the cubs into the pride, they are treated with affection by both adult males and females. Cubs go through an 18- to 24-month apprenticeship learning how to hunt and then are frequently driven from the pride to managethemselves. Order Carnivora, family Felidae.

Reproduction

Reproductive Cycles

Most mammals have definite mating seasons, usually in winter or spring and timed to coincide with the most favorable time of the year for rearing young after birth. Many male mammals are capable of fertile copulation at any time, but female mating function is restricted to a time during a periodic cycle, known as the estrous cycle. Females only copulate with males during a relatively brief period known as heat or estrus.

The estrous cycle is divided into stages marked by characteristic changes in the ovaries, uterus, and vagina. Proestrus, or period of preparation, when new ovarian follicles grow, is followed by estrus, when mating occurs. Almost simultaneously ovarian follicles burst, releasing eggs (ovulation), which are fertilized. In all placental mammals, the fertilized egg implants itself in the uterine wall with pregnancy following. However, should mating and fertilization not occur, estrus is followed by metestrus, a period of repair. This stage is followed by diestrus, during which the uterus becomes small and anemic. The cycle then repeats itself, beginning with proestrus.

How often females are in estrus varies greatly among different mammals. Animals that have only a single estrus during their breeding season are called monestrous; those that have a recurrence of estrus during their breeding season are called polyestrous. Dogs, foxes, and bats belong to the first group; field mice and squirrels are all polyestrous as are many mammals living in the more tropical regions of the earth. Old World monkeys and humans have a somewhat different cycle in which the postovulation period is terminated by menstruation, during which the, endometrium (lining of the uterus) collapses and is discharged with some blood.

Reproductive Patterns

There are three different patterns of reproduction in mammals. One pattern is represented by egg-laying (oviparous) mammals, the monotremes. The duck-billed platypus has one breeding season each year. Ovulated eggs, usually two, are fertilized in the oviduct. As they continue down the oviduct, various glands add albumin and then a thin, leathery shell to each egg. When laid, the eggs are about the size of a robin's egg. The platypus lays its eggs in a burrow nest where they are incubated for about 12 days. After hatching, the young suck milk from the fur of the mother around openings of the mammary glands. Thus in monotremes there is no gestation (period of pregnancy) and the developing embryos draw nutrients stored in their eggs, much as do the embryos of reptiles and birds. But in common with all other mammals, monotremes rear their young on milk.

Marsupials are pouched, viviparous mammals that exhibit a second pattern of reproduction. Although only the eutherians are called "placental mammals," the marsupials do have a primitive type of placenta, called a choriovitelline (or yolk sac) placenta. The embryo (blastocyst) of a marsupial is at first encapsulated by shell membranes and floats free for several days in the uterine fluid. After "hatching" from the shell membranes, the embryo does not implant, or "take root" in the uterus as it would in eutherians, but it does erode a shallow depression in the uterine wall in which it lies and absorbs nutrient secretions from the mucosa by way of the vascularized yolk sac. Gestation (the intrauterine period of development) is brief in marsupials, and therefore all marsupials give birth to tiny young that are effectively still embryos, both anatomically and physiologically. However, early birth is followed by a prolonged interval of lactation and parental care.

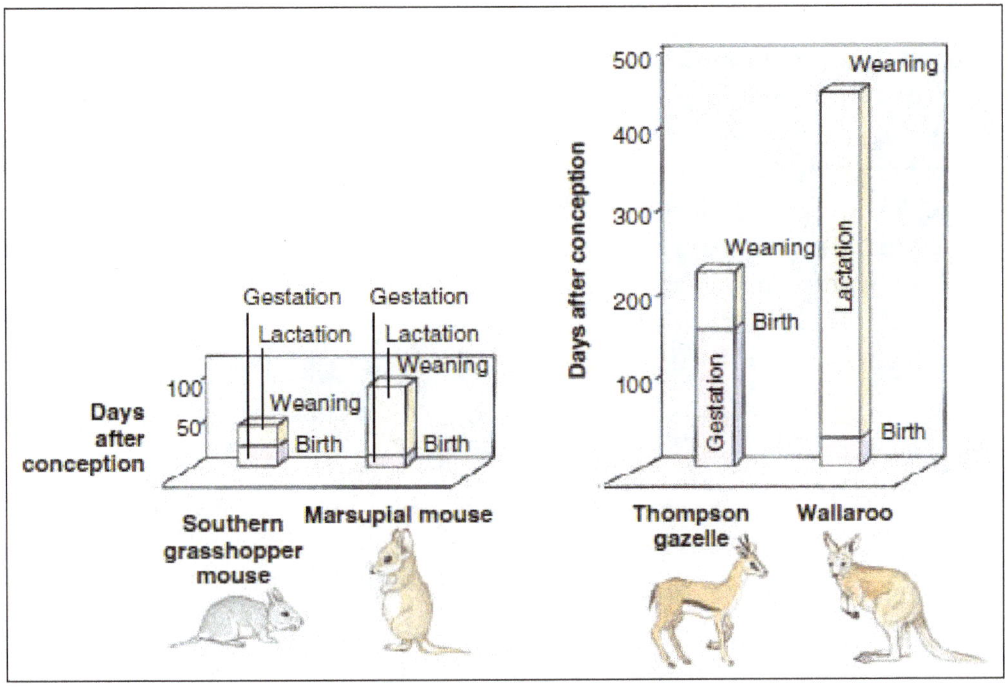

Comparison of gestation and lactation periods between matched pairs of ecologically similar species of marsupial and placental mammals. The graph shows marsupials have shorter intervals of gestation and much longer intervals of lactation than in similar species of placentals.

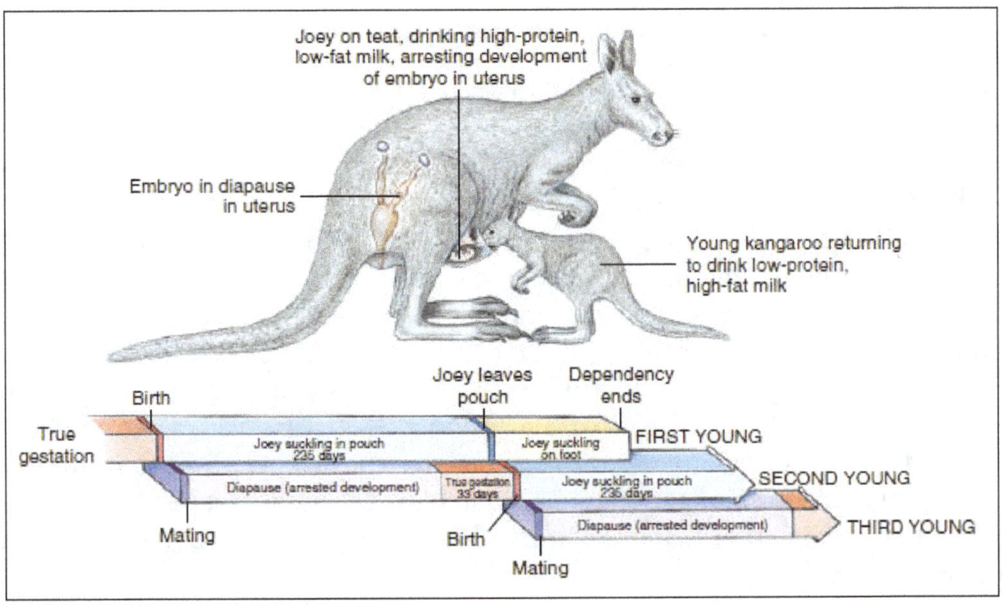

Kangaroos have a complicated reproductive pattern in which the mother may have three young in different stages of development dependent on her at once. Order Diprotodontia, family Macropodidae.

In red kangaroos the first pregnancy of the season is followed by a 33-day gestation, after which the young (joey) is born, crawls to the pouch without assistance from the mother, and attaches to a nipple. The mother immediately becomes pregnant again, but the presence of a suckling young in the pouch arrests development of the new embryo in the uterus at about the 100-cell stage. This period of arrest, called embryonic diapause, lasts approximately 235 days during which time the first joey is growing in the pouch. When the joey leaves the pouch, the uterine embryo resumes development and is born about a month later. The mother again becomes pregnant, but because the second joey is suckling, once again development of the new embryo is arrested. Meanwhile, the first joey returns to the pouch from time to time to suckle. At this point the mother has three young of different ages dependent upon her for nourishment: a joey on foot, a joey in the pouch, and a diapause embryo in the uterus. There are variations on this remarkable sequence—not all marsupials have developmental delays like kangaroos, and some do not even have pouches—but in all, the young are born at an extremely early stage of development and undergo prolonged development while dependent on a teat.

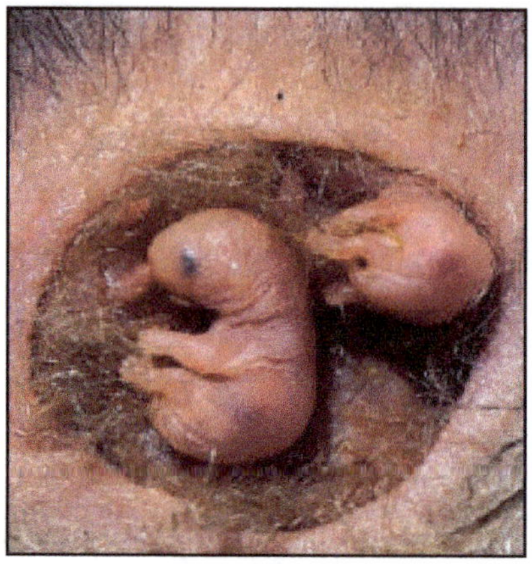

Opossums, Didelphis marsupialis, 15 days old, fastened to teats in mother's pouch. When bornafter a gestation period of only 12 days, they are the size of honeybees. They remain attached to the nipples for 50 to 60 days. Order Marsupialia, family Didelphidae.

The third pattern of reproduction is that of viviparous placental mammals, the eutherians. In placentals, the reproductive investment is in prolonged gestation, unlike marsupials in which the reproductive investment is in prolonged lactation. The embryo remains in the uterus, nourished by food supplied through a chorioallantoic type of placenta, an intimate connection between mother and young. Length of gestation is longer in placentals than marsupials, and in large mammals it is much longer. For example, mice have a gestation period of 21 days; rabbits and hares, 30 to 36 days; cats and dogs, 60 days; cattle, 280 days; and elephants, 22 months (the longest). But there are important exceptions (nature seldom offers perfect correlations). Baleen whales, the largest mammals, carry their young for only 12 months, while bats, no larger than mice, have gestation periods of 4 to 5 months. The condition of the young at birth also varies. An antelope bears its young well furred, eyes open, and able to run about. Newborn mice, however, are blind, naked, and helpless. We all know how long it takes a human baby to gain its footing. Human growth is in fact slower than that of any other mammal, and this is one of the distinctive attributes that sets us apart from other mammals.

The number of young produced by mammals in a season depends on mortality rate, which, for some mammals such as mice, may be high at all age levels. Usually, the larger the animal, the smaller the number of young in a litter. Small rodents, which serve as prey for many carnivores, usually produce more than one litter of several young each season. Meadow mice are known to produce as many as 17 litters of four to nine young in a year. Most carnivores have but one litter of three to five young per year. Large mammals, such as elephants and horses, give birth to a single young with each pregnancy. An elephant produces, on average, four calves during her reproductive life of perhaps 50 years.

Territory and Home Range

Many mammals have territories—areas from which individuals of the same species are excluded. In fact, many wild mammals, like some humans are basically unfriendly to their own kind, especially so to their own sex during the breeding season. If the mammal dwells in a burrow or den, this area forms the center of its territory. If it has no fixed address, the territory is marked out, usually with the highly developed scent glands. Territories vary greatly in size depending on the size of the animal and its feeding habits. Grizzly bears have territories of several square miles, which they guard zealously against all other grizzlies.

Mammals usually use natural features of their surroundings in staking their claims. These are marked with secretions from the scent glands or by urinating or defecating. When an intruder knowingly enters another's marked territory, it is immediately placed at a psychological disadvantage. Should a challenge follow, the intruder almost invariably breaks off the encounter in a submissive display characteristic for the species.

A beaver colony is a family unit, and beavers are among several mammalian species in which the male and female form a strong monogamous bond that lasts a lifetime. Because beavers invest considerable time and energy in constructing a lodge and dam and storing food for winter, the family, especially the adult male, vigorously defends its real estate against intruding beavers. Most of the work of building dams and lodges is undertaken by male beavers, but the females assist when not occupied with their young.

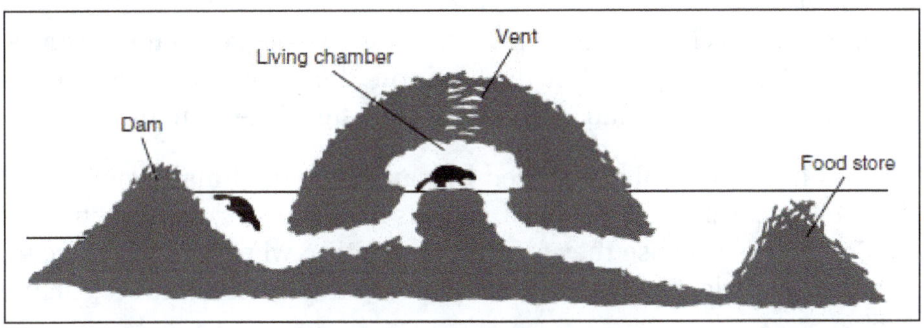

Each beaver colony constructs its own lodge in a pond created by damming a stream. Each year the mother bears four or five young; when the third litter arrives, the 2-year-olds are driven out of the colony to establish new colonies elsewhere. Order Rodentia, family Castoridae.

Immature black-tailed prairie dogs, Cynomys ludovicianus, greeting adult.

These highly social prairie dwellers are herbivores that serve as an important prey to many animals. They live inelaborate tunnel systems so closely interwoven that they form "towns" of as many as 1000 individuals. Towns are subdivided into family units,each with one or two males, severalfemales, and their litters. Although prairie dogs display ownership of burrows with territorial calls, they are friendly with inhabitants of adjacent burrows. The name "prairie dog"derives from the sharp, doglike bark they make when danger threatens. Order Rodentia, family Sciuridae.

An interesting exception to the strong territorial nature of most mammals is the prairie dog, which lives in large, friendly communities called prairie dog "towns". When a new litter has been reared, adults relinquish the old home to the young and move to the edge of the community to establish a new home. Such a practice is totally antithetical to the behavior of most mammals, which drive off the young when they are self-sufficient.

The home range of a mammal is a much larger foraging area surrounding a defended territory. Home ranges are not defended in the same way as are territories; home ranges may, in fact, overlap, producing a neutral zone used by owners of several territories for seeking food.

Mammalian Populations

A population of animals includes all members of a species that share a particular space and potentially interbreed (Animal Ecology). All mammals (like other organisms) live in ecological communities, each composed of numerous populations of different animal and plant species. Each species is affected by the activities of other species and by other changes, especially climatic, that occur. Thus populations are always changing in size. Populations of small mammals are lowest before the breeding season and greatest just after the addition of new members. Beyond these expected changes in population size, mammalian populations may fluctuate from other causes.

Irregular fluctuations are commonly produced by variations in climate, such as unusually cold, hot, or dry weather, or by natural catastrophes, such as fires, hailstorms, and hurricanes. These are density-independent factors because they affect a population whether it is crowded or dispersed. However, the most spectacular fluctuations are density dependent; that is, they correlate with population crowding.

Collared lemming, Dicrostonyx sp., a small rodent of the far north.
Populations of lemmings fluctuate widely. Order Rodentia, family uridae.

Cycles of abundance are common among many rodent species. One of the best-known examples is the mass migrations of the Scandinavian and arctic North American lemmings following population peaks. Lemmings breed all year, although more in summer than in winter. The gestation period is only 21 days; young born at the beginning of the summer are weaned in 14 days and are capable of reproducing by end of summer. At the peak of their population density, having devastated the vegetation by tunneling and grazing, lemmings begin long, mass migrations to find new undamaged habitats for food and space. They swim across streams and small lakes as they go but cannot distinguish these from large lakes, rivers, and the sea, in which they drown. Since lemmings are the main diet of many carnivorous mammals and birds, any change in lemming population density affects all their predators as well.

Varying hares (snowshoe rabbits) of North America show 10- year cycles in abundance. The well-known fecundity of rabbits enables them to produce litters of three or four young as many as five times per year. The density may increase to 4000 hares competing for food in each square mile of northern forest. Predators (owls, minks, foxes, and especially lynxes) also increase. Then the population crashes precipitously for reasons that have long been a puzzle to scientists. Rabbits die in great numbers, not from lack of food or from an epidemic disease (as was once believed) but evidently from some density-dependent psychogenic cause. As crowding increases, hares become more aggressive, show signs of fear and defense, and stop breeding. The entire population reveals symptoms of pituitary-adrenal gland exhaustion, an endocrine imbalance called "shock disease," which results in death. These dramatic crashes are not well understood. Whatever the causes, population crashes that follow superabundance, although harsh, permit the vegetation to recover, providing survivors with a much better chance for successful breeding.

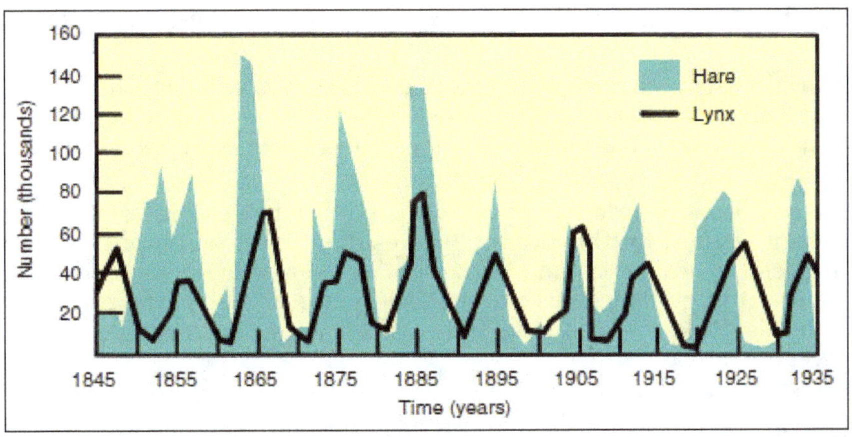

Changes in population size of varying hare and lynx in Canada as indicated by pelts received by the Hudson's Bay Company over a 90-year period. The abundance of lynx (predator) follows that of the hare (prey).

References

- Structural-and-functional-adaptations-of-fishes, general-zoology: biocyclopedia.com, Retrieved 28 June, 2019
- Behavior-and-morphological-adaptations-reptiles-proceedings: veterinarycalendar.dvm360.com, Retrieved 05 March, 2019
- Flight-adaptations, biology: byjus.com, Retrieved 18 July, 2019
- Structural-and-functional-adaptations-of-mammals, general-zoology: biocyclopedia.com, Retrieved 09 April, 2019

PERMISSIONS

All chapters in this book are published with permission under the Creative Commons Attribution Share Alike License or equivalent. Every chapter published in this book has been scrutinized by our experts. Their significance has been extensively debated. The topics covered herein carry significant information for a comprehensive understanding. They may even be implemented as practical applications or may be referred to as a beginning point for further studies.

We would like to thank the editorial team for lending their expertise to make the book truly unique. They have played a crucial role in the development of this book. Without their invaluable contributions this book wouldn't have been possible. They have made vital efforts to compile up to date information on the varied aspects of this subject to make this book a valuable addition to the collection of many professionals and students.

This book was conceptualized with the vision of imparting up-to-date and integrated information in this field. To ensure the same, a matchless editorial board was set up. Every individual on the board went through rigorous rounds of assessment to prove their worth. After which they invested a large part of their time researching and compiling the most relevant data for our readers.

The editorial board has been involved in producing this book since its inception. They have spent rigorous hours researching and exploring the diverse topics which have resulted in the successful publishing of this book. They have passed on their knowledge of decades through this book. To expedite this challenging task, the publisher supported the team at every step. A small team of assistant editors was also appointed to further simplify the editing procedure and attain best results for the readers

Apart from the editorial board, the designing team has also invested a significant amount of their time in understanding the subject and creating the most relevant covers. They scrutinized every image to scout for the most suitable representation of the subject and create an appropriate cover for the book.

The publishing team has been an ardent support to the editorial, designing and production team. Their endless efforts to recruit the best for this project, has resulted in the accomplishment of this book. They are a veteran in the field of academics and their pool of knowledge is as vast as their experience in printing. Their expertise and guidance has proved useful at every step. Their uncompromising quality standards have made this book an exceptional effort. Their encouragement from time to time has been an inspiration for everyone.

The publisher and the editorial board hope that this book will prove to be a valuable piece of knowledge for students, practitioners and scholars across the globe.

INDEX

www.ingramcontent.com/pod-product-compliance
Lightning Source LLC
LaVergne TN
LVHW080923210926
848485LV00004B/716